# Particle Physics: Concepts, Technology and Applications

# Particle Physics: Concepts, Technology and Applications

Editor: Daniel Santiago

**NY** RESEARCH
P R E S S

New York

Published by NY Research Press
118-35 Queens Blvd., Suite 400,
Forest Hills, NY 11375, USA
www.nyresearchpress.com

Particle Physics: Concepts, Technology and Applications
Edited by Daniel Santiago

International Standard Book Number: 978-1-63238-888-9 (Hardback)

**Cataloging-in-Publication Data**

Particle physics : concepts, technology and applications / edited by Daniel Santiago.
p. cm.
Includes bibliographical references and index.
ISBN 978-1-63238-888-9
1. Particles (Nuclear physics). 2. Nuclear physics. 3. Physics. I. Santiago, Daniel.
QC793.2 .P37 2022
539.72--dc23

# Contents

# Preface

The branch of physics which focuses on the study of the nature of particles which comprise radiation and matter is known as particle physics. It also studies the fundamental interactions which are necessary to explain the behavior of the irreducibly small particles. Some of the subatomic particles studied within this field are protons, electrons, neutrons, quarks and leptons. The classification of these particles is done using the theory of the Standard Model of particle physics. There are numerous other theories which are also studied within this field like quantum field theory, effective field theory and lattice field theory. Particle physics is applied in varied sectors such as medicine, computing and national security. This book includes some of the vital pieces of work being conducted across the world, on various topics related to particle physics. It is an upcoming field of science that has undergone rapid development over the past few decades. This book will help the readers in keeping pace with the rapid changes in this field.

This book is a result of research of several months to collate the most relevant data in the field.

When I was approached with the idea of this book and the proposal to edit it, I was overwhelmed. It gave me an opportunity to reach out to all those who share a common interest with me in this field. I had 3 main parameters for editing this text:

1. Accuracy – The data and information provided in this book should be up-to-date and valuable to the readers.

2. Structure – The data must be presented in a structured format for easy understanding and better grasping of the readers.

3. Universal Approach – This book not only targets students but also experts and innovators in the field, thus my aim was to present topics which are of use to all.

Thus, it took me a couple of months to finish the editing of this book.

I would like to make a special mention of my publisher who considered me worthy of this opportunity and also supported me throughout the editing process. I would also like to thank the editing team at the back-end who extended their help whenever required.

**Editor**

# Inhibition of *Staphylococcus aureus* growth in fresh calf minced meat using low density Polyethylene films package promoted by titanium dioxide and zinc oxide nanoparticles

**Arin Marcous[1], Susan Rasouli[2], Fatemeh Ardestani[3,*]**

[1]Department of Food Engineering, Sari Branch, Islamic Azad University, Sari, Iran

[2]Department of Nano Materials and Nano Coating, Faculty of Surface Coating and Modern Technologies, Institute for Color Science and Technology, Tehran, Iran

[3]Department of Chemical Engineering, Qaemshahr Branch, Islamic Azad University, Qaemshahr, Iran, Iran

## HIGHLIGHTS

- ZnO nanoparticle showed 100% bactericidal effect against *Staphylococcus aureus*.

- TiO$_2$ nanoparticle showed 96% bactericidal effect against *Staphylococcus aureus*.

- Mixed TiO$_2$-ZnO showed 98% bactericidal effect against *Staphylococcus aureus*.

## GRAPHICAL ABSTRACT

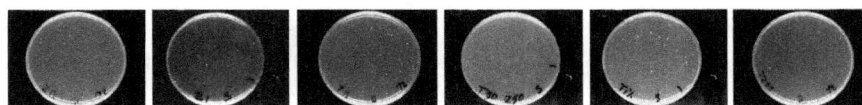

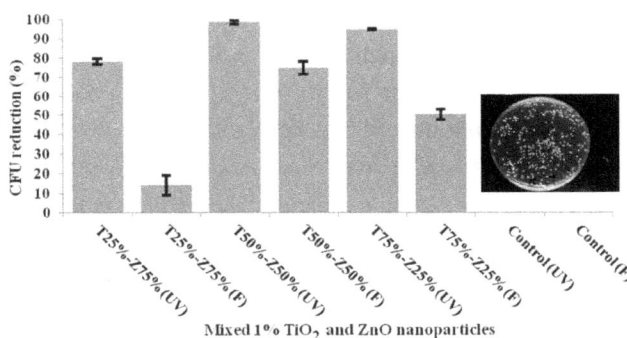

## ARTICLE INFO

*Keywords:*
Antibacterial agents
Fresh minced meat
LDPE flms
Nanoparticles
*Staphylococcus aureus*

## ABSTRACT

Antibacterial properties of TiO$_2$, ZnO as well as mixed TiO$_2$-ZnO nanoparticles coated low density polyethylene films on *Staphylococcus aureus* PTCC1112 were investigated. Bactericidal efficiency of 0.5, 1 and 2 wt% for TiO$_2$ and ZnO nanoparticles and also 1 wt% mixed TiO$_2$-ZnO nanoparticles with TiO$_2$:ZnO ratios of 25:75, 50:50 and 75:25 were tested under UV and fluorescent lights exposure at two different states: films alone (Direct effect) and fresh calf minced meat packed inside the films. The ZnO nanoparticle showed good antibacterial properties against *Staphylococcus aureus* PTCC1112. Maximum CFU reduction of 99.59% and 97.07% were obtained using 2 and 1 wt% ZnO nanoparticle coated LDPE film under UV light for films alone as well as 62.43% and 59.57% for fresh calf minced meat packed. The best antibacterial functionalities of 96.25% and 77.11% CFU reduction were recorded for 1 wt% TiO$_2$ nanoparticle coated LDPE films in the presence of UV light at direct contact with bacteria and fresh calf minced meat packed, respectively. In the case of mixed TiO$_2$-ZnO, maximum CFU reductions of 98.37% and 97.84% were obtained using 50:50 ratio of TiO$_2$: ZnO nanoparticles at the presence of UV light for direct effect and fresh calf minced meat packed, respectively. The 2 wt% ZnO nanoparticle as well as 1 wt% mixed TiO$_2$-ZnO nanoparticles in ratio of 50:50 coated LDPE films were identified as the best case to improve shelf life and prevent *Staphylococcus aureus* PTCC1112 growth in fresh calf minced meat.

* Corresponding author. ; E-mail address: f.ardestani@qaemiau.ac.ir

# 1. Introduction

Today, the presence and propagation of pathogenic bacteria in food industrial products are known as one of the most worldwide serious menaces for public human health. Due to antibacterial resistant subject, development of new methods in bacterial infectious diseases control seems necessary.

Use of active and antibacterial packaging technologies to better protection of spoilable food products has been attended in the recent years [1]. Adding an antimicrobial agent to packaging material causes to extended food shelf life and accepted microbial quality. In this field of knowledge, some metal nanoparticles (NPLs) such as zinc oxide (ZnO), silver doped titanium dioxide (Ag-TiO$_2$), copper oxide (CuO) and silver (Ag) are introduced as applicable agents with good antibacterial characteristics [2-4]. Metal NPLs have higher diffusivity and increased biological impacts such as bactericidal properties in compare to normal size ones [5]. For ZnO NPLs, it may be relating to the dependence of produced H$_2$O$_2$ amount on ZnO surface area [6,7].

Zinc oxide NPLs showed high antibacterial efficiency against *Salmonella typhimurium* and *Staphylococcus aureus*, two important foodborne pathogenic bacteria [8,9]. Panea et al. (2014) reported highly accepted sensorial and microbial characteristics of chicken breast meat packed in low density polyethylene (LDPE) films loaded by ZnO NPLs [10]. Different mechanisms were proposed for antibacterial activity of ZnO NPLs, though the exact mechanism is unknown yet. Direct impact of ZnO on the bacterial cell wall and its destructive effect is one of the most accepted theories [11, 12].

Titanium dioxide (TiO$_2$) has identified photocatalitic activities upon UV irradiation caused to break down some organic vital molecules such as polyunsaturated phospholipids of the microbial cell membrane and finally cell death [13]. Also, dependence of the bactericidal properties of nitrogen-doped TiO$_2$ films on temperature and time duration of exposure to visible artificial light has been proved [14]. However, so far no study has been done to evaluate the antibacterial effects of ZnO, TiO$_2$ and especially mixed ZnO-TiO$_2$ NPLs on *Staphylococcus aureus* in fresh calf minced meat.

In the present work, for the first time, antibacterial effects of low density polyethylene (LDPE) films loaded with ZnO, TiO$_2$ and mixed ZnO-TiO$_2$ NPLs were evaluated against pathogenic *Staphylococcus aureus* PTCC1112 in fresh calf minced meat under UV and fluorescent lights exposure.

# 2. Experimental

## 2.1. Materials

Lyophilized ampoule of *Staphylococcus aureus* PTCC1112 (ATCC 6538) was purchased from Iranian Research Organization for Science and Technology (Persian Type Culture Collection, PTCC, IRAN). Baird-Parker Agar as the selective culture medium for isolation, identification and counting *Staphylococcus aureus* colonies was prepared from Merck Millipore Co. (Germany). All chemicals for analysis and culture media were prepared from Merck Millipore Co. (Germany) too.

Low density polyethylene (LDPE) 020 granules (melt flow rate of 2 g per 10 min, density of 0.92 g ml-1) were funded from Bandar Imam Petrochemical Co. (IRAN).

Zinc oxide (ZnO) nanopowder (molecular weight of 81.39 g mol-1, 15-25 m2 per gram surface area, about 80% zinc basis and <100 nm particle size) and Titanium(IV) oxide (TiO$_2$) nanopowder, anatase (molecular weight of 79.87 g mol-1, 45-55 m$^2$ per gram specific surface area, 99.7% trace metal basis and < 25 nm particle size) were prepared from Sigma Aldrich Co. (USA).

Fresh calf minced meat certified by Iran Veterinary Organization was purchased as without fat and bone meat from Refah Chain Store in Sari, Iran.

## 2.2. NPLs loaded LDPE films manufacturing

LDPE granules were mixed separately with required amounts of TiO$_2$ and ZnO NPLs to make LDPE containing 0.5, 1 and 2 wt% ratios of each NPL and also 1% of the combined two NPLs with ZnO-TiO$_2$ ratios of 25:75, 50:50 and 75:25. Twin screw Extruder (Coperion, ZSk, Screw diameter of 50 mm, Germany) at different temperature profiles applied for melted LDPE nanogranules preparation from each sample, separately. At the next stage, LDPE films coated by different composition of NPLs were made by using Film Blowing Apparatus (Brabender, Screw diameter of 45 mm, Germany). Thickness assessment of the films was performed by using Digital micrometer (Mitutoyo, JAPAN) and determined lower than 50 μm. The images of NPLs coated LDPE films were prepared using Scanning Electron Microscope (SEM, KYKY Technology Development LTD, China).

## 2.3. Initial culturing of S. aureus

The lyophilized PTCC1112 was revived in Trypticase Soy broth (containing g. L-1 of: pancreatic digest of casein, 17; NaCl, 5; papaic digest of soybean meal, 3; $K_2HPO_4$, 2.5; and glucose, 2.5) with pH= 7.2 at 25°C for and 150 rpm and 24 h. Then, remained bacterial cake from centrifugation at 4000×g for 20 min, was suspended in sterile distilled water. Number of S. aureus cells in suspension was counted through colony count method using Baird-Parker Agar [15]. Initial cell number of S. aureus in prepared suspension was estimated equal to 5.95 log CFU/ml.

## 2.4. Evaluation of bactericide effects of NPLs loaded LDPE films

NPL loaded films were sterilized by using alcohol 80% and placed inside the sterile testing tubes. Then, 1 ml of initial culture containing S. aureus was added to each testing tube. For each NPL dosage, six separate testing tubes were prepared. All tests were repeated for three times and the mean values were determined as final result. All inoculated testing tubes were incubated for 72 hr at 25°C in a dark cabinet under UV light (Spectroline, 8 W, 365 nm) or fluorescent light (Osram, Germany). Actually, for each NPL dosage, three testing tube were evaluated under UV light and other three testing tubes investigated under fluorescent light. Counting of viable cells was conducted after 72 hr incubation. For this, 10 ml sterile distilled water was added to each testing tube, and then the tubes were shacked strongly and diluted if necessary. Cell counting was conducted through Most Probable Number (MPN) method by using Petri dishes containing Baird-Parker Agar medium at 25°C for 24 hr [16]. The percent of cell growth inhibition for each testing tube was calculated based on the mean cell numbers in control. Control was LDPE film inside three testing tube without any types of NPL.

## 2.5. Antibacterial effects of NPL loaded LDPE films on fresh calf minced meat

Similar to the previous tests, each NPL dosage

tested in three separate tests, too. Fresh calf minced meat in 100 g samples were packed inside 15×15 cm² LDPE films (containing different NPL dosage) at clean and sterile condition. The packages were stored at 4°C for 72 hr. Some packages were stored under UV light and some others under fluorescent light. Then, after incubation time of 72 hr, number of viable cells was counted and the percent of cell growth inhibition calculated as mentioned in the previous section [16].

## 2.6. Statistical method for results analysis

Obtained experimental results were analyzed by using SPSS 22 software. The mean calculated percent of cell growth inhibition (bactericidal effects of NPLs) for all tests were compared by two ways analysis of variance (ANOVA) and Fishers Least Significant Difference (LSD) test. The value of P≤0.05 was considered as significant difference between each test and other tests.

## 3. Results

Antibacterial effect of various LDPE films containing different dosages of ZnO and $TiO_2$ NPLs was investigated based on the designed experimental procedure. These tests were including LDPE films without any types of NPL, LDPE films loaded with 0.5, 1 and 2 wt% of ZnO NPLs, LDPE films loaded with 0.5, 1 and 2 wt% of $TiO_2$ NPLs, and LDPE films coated by growth inhibition Antibacterial test of different films (uncoated LDPE films, $TiO_2$ NPLs coated LDPE films with $ZnO-TiO_2$ in ratios of 25:75, 50:50 and 75:25. Antibacterial properties of each type of films were evaluated both in the presence of UV and fluorescent light. Also, all experiments were repeated both for free S. aureus inside the testing tubes and S. aureus in fresh calf minced meat.

## 3.1. NPLs loaded verification

To verify the exact loading of ZnO and $TiO_2$ NPLs in LDPE films, SEM analyses was conducted (Figure 1). This image shows the presence of ZnO and $TiO_2$ NPLs, separately at 1 wt% dosages (b,c) and also mixed 1 wt% of combined $ZnO-TiO_2$ NPLs (d) in compare to control (LDPE film without any NPL). The part "a" of Figure 1 presents an image of control sample.

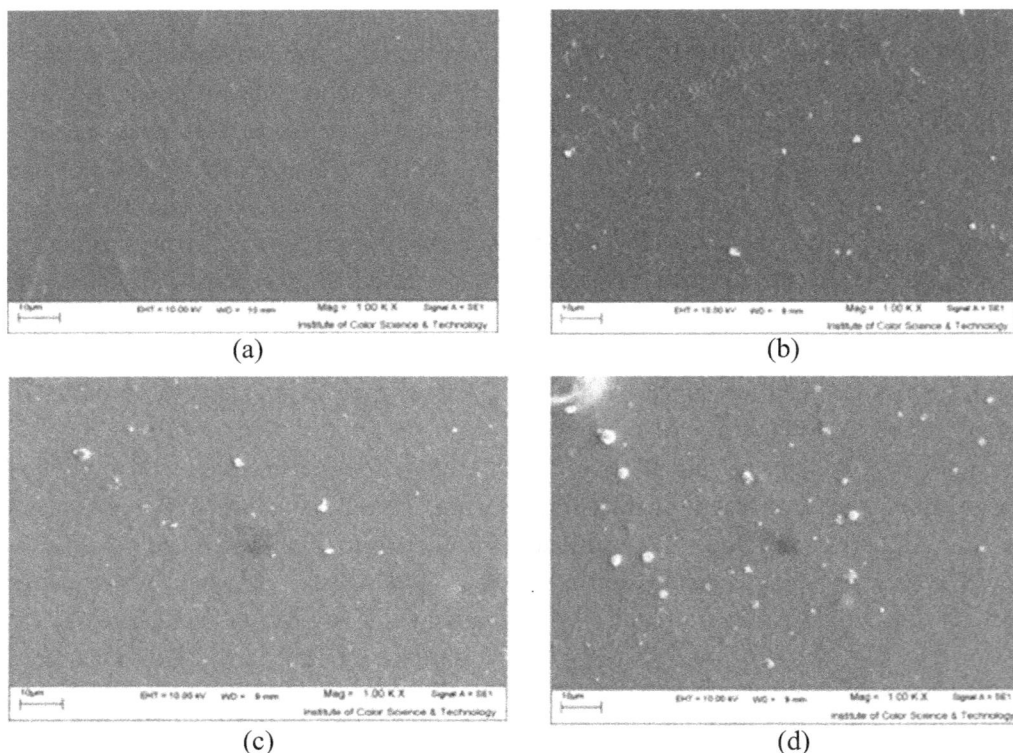

**Fig. 1.** SEM of LDPE film without any NPL (a), LDPE film loaded by 1% ZnO NPLs (b), LDPE film loaded by 1% TiO$_2$ NPLs (c), LDPE film loaded by 1% mixed of TiO$_2$ and ZnO NPLs (d)

### 3.2. Direct inhibition of S. aureus growth by LDPE films coated with TiO$_2$ NPLs

Results showed unconsidered direct bactericidal impacts of LDPE films loaded by TiO$_2$ NPLs on *S. aureus* in the presence of fluorescent light. Increase the dosage of loaded TiO$_2$ had not any perceptible effect on antibacterial properties of the LDPE film (Table 1).

However, a remarkable direct antibacterial potency of 96.25% was observed for LDPE film loaded by 1 wt% TiO$_2$ NPLs in the presence of UV light. This value was obtained very lower and just 26.05% reduction in mean colony counts of *S. aureus* with using of LDPE film loaded by 0.5% TiO$_2$ NPLs. In the case of LDPE film loaded by 2% TiO$_2$ NPLs, a relatively acceptable bactericidal impact equal to 90.65% CFU reduction was obtained under UV light exposure.

### 3.3. S. aureus growth inhibition in fresh calf minced meat by LDPE films coated with TiO$_2$ NPLs

Fairly similar findings to direct antibacterial effects of these types of film on S. aureus were obtained here in the case of fresh calf minced meat (Table 2). Antibacterial properties of TiO$_2$ NPLs were observed significant when UV radiation is used. The best CFU reduction value of 77.11% was obtained under UV light exposure when meat packed inside an LDPE film loaded by 1% TiO$_2$ NPLs, while bactericidal of the same film under fluorescent light was recorded only 18.87% (Figure 3).

### 3.4. Direct inhibition of S. aureus growth by LDPE films coated with ZnO NPLs

Based on the obtained results, a very high CFU reduction of 99.59% was obtained using LDPE film loaded by 2 wt% ZnO NPLs under UV light (Figure 4). Increase ZnO concentration from 0.5 to 2%, showed a positive effect on antibacterial characteristics of the LDPE film. Unlike TiO$_2$ NPLs, LDPE films loaded by ZnO NPLs showed an acceptable bactericidal activity in the absence of UV light. In this case, the higher CFU reduction of 74.51% was recorded for direct effect of LDPE film contained 2% ZnO NPLs on *S. aureus* (Figure 4).

### 3.5. S. aureus growth inhibition in fresh calf minced meat by LDPE films coated with ZnO NPLs

A mediocre CFU reduction of 62.43% was obtained

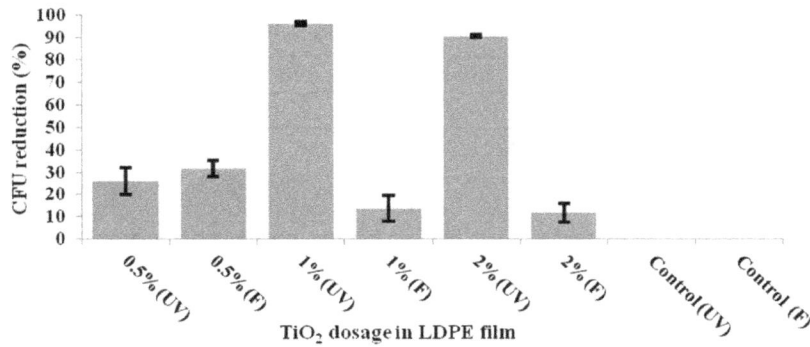

**Fig. 2.** Direct bactericidal effects of TiO$_2$ NPLs loaded LDPE films on *S. aureus* after 72 hr in the presence of fluorescent and UV lights at 20°C

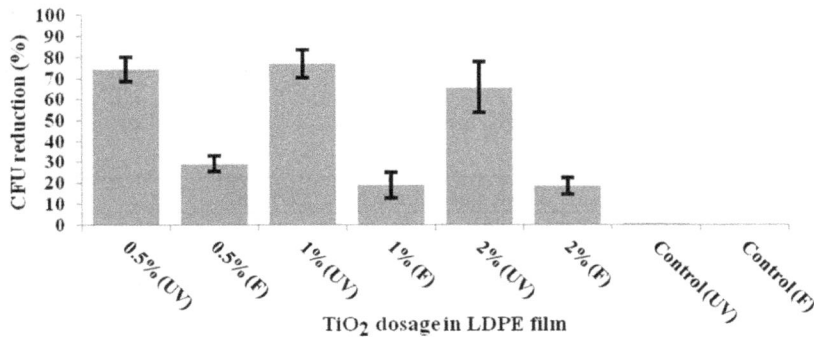

**Fig. 3.** Bactericidal effects of TiO$_2$ NPLs loaded LDPE films on *S. aureus in* fresh calf minced meat after 72 hr in the presence of fluorescent and UV lights at 4°C

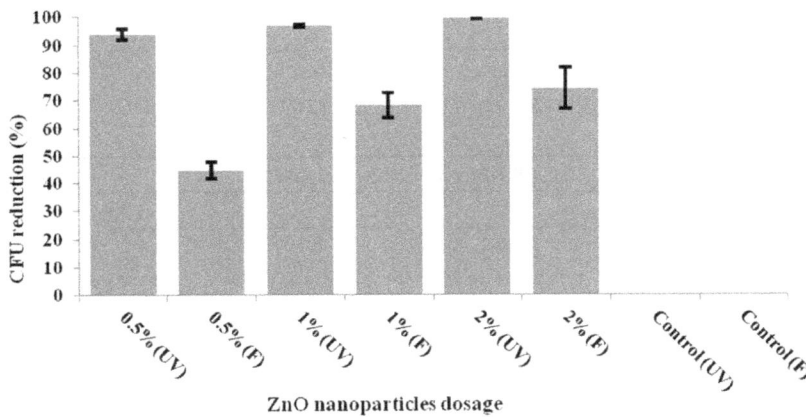

**Fig. 4.** Direct bactericidal effects of ZnO NPLs loaded LDPE films on *S. aureus* after 72 hr in the presence of fluorescent and UV lights at 20°C

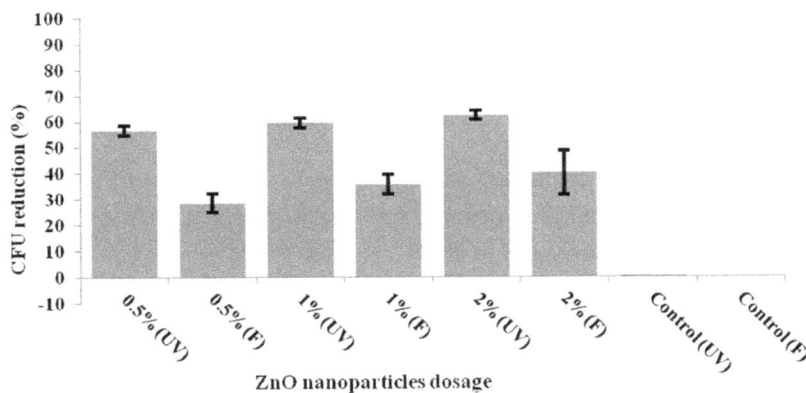

**Fig. 5.** Bactericidal effects of ZnO NPLs loaded LDPE films on *S. aureus* in fresh calf minced meat after 72 hr in the presence of fluorescent and UV lights at 4°C

**Table 1.** The mean values of *S. aureus* colony counts (log CFU/g) for direct effect of LDPE films loaded by various dosages of ZnO and TiO$_2$ NPLs after 72 hr exposed under fluorescent and UV light at 20°C inside the testing tubes.

| LDPE films loaded by TiO$_2$ NPLs | | | | | |
|---|---|---|---|---|---|
| UV | | | Fluorescent | | |
| 0.5% | 1% | 2% | 0.5% | 1% | 2% |
| 5.8143 ±0.0308 | 4.5018±0.1401 | 4.9167 ±0.0277 | 5.8418 ±0.0289 | 5.9426 ±0.0505 | 5.9533 ±0.0289 |

| LDPE films loaded by ZnO NPLs | | | | | |
|---|---|---|---|---|---|
| UV | | | Fluorescent | | |
| 0.5% | 1% | 2% | 0.5% | 1% | 2% |
| 4.6969±0.2198 | 4.4025±0.1473 | 3.3004 ±0.3973 | 5.7493 ±0.0504 | 5.5007 ±0.1121 | 5.3687 ±0.2896 |

| LDPE films loaded by 1% mixed TiO$_2$ and ZnO NPLs | | | | | |
|---|---|---|---|---|---|
| UV | | | Fluorescent | | |
| 25:75 | 50:50 | 75:25 | 25:75 | 50:50 | 75:25 |
| 5.2846 ±0.0542 | 3.9469 ±0.0984 | 4.6556 ±0.0758 | 5.9412 ±0.041 | 5.4010±0.0761 | 5.7033 ±0.0448 |

| LDPE film without any NPL (Control) | | |
|---|---|---|
| UV | | Fluorescent |
| 5.9848 ±0.0367 | | 6.0086±0.0003 |

**Table 2.** The mean values of *S. aureus* colony counts (log CFU/g) for fresh calf minced meat packed in LDPE films loaded by various dosages of ZnO and TiO$_2$ NPLs after 72 hr exposed under fluorescent and UV light at 4°C.

| LDPE films loaded by TiO$_2$ NPLs | | | | | |
|---|---|---|---|---|---|
| UV | | | Fluorescent | | |
| 0.5% | 1% | 2% | 0.5% | 1% | 2% |
| 3.3271 ±0.1900 | 3.2632±0.2495 | 3.4029±0.3713 | 2.0182±0.0418 | 2.0769±0.0494 | 2.0795±0.0362 |

| LDPE films loaded by ZnO NPLs | | | | | |
|---|---|---|---|---|---|
| UV | | | Fluorescent | | |
| 0.5% | 1% | 2% | 0.5% | 1% | 2% |
| 3.7947 ±0.0224 | 3.7206 ±0.0265 | 3.7805±0.0945 | 1.8069±0.0346 | 1.7759±0.0336 | 1.7439±0.0422 |

| LDPE films loaded by 1% mixed TiO$_2$ and ZnO NPLs | | | | | |
|---|---|---|---|---|---|
| UV | | | Fluorescent | | |
| 25:75 | 50:50 | 75:25 | 25:75 | 50:50 | 75:25 |
| 3.1017 ±0.0990 | 2.2746±0.0571 | 3.3264 ±0.217 | 2.0378±0.0184 | 2.0959±0.0226 | 2.1138 ±0.0405 |

| LDPE film without any NPL (Control) | | |
|---|---|---|
| UV | | Fluorescent |
| 4.1760 ±0.0002 | | 3.9426±0.0003 |

for *S. aureus* in fresh calf minced meat when packed into LDPE film loaded by 2% ZnO NPLs under UV exposure (Figure 5). Increase NPL concentration from 0.5 to 2% caused an increase in bactericidal characteristics of films both in the presence of UV and fluorescent lights (Table 2). Under fluorescent light exposure, at the best condition, 40.12% CFU reduction was obtained using LDPE film loaded by 2% ZnO NPLs (Figure 5).

### 3.6. Direct inhibition of S. aureus growth by LDPE films coated with mixed TiO$_2$-ZnO NPLs

For LDPE films contained a mixed of TiO$_2$ and ZnO NPLs, the best results on CFU reduction of *S. aureus* was obtained when an equal mixing of two NPLs was applied. In this regard, a good antibacterial effect equal to 98.37% direct CFU reduction was recorded by using

LDPE film loaded by 1% mixed $TiO_2$-ZnO NPLs in ratio of 50:50 under UV light (Figure 6). A meaningful difference was observed between treatments under UV light and the ones under fluorescent lights (Table 1). Under UV light exposure, adding more amounts of $TiO_2$ NPLs in mixture had positive antibacterial impacts on *S. aureus* in compare to treatments with more portions of ZnO NPLs. So that, CFU reductions of 94.86% and 78.23% were recorded for LDPE films loaded by mixed $TiO_2$-ZnO NPLs in ratios of 75:25 and 25:75, respectively (Figure 6). Under fluorescent light exposure, the reverse results were observed. So that CFU reductions of 14.07% and 50.34% were recorded for LDPE films loaded by mixed $TiO_2$-ZnO NPLs in ratios of 75:25 and 25:75, respectively (Figure 6).

*3.7. S. aureus growth inhibition in fresh calf minced meat by LDPE films coated with mixed $TiO_2$-ZnO NPLs*

A manifest difference was observed between bactericidal impacts of LDPE film coated with mixed $TiO_2$-ZnO NPLs under UV and fluorescent lights. The higher obtained CFU reduction in the presence of fluorescent light recorded only 26.20% using 1% mixed $TiO_2$-ZnO NPLs in ratio of 25:75 (Figure 7). This effect has been decreased even to 11.97% with applying reverse ratio of 75:25. However, a very desirable antibacterial property equal to CFU reduction of 97.84% was recorded for LDPE film loaded by 1% mixed $TiO_2$-ZnO NPLs in ratio of 50:50 under UV light exposure (Figure 7).

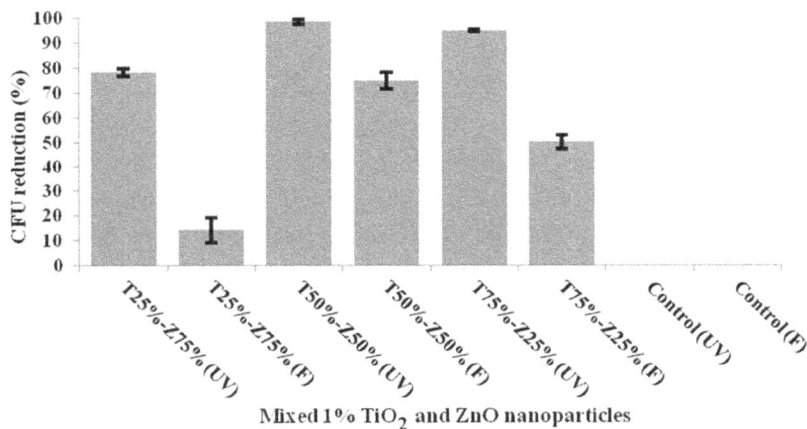

**Fig. 6.** Direct bactericidal effects of 1% mixed of $TiO_2$ and ZnO NPLs loaded LDPE films on *S. aureus* after 72 hr in the presence of fluorescent and UV lights at 20°C

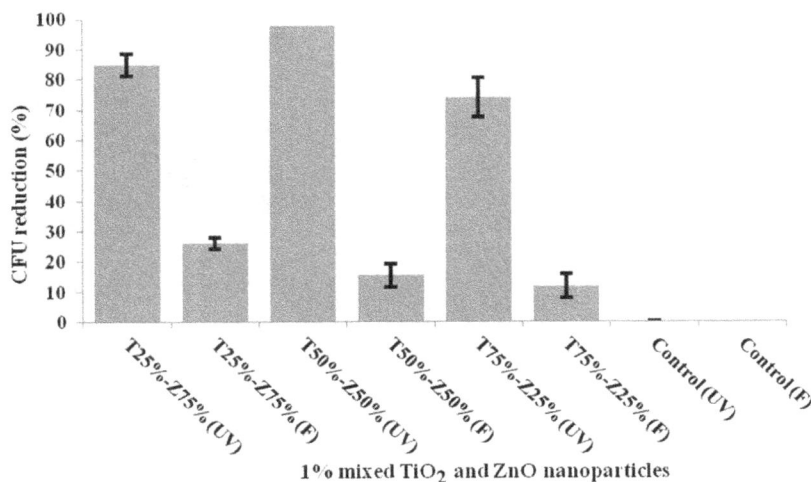

**Fig. 7.** Bactericidal effects of 1% mixed of $TiO_2$ and ZnO NPLs loaded LDPE films on *S. aureus* in fresh calf minced meat after 72 hr in the presence of fluorescent and UV lights at 4°C

## 4. Discussion

### 4.1. Direct bactericidal effect of LDPE films coated with TiO₂ NPLs

Figure 8 showed *Staphylococcus aureus* PTCC 1112 colonies in direct treatments with LDPE films loaded by ZnO, TiO₂ and also mixed ZnO-TiO₂ NPLs in the present work.

Ibrahim (2015) reported high bactericidal efficiency of 95% for Ag-TiO₂ films against *Staphylococcus aureus* after 3 hours incubation, a little less than our result (96.25% for LDPE film loaded by 1% TiO₂ NPLs in the presence of UV light) [13]. Also, Xing et al. (2012) findings showed 95.2% inhibition of *S. aureus* with using TiO₂ NPLs loaded poly ethylene films irradiated by ultraviolet light for 1 hour [17]. As respects to antibacterial standards, a powerful antibacterial agent must cause a bacterial CFU reduction more than 70% [18]. Then, based on our results, TiO₂ NPLs at 1 wt% under UV light is known as a good antibacterial substance against to *S. aureus* PTCC1112. However this NPL showed low bactericidal properties under visible light irradiation (maximum about 32% obtained CFU reduction).

Based on our obtained results as well as some other reported researches, TiO₂ requires photo-activation to display antibacterial properties [19-20]. Antibacterial feature of UV light by itself isn't enough to prevent microbial growth with an acceptable efficiency [21]. Under ultraviolet light exposure, TiO₂ NPLs are taking part in some photo-related reactions lead to excited electrons and hallow pair formation and then, their diffusion to the surface of TiO₂ NPLs. In the following, hydroxyl radicals and superoxide ions with high performance in destroying the bacterial cell membrane and bactericidal properties are formed [22].

### 4.2. Bactericidal effect of LDPE films coated with TiO₂ NPLs in fresh calf minced meat

Evaluation of antibacterial effects of this category of films confirmed again the important role of ultraviolet rays in bactericidal properties of TiO₂ NPLs. The same as direct bactericidal studies, in the case of fresh calf minced meat, results showed significant difference between *S. aureus* growth inhibition by TiO₂ NPLs in the presence of UV and fluorescent lights, too. The antibacterial efficiency of TiO₂ NPLs on *S. aureus* in fresh calf minced meat was obtained lower than its direct effect in testing tubes. This may be related to the faster and easier diffusion of hydroxyl radicals and superoxide ions to bacterial cell membrane in direct status. However, according to the standard, an agent

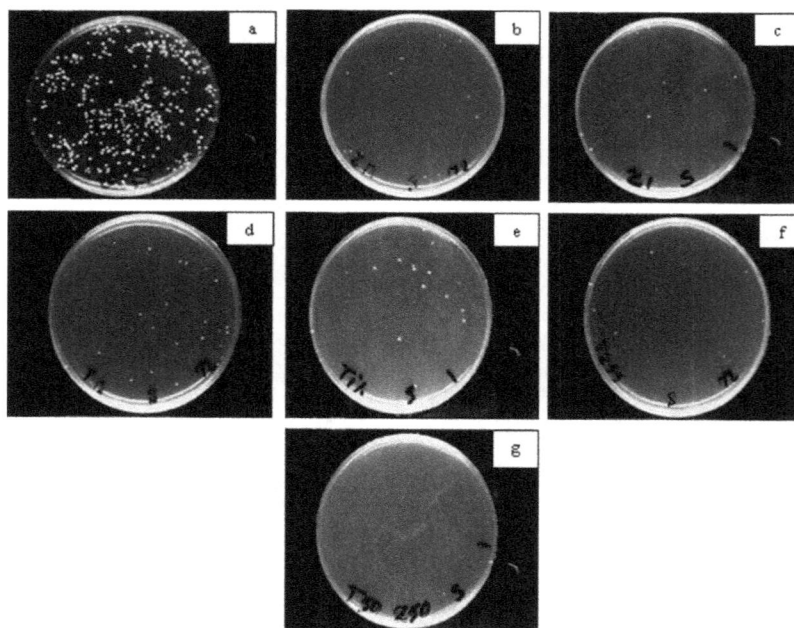

**Fig. 8.** *S. aureus* colonies in direct treatments with LDPE films loaded by any NPLs in control sample (a), 1% ZnO NPLs under fluorescent light (b), 1% ZnO NPLs under UV light (c), 1% TiO₂ NPLs under fluorescent light (d), 1% TiO₂ NPLs under UV light (e), 1% mixed of 50-50 TiO₂ and ZnO NPLs under fluorescent light (f), 1% mixed of 50-50 TiO₂ and ZnO NPLs under UV light (g).

with CFU reduction less than 20% isn't known as a bactericide material. The values between 20-50% indicate low antibacterial property. CFU reduction between 50-70% and more than 70% show expressive and high bactericidal properties, respectively [18]. Therefore, the results showed that $TiO_2$ NPLs at dosage of 1 wt% have high bactericidal efficiency of 77% CFU reduction against *S. aureus* PTCC1112 in fresh calf minced meat under UV light exposure. While, isn't introduced as a bactericidal agent in the presence of fluorescent light.

### 4.3. Direct bactericidal effect of LDPE films coated with ZnO NPLs

Antibacterial effects of ZnO NPLs on *S. aureus* have been reported by some researchers [23]. This agent showed better antibacterial properties against gram positive bacteria such as *S. aureus* than gram negative bacteria such as *E. coli*. Different internal antioxidant content and detoxification agents in gram positive and gram negative bacteria have been mentioned as the main reason of this phenomenon [24]. The actual mechanism of bactericidal properties of ZnO is unknown. However some mechanisms includes release $Zn^{2+}$, production of reactive oxygen species such as hydroxyl radicals and hydrogen peroxide, as well as destruct the bacterial cell wall have been proposed [25-27]. Results showed about 100% bactericidal efficiency for ZnO NPLs at 2 wt% against *S. aureus* PTCC1112 under UV light (Figure 4). Also, ZnO NPLs at 2 wt% is introduced as a powerful antibacterial agent in the absence of UV light with about 75% CFU reduction. Thus, bactericidal function of ZnO seems to not be depended to ultraviolet rays. Emami-Karvani and Chehrazi (2011) reported compelet inhibition of *S. aureus* growth applying 1.56 mg ml$^{-1}$ ZnO NPLs [28]. They also understood that ZnO NPLs is more effective against gram-positive bacteria than gram-negative ones, may be due to different cell wall structure and metabolism [28, 29].

### 4.4. Bactericidal effect of LDPE films coated with ZnO NPLs in fresh calf minced meat

The results of Akbar and Anal (2014) showed that use an active package containing calcium alginate film loaded by ZnO NPLs decreased the number of *Salmonella typhimurium* and *Staphylococcus aureus* in

ready-to-eat poultry meat from log seven to zero at a period of 10 days incubation at 8±1°C [8]. Panea et al. (2014) reported good antibacterial effects of LDPE films blended with a combination of 5% ZnO and 10% Ag NPLs applied for chicken breast meat packaging [10]. However, our results indicated expressive and low antibacterial effect of ZnO NPL on *S. aureus* PTCC1112 in fresh calf mined meat under UV and fluorescent lights exposure, respectively (Figure 5).

### 4.5. Direct bactericidal effect of LDPE films coated with mixed $TiO_2$-ZnO NPLs

An equal parts mixture of $TiO_2$ and ZnO NPLs under UV light showed high direct antibacterial functionality (near to 100% CFU reduction) at 1 wt% against *S. aureus* PTCC1112 (Figure 6). According to our knowledge, until now, there is not any research work noted the antibacterial characteristics of this mixture on *S. aureus*.

### 4.6. Bactericidal effect of LDPE films coated with $TiO_2$-ZnO NPLs in fresh calf minced meat

Mixture of $TiO_2$-ZnO NPLs in ratio of 50:50 is a powerful bactericidal option with high bacterial growth inhibitory function against *S. aureus* PTCC1112 in fresh calf minced meat. With a CFU reduction of 98%, this mixture is introduced as a favorable antibacterial agent for *S. aureus*. Results showed significant improve in safety and shelf life of fresh calf minced meat by using mixed $TiO_2$-ZnO NPLs instead of each of these two NPLs separately.

## 5. Conclusion

Based on the results of this work, antibacterial activity of $TiO_2$ NPLs is depend to UV light presence while ZnO NPLs are independent from this. Under UV light exposure, $TiO_2$ NPLs showed high antibacterial properties against *S. aureus* in fresh calf minced meat. ZnO NPLs showed a complete direct bactericidal efficiency against *S. aureus* under and also in the absence of UV. However, low antibacterial effect of ZnO NPL on *S. aureus* in fresh calf mined meat was recorded. $TiO_2$-ZnO NPLs (50:50) under UV light is introduced as a powerful direct and also in fresh calf mined meat antibacterial agent against *S. aureus*.

## Acknowledgments

The authors have special thanks from Institute for Color Science and Technology as well as Offices of Vice Chancellor for Research of Islamic Azad University, Qaemshahr and Sari Branches.

## References

[1] J. Chen, A.L. Brody, Use of active packaging structures to control the quality of a ready-to-eat meat product, Food Control. 30 (2013) 306-310.

[2] P. Swain, S.K. Nayak, A. Sasmal, T. Behera, S.K. Barik, S.K. Swain, S.S. Mishra, A.K. Sen, J.K. Das, Antimicrobial activity of metal based NPLs against microbes associated with diseases in aquaculture, World J. Microb. Biot. 30 (2014) 2491-2502.

[3] L. Ozimek, E. Pospiech, S. Narine, Nanotchenologies in food and meat processing, Acta. Sci. Pol. Technol. Aliment. 9 (2010) 401-412.

[4] R. Tankhiwale, S.K. Bajpai, Preparation, characterization and antibacterial applications of ZnO-NPLs coated polyethylene films for food packaging, Colloids. Surf. B. Biointerfaces. 90 (2012) 16-20.

[5] R.K. Raghupati, R.T. Koodali, A.C. Manna, Size-dependent bacterial growth inhibition and mechanism of antibacterial activity of zinc oxide NPLs, Langmuir. 27 (2011) 4020–4028.

[6] T. Ohira, O. Yamamoto, Y. Iida, Z.E. Nakagawa, Antibacterial activity of ZnO powder with crystallographic orientation, J. Mater. Sci. -Mater. M. 19 (2008) 1407–1412.

[7] N. Padmavathy, R. Vijayaraghavan, Enhanced bioactivity of ZnO NPLs-an antibacterial study, Sci. Technol. Adv. Mat. 9 (2008) 432-438.

[8] A. Akbar, A.K. Anal, Zinc oxide NPLs loaded active packaging, a challenge study against *Salmonella typhimurium* and *Staphylococcus aureus* in ready-to-eat poultry meat, Food Control. 38 (2014) 88-95.

[9] P.J.P. Espitia, N.F.F. Soares, R.F. Teofilo, J.S.R. Coimbra, D.M. Vitor, R.A. Batista, S.O. Ferreira, N.J. Andrade, A.A. Medeiros, Physical-mechanical and antimicrobial properties of nanocomposite films with pediocin and ZnO NPLs, Carbohyd. Polym. 94 (2013) 199-208.

[10] B. Panea, G. Ripoll, J. Gonzalez, A. Fernandez-Cuello, A. Alberti, Effect of nanocomposite packaging containing different proportions of ZnO and Ag on chicken breast meat quality, J. Food. Eng. 123 (2014) 104-112.

[11] L. Zhang, Y. Jiang, Y. Ding, M. Povey, D. York, Investigation into the antibacterial behavior of suspension of ZnO NPLs (ZnO nanofluids), J. Nanopart. Res. 9 (2007) 479-489.

[12] R. Brayner, R. Ferrari-lliou, N. Brivois, S. Djediat, M. F. Benedetti, F. Fievet, Toxicological impact studies based on Escherichia coli bacteria in ultrafine ZnO NPLs colloidal medium, Nano. Lett. 6 (2006) 866-870.

[13] H.M.M. Ibrahim, Photocatalytic degradation of methylene blue and inactivation of pathogenic bacteria using silver NPLs modified titanium dioxide thin films, World J. Microb. Biot. 31 (2015) 1049-1060.

[14] C. Vacaroiu, M. Enache, M. Gartner, G. Popescu, M. Anastasescu, A. Brezeanu, N. Todorova, T. Giannakopoulou, C. Trapalis, The effect of thermal treatment on antibacterial properties of nanostructured TiO2(N) films illuminated with visible light, World J. Microb. Biot. 25 (2009) 27-31.

[15] S.H. Othman, N.R. Abd Salam, N. Zainal, R.K. Basha, R.A. Talib, Antimicrobial activity of $TiO_2$ NPL-coated film for potential food packaging applications, Int. J. Photoenerg. 2014 (2014) 1-6.

[16] R. Mehrpour, Microbiology of food and animal feeding stuffs-Horizontal method for the enumeration of positive Staphylococci- coagulase (*Staphylococcus aureus* and other species)-Part 3: Detection and MPN technique for low numbers, Inst. Standard. Ind. Res. Iran. 6806-3 (2006) 3-11.

[17] Y. Xing, X. Li, Z. Li, Q. Xu, Z. Che, W. Li, Y. Bai, K. Li, Effect of $TiO_2$ NPLs on the antibacterial and physical properties of polyethylene-based film, Prog. Org. Coat. 73 (2012) 219-224.

[18] D. Prasad, C.R. Girija, A.J. Reddy, H. Nagabhushana, B.M. Nagabhushana, T.V. Venkatesha, S.T. Arun Kumar, A Study on the antibacterial activity of ZnO NPLs prepared by combustion method against E. coli, Int. J. Eng. Res. Appl. 4 (2014) 84-89.

[19] M. Fang, J.H. Chen, X.L. Xu, P.H. Yang, H.F. Hildebrand, Antibacterial activities of inorganic agents on six bacteria associated with oral infections by two susceptibility tests, Int. J. Antimicrob. Agents. 27 (2006) 513-517.

[20] J.R. Villalobos-Hernandez, G.G. Muller-Goymann, Sun protection enhancement of titanium dioxide crystals by the use of carnauba wax NPLs: the synergistic interaction between organic and inorganic sunscreens at nanoscale, Int. J. Pharm. 322 (2006) 161–170.

[21] P.J. Meechan, Ch. Wilson, Use of ultraviolet lights in biological safety cabinets: A contrarian view, Appl. Biosafety. 11(2006) 222-227.

[22] R.J. Watts, D. Washington, J. Howsawkeng, A.L. Teel, Comparative toxicity of hydrogen peroxide, hydroxyl radicals, and superoxide anion to Escherichia coli, Adv. Environ. Res. 7 (2003) 961-968.

[23] N. Jones, B. Ray, R.T. Koodali, A.C. Manna, Antibacterial activity of ZnO NPLs suspensions on a broad spectrum of microorganisms, FEMS. Microbiol. Lett. 279 (2008) 71–76.

[24] G. Applerot, N. Perkas, G. Amirian, O. Girshevitz, A. Gedanken, Coating of glass with ZnO via ultrasonic irradiation and a study of its antibacterial properties, Appl. Surf. Sci. 256 (2009) 3-8.

[25] R. Jalal, E.K. Goharshadi, M. Abareshi, M. Moosavi, A. Yousefi, P. Nancarrow, ZnO nanofluids: green synthesis, characterization, and antibacterial activity, Mater. Chem. Phys. 121 (2010) 198–201.

[26] Y. Xie, Y. He, P.L. Irwin, T. Jin, X. Shi, Antibacterial activity and mechanism of zinc oxide NPLs on Campylobacter jejuni, Appl. Environ. Microb. 77 (2011) 2325–2331.

[27] T. Gordon, B. Perlstein, O. Houbara, I. Felner, E. Banin, S. Margel, Synthesis and characterization of zinc/iron oxide composite NPLs and their antibacterial properties, Colloids. Surf. A. 374 (2011) 1-8.

[28] Z. Emami-Karvani, P. Chehrazi, Antibacterial activity of ZnO NPL on gram positive and gram-negative bacteria, Afr. J. Microbiol. Res. 5 (2011) 1368-1373.

[29] T. Jin, D. Sun, J. Y. Su, H. Zhang, H. J. Sue, Antimicrobial efficacy of zinc oxide quantum dots against Listeria monocytogenes, Salmonella enteritidis, and Escherichia coli O157:H7, J. Food. Sci. 74 (2009) 46-52.

# 2

# Pool boiling heat transfer coefficient of pure liquids using dimensional analysis

Ahmadreza Zahedipoor, Mehdi Faramarzi*, Shahab Eslami, Asadollah Malekzadeh

*Department of Chemical Engineering, Gachsaran Branch, Islamic Azad University, Gachsaran, Iran*

## HIGHLIGHTS

- Dimensionless groups were created linking new boiling heat transfer coefficient to physical properties of boiling liquids.

- Boiling heat transfer coefficient of liquids increased slowly through an increase in heat flux.

- A precise correlation was achieved to link dimensionless groups by optimizing the model using genetic algorithm.

## GRAPHICAL ABSTRACT

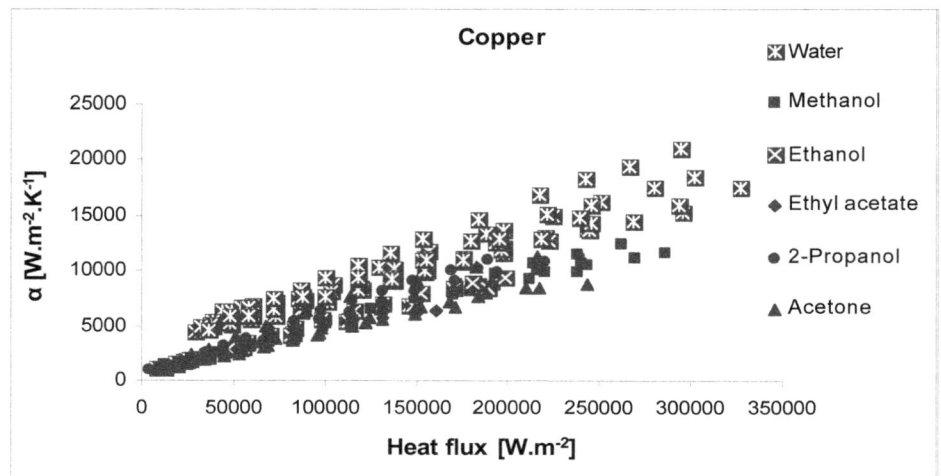

## ARTICLE INFO

*Keywords:*
Pool boiling
Heat transfer
Atmospheric pressure
Heat transfer coefficientay

## ABSTRACT

The pool boiling heat transfer coefficient of pure liquids were experimentally measured on a horizontal bar heater at atmospheric pressure. These measurements were conducted for more than three hundred data in thermal currents up to 350 kW.m$^{-2}$. Original correlations and the unique effect of these correlations on experimental data were discussed briefly. According to the analysis, a new empirical relationship implying a performance superior to other available correlations is presented.

* Corresponding author.  ; E-mail address: Faramarzi.iaug@gmail.com

# 1. Introduction

Free movement mechanism plays an important role in industrial heat transfer processes. In recent decades, a lot of free convections have been presented for predicting heat transfer coefficient developed in different situations such as various geometrical situations. Boiling plays an important role in many chemical engineering issues such as cooling cycle, strength, and distillation processes. Proper design of the equipment requires understanding of the boiling process and predicting the boiling heat transfer coefficient. Due to the high boiling heat transfer coefficient in the nuclear area, this area is very important. In general, nuclear boiling is defined as the formation of vapor bubbles in active positions of the surface, when the surface temperature is higher than the liquid saturation temperature in contact with the surface.

Movement processes which are accompanied with a fluid phase change involve boiling, too. Boiling is one of the most popular and very complicated processes in engineering science because of its multiple sub-processes. These sub-processes are realized as dynamics of bubbles including the diameter of bubbles separation, compression of bubble generator sources per unit area, and bubble generation frequency. Understanding and more accurate modeling of the boiling process requires modeling of these sub-processes.

## 1.1. Literature review

Despite extensive research in the field of nuclear pool boiling, fundamental mechanisms of this process, including sub-processes such as bubble diameter, density of bubble generation points and bubble separation frequency, have not been fully understood. A lot of relationships have been presented to predict the boiling heat transfer coefficient of pure liquids that are generally empirical or semi-empirical.

Pool boiling proposed by Kutateladze in 1952 is one of the oldest equations for pure liquids [1]. This experimental model, with two dimensionless parameters, calculates the heat transfer coefficient without considering the heat transfer surface roughness. By the end of 1962 McNelly's model [2] was used as the most reliable predictor of pure liquids heat transfer coefficient. The model was only reliable for the prediction of pure and single-component materials and

includes physical properties of the liquid phase while boiling as well as vapor phase.

In early 1963 Mostinski [3] carried out numerous tests on the critical heat flux using corresponding states. In the same year, he presented results of his experiments using a mathematical model. Boyko-KruzhilinIn [4] introduced a more complete model in 1967 in which physical properties of heat transfer coefficient were used for calculation and prediction of heat transfer coefficient. This model, obtained by dimensional analysis, calculates more accurate values for pure liquid heat transfer coefficient. After numerous tests on various fluids, Labantsov [5] proposed his empirical relationship in 1972. Since his relationship is based on experimental data for more than 200 materials, it provides a good overlap for this data. One of the most accurate equations to predict the boiling heat transfer coefficient of pure liquids was offered by Stephan and Abdelsalam [6]. Introducing this exact relationship, Stephan has sharply declined the deviation of heat transfer coefficient from experimental data. He used groups of six fold thermo-physical properties which facilitated calculations of heat transfer coefficient prediction due to sorting values of these groups and creating linear relationship for the multiplication of these properties by each other. The diversion rate of the model, used for more than 500 materials, is about 5-16%.

Gorenflo [7] is the innovator of a new experimental system for estimating the heat transfer coefficient. By doing numerous tests he offered a new multi-parameter model in 1984 and again in 1993. The relationship presented by Alavi Fazel et al. [8] was based on measured data (of water, ethanol, methanol, acetone and 2-propanol) and dimensional analysis. Simplicity is one of the benefits of this relationship. In addition, via principle of corresponding states and dimensional analysis, Sarfaraz introduced his model for the pool boiling heat transfer coefficient of pure liquids in 2013 [9]. Table 1 shows equations to predict the heat transfer coefficient of pure liquids.

# 2. Experimental device

The instrument applied for data collection and measurement is known as Gorenflo's pool boiling device. The device is made of a smooth stainless steel surface cylinder located inside a glass enclosure with high thermal tolerance. There are four thermocouples

**Table 1.** Relationships for estimating the boiling Hhat transfer coefficient of pure liquids.

| Researcher | Model |
|---|---|
| Kutateladze [1] | $\alpha = \left[ 3.37E - 9\dfrac{k_l}{l^*}\left(\dfrac{H_{fg}}{C_{pl}q}\right)^{-2} M_*^{-4} \right]^{\frac{1}{3}}$ |
| McNelly [2] | $\alpha = 0.225\left(\dfrac{C_{pl}q}{H_{fg}}\right)^{0.69}\left(\dfrac{Pk_l}{\sigma}\right)^{0.31}\left(\dfrac{\rho_l}{\rho_v} - 1\right)^{0.33}$ |
| Mostinsk [3] | $\alpha = (3.596E - 5)P_c^{0.69}q^{0.7}[1.8(\dfrac{P}{P_c})^{0.17} + 4\left(\dfrac{P}{P_c}\right)^{1.2} + 10\left(\dfrac{P}{P_c}\right)^{10}$ |
| Boyko-Kruzhiline [4] | $\alpha = 0.082\dfrac{k_l}{l^*}\left[\dfrac{H_{fg}q}{g(T_s + 273.15)k_l}\left(\dfrac{\rho_v}{\rho_l-\rho_v}\right)\right]^{0.7}\left[\dfrac{(T_s + 273.15)C_{pl}\sigma P}{H_{fg}\rho_v l^*}\right]^{0.33}$ |
| Labantsov [5] | $\alpha = 0.075\left[1 + 10\left(\dfrac{\rho_v}{\rho_l - \rho_v}\right)^{0.67}\right]\left[\dfrac{k_l^2}{v\sigma(T_s + 273.15)}\right]^{0.33}q^{0.67}$ |
| Stephan and Abdelsalam [6] | $\alpha = 0.23\dfrac{k_l}{d_b}\left(\dfrac{qd_b}{k_lT_s}\right)^{0.674}\left(\dfrac{\rho_v}{\rho_l}\right)^{0.297}\left(\dfrac{H_{fg}d_b^2}{\hat{\alpha}_l^2}\right)^{0.371}\left(\dfrac{\hat{\alpha}_l^2\rho_l}{\sigma d_b}\right)^{0.35}\left(\dfrac{\rho_l - \rho_v}{\rho_l}\right)^{-1.73}$ |
| Gorenflo [7] | $F_p = 1.73Pr^{0.27} + 6.1Pr + 0.68Pr/(1-P_r) \qquad \alpha = \alpha_0 F_q F_p F_{WR} F_{WM}$ |
| Alavi Fazel [8] | $\alpha = \dfrac{3.253\sigma^{0.125}H_{fg}^{0.125}(q/A)^{0.876}}{T_{sat}\hat{\alpha}^{0.145}}$ |
| Sarafraz [9] | $\alpha = \dfrac{3.0219\sigma^{0.12}\Delta H_{fg}^{0.1107}q^{0.8045}}{T_{sat}\hat{\alpha}^{0.1398}}$ |
| Rohsenow [10] | $\dfrac{C_{pl}\Delta T_{sat}}{H_{fg}P_r^s} = C_{sf}\left[\dfrac{\left(\dfrac{q}{A}\right)}{\mu l H_{fg}}\sqrt{\dfrac{\sigma g_c}{g(\rho_l - \rho_v)}}\right]^{0.33}$ |
| Nishikawa [11] | $\alpha = \dfrac{31.4P_c^{0.2}}{M_w^{0.1}T_c^{0.9}}(8R_p)^{0.2(1-P_r)}\dfrac{(P_r^{0.23})q^{0.8}}{(1 - 0.99P_r)^{0.9}}$ |
| Fujita [12] | $\alpha = 1.21\,(q/A)^{0.83}$ |
| Cooper [13] | $\alpha = 55P_r^{0.12-0.443Rp}(-\log P_r)^{-0.55}MW^{-0.5}(q/A)^{0.67}$ |

on the cylinder surface. The thermocouples arithmetic mean at any moment and after the final correction indicates cylinder surface temperature. The main heater of the experiment is placed at the center of the cylinder which operates near the boiling point. The pool boiling phenomenon occurs by contribution of this heater. Figures (1) and (2) display full details of the device [14].

A silicone paste is used to reduce the thermal contact resistance between the thermocouple and holes. The required voltage is provided by urban power supply and is regulated by an autotransformer. In order to remove roughness, cylinder surface is polished by smooth sandpaper with roughness of 400 micrometers and is

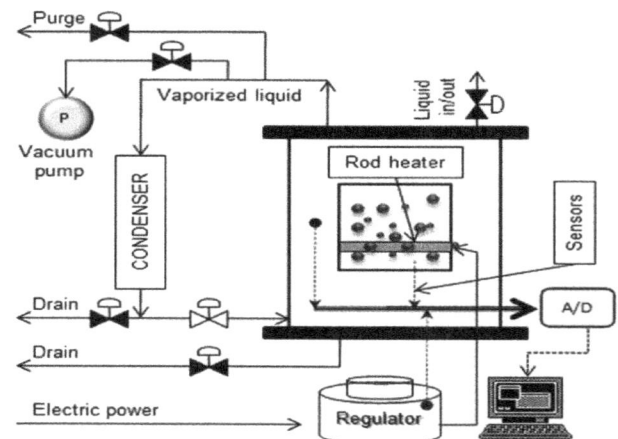

**Fig. 1.** Laboratory device details [14].

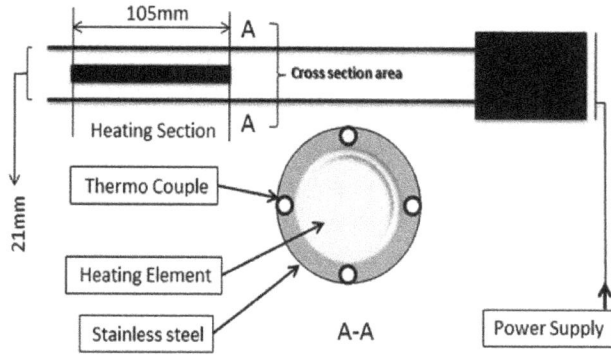

**Fig. 2.** Laboratory main heater details [14].

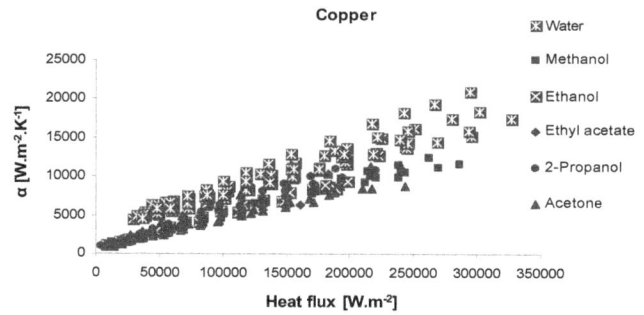

**Fig. 3.** The effect of heat transfer level on the heat transfer coefficient rate of pure liquids.

refined by fine polishing oil. Moreover, in order that temperature sensors operate precisely, sensors diameter was 2 mm with a length of 100 mm. The container volume is about 4 liters. A condense is placed on top of the container in order to restore the steam produced from the boiling process. This experiment is based on Newton's cooling law as seen in equation (1).

$$q = h \ (T_w\text{-}T_{sat}) \tag{1}$$

This article examines some of the most important physical properties of pure liquids including liquid phase density, surface tension, liquids' specific heat capacity and vaporization enthalpy.

*2.1. Experimental results*

Figure 3 shows row experimental values of boiling

heat transfer for the tested liquids. According to the graph, boiling heat transfer coefficient of any liquid increases slowly as the heat flux increases. However, there are small fluctuations that are primarily related to experimental error and residual effect. Note that A/D function (analog to digital convertor) is sensitive to environmental conditions.

## 3. Available correlation functions

A purely theoretical and predictive model has not been offered due to the complexity of the boiling phenomenon. There are a lot of parameters impacting pool boiling heat transfer which require extensive research in order to be correlated with boiling heat transfer coefficient without any empirical tuning parameter. Table 2 shows the ordinary physical constants of test pure liquids. Some important physical properties of test pure liquids during experiments are presented in Table 3. Table 4 compares

**Table 2.** Summarization of the boiling point and critical important constants of pure liquids.

| Chemical component | $T_b$ (°C) | $V_c$ (m³/kg.mol) | $P_c$ (kPa) | $T_c$ (K) | Mw (g/mol) | $\omega$ |
|---|---|---|---|---|---|---|
| Water | 100 | 0.056 | 22055 | 647.069 | 18.153 | 0.3449 |
| Methanol | 64.7 | 0.117 | 8084 | 512.5 | 32.043 | 0.5658 |
| Ethanol | 78.4 | 0.168 | 6120 | 514 | 46.069 | 0.6436 |
| Ethyl acetate | 77.1 | 0.286 | 3880 | 523.3 | 88.105 | 0.3664 |
| 2-Propanol | 82.5 | 0.222 | 4765 | 508.3 | 60.095 | 0.6544 |
| Acetone | 56.5 | 0.209 | 4701 | 508.2 | 58.08 | 0.3065 |

**Table 3.** Important physical properties of pure liquids.

| Physical properties | Water | Methanol | Ethanol | Ethyl acetate | 2-Propanol | Acetone |
|---|---|---|---|---|---|---|
| liquid phase density (kg.m⁻³) | 968.79 | 749.60 | 743.66 | 828.64 | 720.93 | 749.48 |
| surface tension [N/m] | 0.058 | 0.018 | 0.017 | 0.016 | 0.014 | 0.018 |
| specific heat capacity [J/kg.K] | 4219 | 2823 | 3003 | 2133 | 3414 | 2595 |
| vaporization enthalpy (j.kg⁻¹) | 2263717 | 1096692 | 845032 | 365769 | 664324 | 508699 |

**Table 4.** Percentage of average absolute error for pure liquids.

| Model | Acetone | 2-Propanol | Ethyl acetate | Ethanol | Methanol | Water |
|---|---|---|---|---|---|---|
| Kutateladz | 91.3 | 80 | 94.93 | 62.8 | 95.4 | 42.11 |
| McNelly | 60.14 | 42.59 | 75.93 | 20.76 | 30.13 | 18 |
| Mostinski | 13.22 | 21.9 | 18.98 | 14.38 | 15.82 | 15.73 |
| Labantsov | 14.34 | 43.64 | 14.1 | 37.97 | 14.83 | 14 |
| Boyko-Kruzhiline | 19.71 | 21.79 | 71.44 | 15.97 | 13.84 | 18.61 |
| Stephan-Abdelsalam | 42.88 | 34.48 | 48.85 | 11.43 | 64.16 | 14.45 |
| Nishikawa | 35.42 | 14.17 | 36.46 | 18.67 | 38.1 | 79.6 |
| Fujita | 255.9 | 176 | 246 | 174.7 | 225.3 | 110.3 |
| Cooper | 66.37 | 60.68 | 47 | 52 | 93 | 22.15 |
| Gorenflo | 14.33 | 16.92 | 13.78 | 48 | 14.41 | 15.76 |
| Alavi Fazel | 50.18 | 24.44 | 41.41 | 26 | 58 | 35.19 |

functions of available predictive and major correlations with current experimental data. Absolute relative error referenced in the table is calculated by equation (2).

$$A.D.D.\% = \frac{\text{Estimated Value from correlation - Experimental data}}{\text{Experimental data}} \times 100 \quad (2)$$

## 4. New experimental model

In this study, all groups were created without considering probable dimension. All dimensionless groups were extracted using more than 300 experimental data and the database available in this article and employing Buckingham theory. The model is optimized via genetic algorithm. A precise correlation is achieved to link dimensionless groups by the following:

$$Nu = (Pr)^{g0} (Re)^{g1} (p/p_c)^{g2} (\rho_v/\rho_l)^{g2} \quad (3)$$

In this equation, $Nu$, $Pr$, $Re$, $p/p_c$, and $\rho_v/\rho_l$ represent Nusselt dimensionless number, Prandtl dimensionless number, Reynolds dimensionless number, the ratio of atmospheric pressure to critical pressure in [Pa], and vapor density to liquid density in [kg/m$^3$], respectively. Parameters of g0 to g3 are calculated by genetic algorithm such that the boiling heat transfer coefficient error of pure liquids reaches the minimum value (Table

**Table 5.** Parameters of the proposed model

| Genome | g0 | g1 | g2 | g3 |
|---|---|---|---|---|
| Optimum | 0.907 | 0.755 | -0.743 | 0.849 |

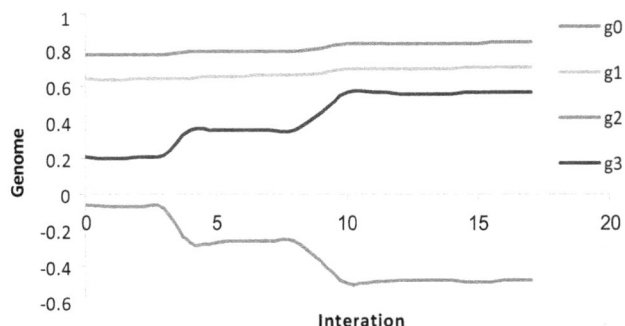

**Fig. 4.** Gene convergence of the genetic algorithm with optimal reply.

5). As seen in Figure 4, the output of the program encoded by the genetic algorithm technique indicates powers as the equation's unknowns.

Figure 5 illustrates the comparison between experimental values and values predicted by the proposed model that represents good overlap of the model with experimental data.

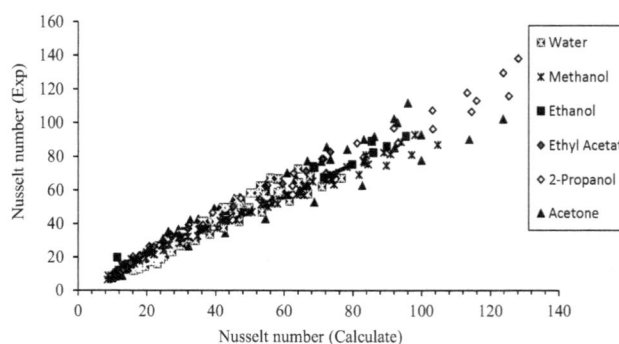

**Fig. 5.** A comparison of the calculated Nusselt's dimensionless number for the experimental results and predicted values.

# 5. Conclusion

Pool boiling heat transfer coefficient for pure liquids, such as water, acetone, isopropanol, methanol and ethanol, were measured experimentally at atmospheric pressure. Correlations of the main prediction for transferring the boiling heat of pure liquids were briefly investigated. The comparison between experimental data and available main correlations imply a significant error. So far, numerous relationships have been proposed to predict heat transfer coefficient in pure liquids. In each of the available models, some parameters are taken into account while others are neglected. Our objective was to find a highly accurate relationship for predicting the boiling heat transfer coefficient of pure liquids, through which this parameter can be related to dimensionless groups.

In this paper, based on dimensional analysis, dimensionless groups were created that can link the new boiling heat transfer coefficient to physical properties of boiling liquids. This new correlation provides more accuracy than other available correlations.

## Nomenclature

| | |
|---|---|
| A | Area, $m^2$ |
| C | Heat capacity, $J.kg^{-1}.°C^{-1}$ |
| $d_b$ | Bubble departing diameter, m |
| $F_P$ | See Gorenflo [7] equation |
| $F_q$ | See Gorenflo [7] equation |
| $F_{WM}$ | See Gorenflo [7] equation |
| $F_{WR}$ | See Gorenflo [7] equation |
| G | Gravitational acceleration, $m^2.s^{-1}$ |
| $\Delta H_{fg}$ | Heat of vaporization, $J.kg^{-1}$ |
| K | Thermal conductivity, $W.m^{-1}.°C^{-1}$ |
| l* | See Boyko-Kruzhilin [4] equation |
| n | See Gorenflo [7] equation |
| N | Number of components |
| Nu | Nusselt number |
| Pr | Prandtl number |
| Re | Reynolds number |
| P | Pressure, Pa |
| q | Heat, W |
| Ra | Roughness, m |
| s | Distance, m |
| T | Temperature, K |
| $\Delta$ | Difference |
| $\alpha$ | Heat transfer coefficient, $W.m^{-2}.°C^{-1}$ |
| $\hat{\alpha}$ | Thermal diffusion, $m^2.s^{-1}$ |
| $\rho$ | Density, $kg.m^{-3}$ |
| $\sigma$ | Surface tension, $Dy/cm$ or $N.m^{-1}$ |

*Subscripts*

| | |
|---|---|
| b | Bulk |
| c | Critical |
| i | Component |
| id | Ideal |
| l | Liquid |
| o | Reference |
| r | Reduced |
| s | Saturated or Surface |
| th | Thermocouples |
| v | Vapor |

## References

[1] S. Kutateladze, Heat Transfer and Hydrodynamic Resistance, Energoatomizdat Publishing HOUSE, (1990).

[2] M.J. McNelly, A Correlation of rates of heat transfer to nucleate boiling of liquids, J. Imperial College Chem. Eng. Soc. 7 (1953) 18–34.

[3] I.L. Mostinski, Application of the rule of corresponding states for calculation of heat transfer and critical heat flux, Teploenergetika (Therm. Eng.+) 4 (1963) 66.

[4] Boyko-Kruzhilin, Perry's Chemical Engineering Handbook, 7th ed., (1997) pp. 419.

[5] D.A. Labantsov, Mechanism of vapor bubble growth in boiling under on the heating surface, J. Eng. Phys. 6 (1963) 33-39.

[6] K. Stephan, K. Abdelsalam, Heat transfer correlation for natural convection boiling, Int. J. Heat Mass Tran. 23 (1980) 73-87.

[7] D. Gorenflo, Pool boiling, VDI Heat Atlas, 1st English ed., Springer, (1993) pp. 776 -926.

[8] S.A. Alavi Fazel, R. Roumana, Pool Boiling Heat Transfer to Pure Liquids, International Conference on Continuum Mechanics Fluids Heat, WSEAS Mech. Eng. Se. (2010) 211-216.

[9] M.M. Sarafraz, Experimental Investigation on Pool Boiling Heat Transfer to Formic Acid, Propanol and 2-Butanol Pure Liquids under the Atmospheric Pressure, J. Appl. Fluid Mech. 6 (2013) 73-79.

[10] W.M. Rohsenow, A method of correlating heat transfer data for surface boiling of liquids, ASME J.

Heat Trans. 74 (1952) 969-976.

[11] K. Nishikawa, Effect of the surface roughness on the Nucleate Boiling Heat Transfer on the Wide Range of Pressure, Proceedings of the 7th International Heat Transfer Conference, Germany, 4 (1982) pp.1-6.

[12] K. Nishikawa, Y. Fujita, H. Ohta, S. Hitaka, Effects of system pressure and surface roughness on nucleate boiling heat transfer, Memoirs of the faculty of engineering, Kyushu University, 42 (1982) 95-123.

[13] M.G. Cooper, Saturation nucleate pool boiling-a simple correlation, Proc. Int. Chem. Eng. Symposium (1984) 786-793.

[14] S.A. Alavi Fazel, A.A. Safekordi, M. Jamialahmadi, Pool boiling heat transfer coefficient in water-amines solutions, Int. J. Eng. Trans. A 21 (2008) 113-130.

# 3

# Gamma irradiation induced surface modification of silk fabrics for antibacterial application

Sahar S. El Sayed, Amal A. El-Naggar*, Sayeda M. Ibrahim

*Department of Radiation Chemistry, National Center for Radiation Research and Technology, P. N. 13759, Cairo, Egypt*

HIGHLIGHTS

- Fabrics were carried out by coating with silver nanoparticles (AgNPs) stabilized with polyvinylpyrrolidone (PVP) through γ-irradiation.

- The AgNPs-coated silk fabrics demonstrated an excellent antibacterial activity against the tested bacteria, *Escherichia coli* and *Staphylococcus aureus*.

- This work offers potentials to produce specific AgNPs-coated antimicrobial silk for various applications in the textile industry.

GRAPHICAL ABSTRACT

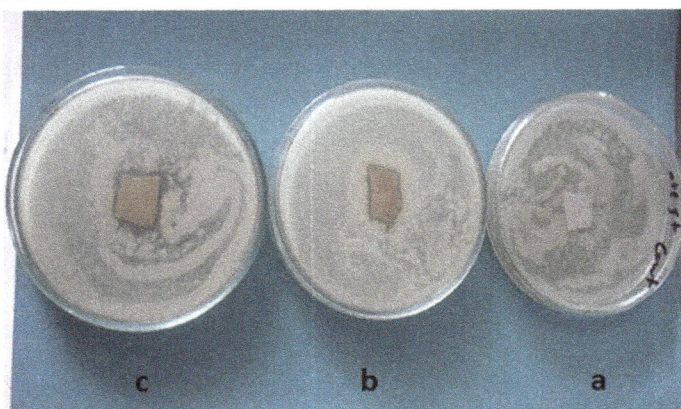

ARTICLE INFO

*Keywords:*
Silk
Silver nanoparticle
Antibacterial activity
Surface modification
γ-Irradiation

ABSTRACT

Silk fabrics were modified by a treatment of silver nitrate solution ($AgNO_3$) and polyvinylpyrrolidone (PVP) as a stabilizer then exposure to γ-irradiation to create antibacterial properties. Effects of the absorbed dose on treated fabrics were investigated. The scanning electron microscopy (SEM) and X-ray diffraction (XRD) patterns were used to confirm the presence of silver nanoparticles (AgNPs) on the fabric. The treated fabrics should have enhanced thermal stability due to the presence of AgNPs. The treated silk fabric was examined for its antibacterial activity toward various types of bacteria. The AgNPs-treated silk fabrics demonstrated excellent antibacterial activity against the tested bacteria, *Escherichia coli* and *Staphylococcus aureus*. This work opens the door for production of specific AgNPs-silk as a type of textile in the antibacterial domain.

* Corresponding author.  ; E-mail address: amalelnaggar@yahoo.com

## 1. Introduction

As a natural protein fiber, silk possesses a structure very similar to human skin with smooth, breathable, soft, non-itching and antistatic characteristics [1]; however, bacteria can easily adhere and grow on silk fabric causing deformation and degradation [2]. So, it is necessary to produce a silk fiber having antimicrobial activity which can be used in a wide range of applications. The antimicrobial finishing can be used for fabrics and clothes used in hospital and crowded public areas or textiles that are left wet for a long period of time between processing steps. Finally, the use of this finishing in intimate apparel, underwear and socks can prevent unpleasant odors [3].

As advances in nanotechnology have been made, numerous nanomaterials with various structures and unique properties have been fabricated [4-9]. Researchers have recently become interesting in the immobilization of fibers with nanomaterials to produce multifunctional textiles. Since the silver nanoparticle (AgNP) is a typical nanomaterial with broad-spectrum antibacterial effects on both Gram-negative and Gram-positive bacteria [10,11], it has been utilized to treat silk fibers for antimicrobial properties [12-15]. The antimicrobial properties of silver particles have been exploited for a long time in biomedical textiles as its broad-spectrum action is particularly significant in preventing polymicrobial colonization associated with hospital-acquired infections [16,17]. Silver has revealed bactericidal activity against a wide range of Gram-positive and Gram-negative bacteria, namely *Pseudomonas aeruginosa*, *S. aureus*, *Staphylococcus epidermidis*, *E. coli* and *Klebsiella pneumonia* [18]. The AgNPs interact with the bacterial membrane and are able to penetrate inside the cell. The mechanisms of interaction involve AgNPs attaching to bacterial cell membranes, which increase permeability and disturb respiration. Another advantage of AgNPs is that they are easily embedded into the fibers' polymeric matrices [19,20].

Various polymers, such as polyamide, polyvinyl-pyrrolidone, and polyacrylic acid, were employed to functionalize the silk surface [21-23] to produce a hydrophilic character to increase the adsorption amount of $Ag^+$, which subsequently reduced to AgNPs [11,24].

Antibacterial fabrics can be used to make bandages, gauze, bed sheets, and surgical clothes [25,26].

Mohammad Mirjalili *et al.* imparted an antimicrobial finishing to cellulose fabric using a nano silver solution [27]. The surface characteristics of these fabrics have been studied by scanning electron microscopy (SEM) which indicated silver nanoparticles were well dispersed on the fabric surface. An antibacterial test, using Gram-negative bacteria (*Escherichia coli*), was used to estimate the biological activity of the treated fabrics.

Wasif A. I. and Laga S. K. studied antimicrobial finishing on cotton woven fabric using a nano silver solution as an antimicrobial agent against Gram positive (*Staphylococcus aureus*) and Gram-negative bacteria (*Escherichia coli*), at various concentrations in the presence of polyvinylalcohol (PVOH) [28]. They found that the higher the concentration of antimicrobial agent the larger the zone of inhibition in the cases of both Gram-positive and Gram-negative bacteria. Various properties like tensile strength, bending length, crease recovery angle, and zone of inhibition were also studied.

In this study, we prepared silk fabrics treated with AgNPs via γ-ray irradiation. SEM, XRD, and TGA were used to confirm the properties of these treated fabrics. The antimicrobial finish was applied to make silk fabrics more suitable for industrial applications.

## 2. Experimental

### 2.1. Materials

Silver nitrate, crystal, and ACS, M.W. 169.87, was obtained from GAMMA, Laboratory Chemicals. Polyvinylpyrolidone (PVP), M.W. 40000, was supplied from Universal Fine Chemicals PVT. LTD, India. All the materials were used without further purification. Silk fabric was kindly supplied by Al Khateeb Company, Akhmen, Egypt.

### 2.2. Treatment of silk fabric

Approximately 60 g of washed silk fabrics were irradiated in 500 ml of 3 mM $AgNO_3$ solution using the stabilizer of 1.0% Polyvinylpyrrolidone (PVP) in the dose range from 5 to 20 kGy of gamma irradiation at the National Center for Radiation Research and Technology, Cairo, Egypt.

Afterwards, samples were rinsed with water and dried at 40°C for 40 min. A similar procedure was applied for 40 repeated washes.

## 3. Characterizations

Samples (silk treated with AgNPs) were digested in conc. Nitric acid 69% (Anular) and $H_2O_2$ in a 5:1 ratio using a Microwave Digester Instrument, Milestone 1200 Miga, Italy [29]. The Ag element was estimated using an Atomic Absorption Spectrometer, Thermo Scientific E3000 series, England. Tensile (tensile strength and % elongation-at-break) properties were determined using a Mecmesin, (Model 10-1) UK, equipped with software and a crosshead speed of 50 mm/min. All mechanical parameters were directly calculated. The samples for tensile measurements were dumbbell shaped with a width of 4 mm and length of 50 mm. The surface morphology of the modified silk fabrics in comparison with unmodified silk fabrics was examined by SEM. The micrographs were taken with a JSM-5400 instrument manufactured by Joel, Japan. An XRD experiment of the samples was performed at room temperature using a Philips PW 1390 diffractometer (30 kV, 10 mA) with copper target irradiation at a scanning rate of 80/min in a $2\Theta$ range of 400-900°C. Thermal characteristics of AgNPs treated silk fabrics were determined from the thermogravimetric analysis (TGA) data, using a TGA-30 apparatus (at Shimadzu, Kyoto, Japan), at a heating rate of 10°C/min. in air, over a temperature range from room temperature to 600°C.

## 4. Antibacterial activity testing

A disk diffusion test, according to the Kirby Baur method [30], was applied to identify the bacterial effect through the measurements of bacterial broth. A known quantity of bacteria was grown overnight on agar (solid growth media) plates. The samples were placed in Petri-dishes and then incubated for 24 hr at 30-32°C. The inhibition zone was then measured in cm from one side of the square sample. Both positive and negative bacteria, *Escherichia Coli* and *Staphylococcus aurous*, were tested.

## 5. Results and discussions

### 5.1. Measurement of silver content on the treated silk fabrics

The amount of silver content on the treated silk fabrics was calculated. Figure 1 shows the relationship between the absorbed dose and silver content. The results obtained from this figure indicate that the content of AgNPs on the silk fabrics increases as the absorbed dose increases up to 10 kGy then tends to level off. A maximal value of the AgNPs content on fabrics was found to be 177 ppm for the silk fabrics in a 3 mM $AgNO_3$ solution irradiated with an absorbed dose of 10 kGy.

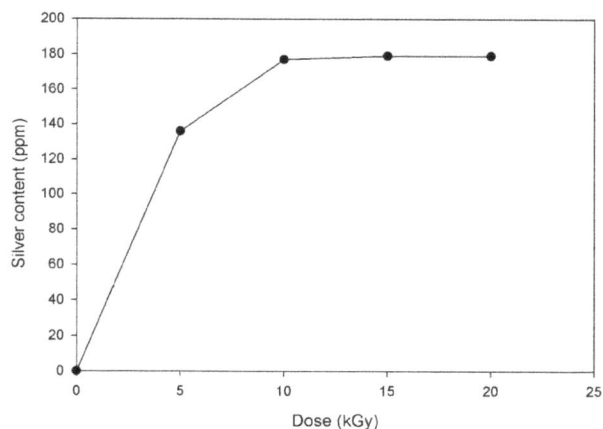

**Fig. 1.** Relationship between the content of AgNPs on the silk fabrics and irradiation dose.

The formation of $Ag^+$ ions was a result of silk fabrics treated with the $AgNO_3$ solution. These $Ag^+$ ions were reduced to Ag atoms and simultaneously deposited on the silk fabrics. The interaction between fibers and metallic AgNPs stabilized by PVP polymer caused the formation of a chemical bond between the silver and alcoholic groups of silk and resulted in the physical adsorption of AgNPs on the fabric surface [31].

### 5.2. Mechanical properties

Mechanical properties of the treated silk fabrics in terms of tensile strength (TS) and elongation (E,%) in the dose range from 0 to 20 kGy were plotted in Figures 2 and 3. The results show that tensile strength and elongation at break (%) of the AgNPs treated silk fabrics are slightly affected at the doses (5-20 kGy) in comparison with the control silk fabrics. The results showed that tensile strength (TS) of the AgNPs treated silk fabrics irradiated with 5 kGy exhibit slightly lower tensile strength than the control silk fabric as shown in Figure 2. Whereas, the strain value of the treated fabric at the same dose was more than that of the untreated fabric as shown in Figure 3. The results obtained from this figure indicate that the treatment process did not alter the mechanical properties of the silk fabric.

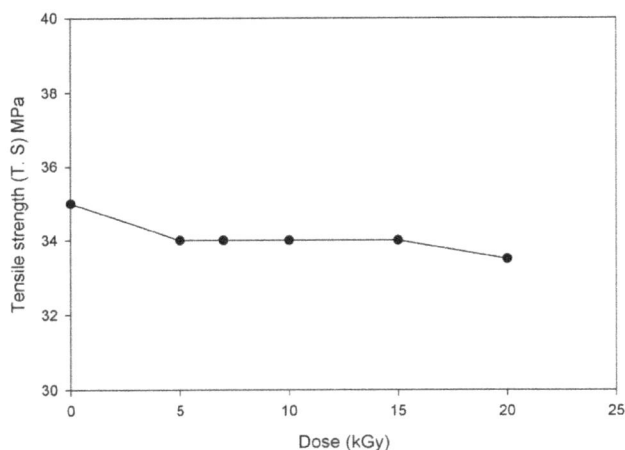

**Fig. 2.** Effect of different irradiation dose on the tensile strength at break of untreated silk fabric and AgNPs-treated ones with different gamma irradiation doses.

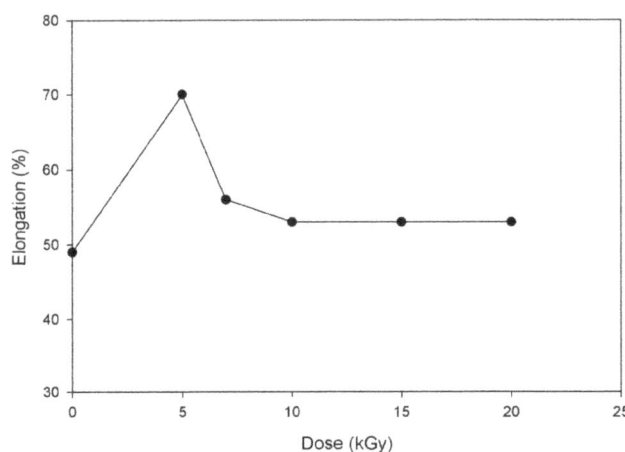

**Fig. 3.** Effect of different irradiation dose on the elongation (%) at break of untreated silk fabric and AgNPs-treated silk fabrics with different gamma irradiation doses.

## 5.3. Scanning electron microscopy (SEM) of untreated and treated silk fabrics

The surface morphology of untreated and treated silk fabrics irradiated to different doses (5 and 10 kGy) was investigated by SEM (Figure 4 a, b, and c). (a) shows the surface morphology of untreated fabric, which appeared smooth and homogeneous. Figure 4 (b, and c) illustrates the surface morphology of silk fabric treated with silver NPs, which appeared rough and full of particle aggregates due to the imbedding of AgNPs. It was noted that after irradiation to 5 kGy in AgNO₃ solution, some of the AgNPs were agglomerated and therefore caused the formation of large nanoparticles dispersed on the fabric surface. As the irradiation dose increased from 5 kGy to 10 kGy, the aggregation of the AgNPs increases [32]. To limit the side effects of AgNPs aggregation, a

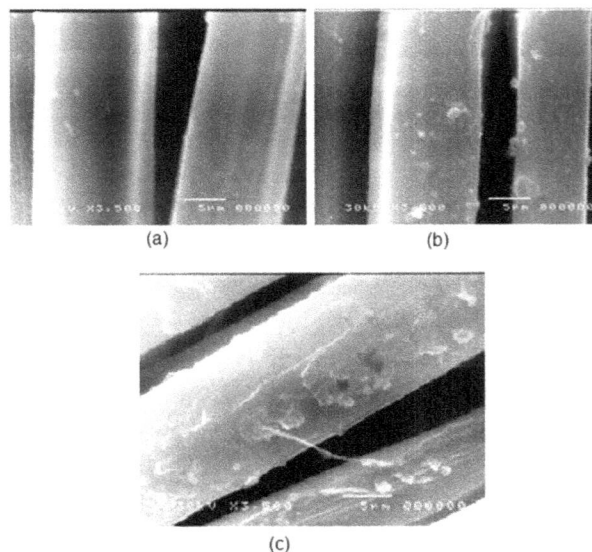

**Fig. 4.** SEM of untreated and AgNPs-treated silk fabrics with different gamma irradiation doses, a) Original silk fabric, b) 5 kGy, c) 10 kGy.

low irradiation dose with efficient AgNPs production is preferred for obtaining homogeneously dispersed AgNPs on the surface.

## 5.4. XRD of untreated and treated silk fabrics

XRD patterns of untreated and treated silk fabrics with AgNPs are shown in Figure 5 (a, b and c). The silk fabrics had a crystalline peak located at $2\Theta=20.28°$ [33]. Figure (5a) shows there is no significant change in this peak after the treatment; this suggests that the gamma irradiation and AgNPs attachment do not alter the structure of the silk fibers. (Figure 5b and c) shows four new peaks at $2\Theta$ values of 38.4°C, 44.6°C, 64.7°C and 77.5°C were detected, which were attributed to the diffraction peaks of the (111), (200), (220) and (311) planes of silver NPs, respectively [34]. The XRD pattern clearly indicates that AgNPs were deposited directly on the silk surface and the silk fabrics treated with AgNPs were successfully prepared using silver nitrate by radiation.

## 5.5. Thermogravimetric analysis (TGA) of untreated and treated silk fabrics

Thermal behavior of untreated and treated silk fabrics in a temperature range from room temperature to 600°C at all gamma irradiation doses (5-20 kGy) are shown in Figure 6. It can be seen that the thermal decomposition

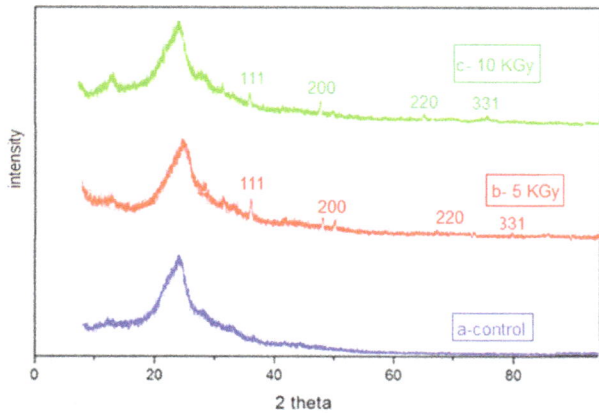

**Fig. 5.** XRD patterns of untreated (a) and AgNPs-treated silk fabric with different gamma irradiation doses (b and c).

of both the untreated and treated silk fabrics passes through three stages upon heating from room temperature to 600°C. All the curves show an initial stage of weight loss below 100°C, which can be ascribed to the elimination of adsorbed water from the fabrics. In the second stage at temperatures from 100-290°C, all samples again show weight loss, which may be attributed to the degradation of side chain groups of silk fibroin proteins [35]. The third stage of weight loss from 300 to 600°C may be caused by the breakdown of the main chains of silk fibers [35]. Moreover, it was found that thermal stability increases as the irradiation dose increases up to 10 kGy. For example, at a constant temperature of 350°C weight loss % is 30% and 19% at an irradiation dose of 5 and 10 kGy, respectively. Since carboxyl groups on the side chains of silk proteins are expected to be the binding sites of AgNPs, its attachment may possibly protect the side chains from thermal degradation [36-38].

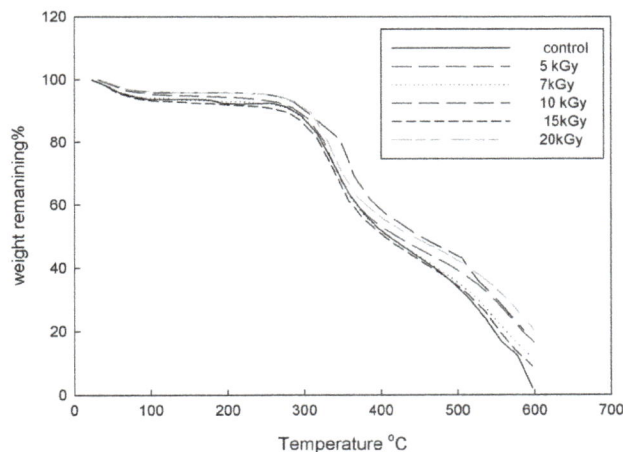

**Fig. 6.** TGA curves of untreated and AgNPs-treated silk fabric with different gamma irradiation doses.

## 5.6. Antibacterial activity testing

Figures 7 and 8 show the antibacterial effects of untreated and Ag NPs-treated silk fabrics (5 and 10 kGy) on *E.coli* (A) Gram-negative and *S. aureus* (B) Gram-positive bacteria. *E. coli* and *S. aureuswere* selected as model bacteria to explore the antimicrobial effects of the modified silk fabric in the zone of inhibition test since they are typical Gram-negative and Gram-positive bacteria, respectively. In this study [39], no clear zone occurs on either the *E. coli* and *S. aureus* agar plates for the untreated silk fabric. In contrast, all AgNPs-treated fabrics have obvious inhibition zones. The results obtained from these figures signified the following features: (a) the untreated silk fabric (control) exhibited no inhibition zone indicating no antibacterial activity; (b) treatment of a constant concentration of silver nitrate (3 mM) irradiated to 5 and 10 kGy rendered the fabric antibacterial irrespective of the bacteria used, (c) resistance of the modified fabric to *E.coli* (A) Gram-negative bacteria is greater than that for the *S. aureus* (B) Gram-positive bacteria as evidenced by comparing the inhibition zone for all used bacteria. These results strongly prove that AgNPs-treated silk fabrics possess the capability to kill both Gram-negative and Gram-positive bacteria.

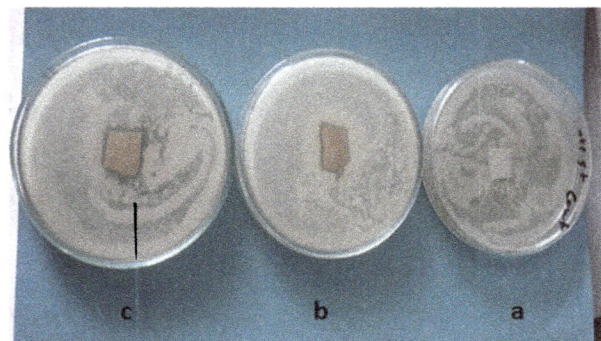

**Fig. 7.** Antibacterial effects of the untreated and AgNPs-treated silk fabrics on *E.coli* (A) Gram-negative bacteria.

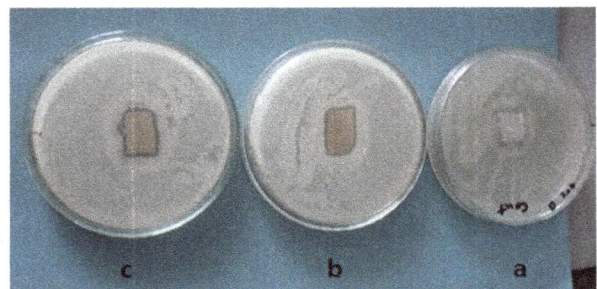

**Fig. 8.** Antibacterial effects of the untreated and AgNPs-treated silk fabrics on *S. aureus* (B) Gram-positive bacteria.

## 6. Conclusions

The preparation of the treated fabric depends on the gamma irradiation doses in the presence of PVP polymer and silver ions. Gamma irradiation doses cause significant aggregation of AgNPs on the fabric. A good distribution of AgNPs on the silk fabric surface, exhibited in SEM images, proves the existing of AgNPs on the silk fabrics. The XRD data reveal that the synthesized treated fabric possess good crystalline structures. The presence of AgNPs greatly changed the pattern of the TGA curve, and may be due to the attached AgNPs protecting the silk fibers from degradation. The residue weight observed in the TGA experiment further proves the successful synthesis of AgNPs on the fiber surface. The antimicrobial activity including bacterial growth inhibition and bactericidal effects of the AgNPs-coated silk is demonstrated in the zone of inhibition. Based on the data, it can be concluded that AgNPs-coated silk, which possesses excellent antibacterial activity, is directly fabricated with a gamma irradiated-assisted in situ synthesis approach using degummed silk fibers and a $AgNO_3$ solution as raw materials. This work offers the potential to produce treated antimicrobial silk for applications including clothing and industrial textiles.

## Acknowledgements

We would like to sincerely thank the Department of Radiation Chemistry National Center for Radiation Research and Technology. Authors also gratefully acknowledge the Atomic Energy Authority Cairo, Egypt for its support of this work and for the cooperation of its staff working in the radiation source.

## References

[1] D.M. Phillips, L.F. Drummy, D.G. Conrady, D.M. Fox, R.R. Naik, M.O. Stone, P.C. Trulove, H.C. De Long, R.A. Mantz, Dissolution and regeneration Bombyx mori Silk fibroin using ionic liquids, J. Am. Chem. Soc. 126 (2004) 14350-14351.

[2] H.J. Jin, J. Park, R. Cebe, P. Valluzzi, D.L. Kaplan, Biomaterial films of Bom-byx mori silk fibroin with poly(ethylene oxide), Biomacromolecules 5 (2004) 711-717.

[3] G. Arai, G.M. Colonna, E. Scotti, A. Boschi, R. Murakami, M.T. Tsukada, Absorption of metal cations by modified *B. mori* silk and preparation of fabrics with antimicrobial activity, J. Appl. Polym. Sci. 80 (2001) 297-303.

[4] V. Scognamiglio, Nanotechnology in glucose monitoring: advances and challenges in the last 10 years, Biosens. Bioelectron. 47 (2013) 12-25.

[5] Z.S. Lu, C.X. Guo, H.B. Yang, Y. Qiao, J. Guo, C.M. Li, One-step aqueous synthesis of graphene-CdTe quantum dot-composed nanosheet and its enhanced photoresponses, J. Colloid Interf. Sci. 353 (2011) 588-592.

[6] Z.S. Lu, W.H. Hu, H.F. Bao, Y. Qiao, C.M. Li, Interaction mechanisms of CdTe quantum dots with proteins possessing different isoelectric points, Med. Chem. Commun. 2 (2011) 283286.

[7] Z.S. Lu, C.M. Li, Quantum dot-based nanocomposites for biomedical applications, Curr. Med. Chem. 18 (2011) 3516-3528.

[8] Z.S. Lu, C.M. Li, H.F. Bao, Y. Qiao, Q.L. Bao, Photophysical mechanism for quantum dots- induced bacterial growth inhibition, J. Nanosci. Nanotechnol. 9 (2009) 3252-3255.

[9] Z.S. Lu, C.M. Li, H.F. Bao, Y. Qiao, Y. Toh, X. Yang, Mechanism of antimicrobial activity of CdTe quantum dots, Langmuir 24 (2008) 5445-5452.

[10] E. Amato, Y.A. Diaz-Fernandez, A. Taglietti, P. Pallavicini, L. Pasotti, L. Cucca, C. Milanese, P. Grisoli, C. Dacarro, J.M. Fernandez-Hechavarria, Synthesis, characterization and antibacterial activity against Gram positive and Gram negative bacteria of biomimetically coated silver nanoparticles, Langmuir 27 (2011) 9165-9173.

[11] L.Y. Guo, W.Y. Yuan, S. Lu, C.M. Li, Polymer/ nanosilver composite coatings for antibacterial applications, Colloid Surface A 439 (2013) 69-83.

[12] W.D. Yu, T. Kuzuya, S. Hirai, Y. Tamada, K. Sawada, T. Iwasa, Preparation of Ag nanoparticle dispersed silk fibroin compact, Appl. Surf. Sci. 262 (2012) 212-217.

[13] L. He, S.Y. Gao, H. Wu, X.P. Liao, Q. He, B. Shi, Antibacterial activity of silver nanoparticles stabilized on tannin grafted collagen fiber, Mater. Sci. Eng. C 32 (2012) 1050-1056.

[14] J.J. Wu, G.J. Lee, Y.S. Chen, T.L. Hu, The synthesis of nano silver/polypropylene plastics for antibacterial application, Curr. Appl. Phys. 12 (2012) S89-S95.

[15] R. Bhattacharya, P. Mukherjee, Biological properties of "naked" metal nanoparticles, Adv. Drug

Deliver. Rev. 60 (2008) 1289-1306.

[16] S. Shahidi and J. Wiener, Antimicrobial Agents-Chapter 19: Antibacterial Agents in Textile Industry; InTech: Rijeka, Crotia, 2012, pp. 387-406.

[17] Y. Gao, R. Cranston, Recent advances in antimicrobial treatments of textiles, Text. Res. J. 78 (2008) 60-72.

[18] J. Hasan, R.J. Crawford, E.P. Ivanova, Antibacterial surfaces: The quest for a new generation of biomaterials, Trends Biotechnol. 31 (2013) 295-304.

[19] B. Simoncic, B. Tomsic, Structures of novel antimicrobial agents for textiles-A review, Text. Res. J. 80 (2010) 1721-1737.

[20] H. Palza, Antimicrobial polymers with metal nanoparticles, Int. J. Mol. Sci. 16 (2015) 2099-2116.

[21] D. Zhang, G.W. Toh, H. Lin, Y.Y. Chen, In situ synthesis of silver nanoparticles on silk fabric with PNP for antibacterial finishing, J. Mater. Sci. 47 (2012) 5721-5728.

[22] S. Tangbunsuk, G.R. Whittell, M.G. Ryadnov, G.W.M. Vandermeulen, D.N. Woolfson, I. Manners, Metallopolymer-peptide hybrid materials: Synthesis and Self-Assembly of Functional, Poly ferrocenylsilane-Tetrapeptide Conjugates, Chem.-Eur. J. 18 (2012) 2524-2535.

[23] X.M. Wang, W.R. Gao, S.P. Xu, W.Q. Xu, Luminescent fibers: in situ synthesis of silver nanoclusters on silk via ultraviolet light-induced reduction and their anti-bacterial activity, Chem. Eng. J. 210 (2012) 585-589.

[24] A.R. Abbasi, A. Morsali, Influence of various reduction reagents on the morphological properties of Ag nanoparticles@silk fiber prepared using sonochemical method, J. Inorg. Organomet. P. 21 (2011) 369-375.

[25] S.T. Dubas, P. Kimlangdudsana, P. Potiyaraj, Layer-by-layer deposition of antimicrobial silver nanoparticles on textile fibers, Colloid. Surface. A 289 (2006) 105-109.

[26] P. Gupta, M. Bajpai, S.K. Bajpai, Investigation of antibacterial properties of silver nanoparticle-loaded poly (acrylamide-co-itaconic acid)-grafted cotton fabric, J. Cotton Sci. 12 (2008) 280-286.

[27] M. Mirjalili, N. Yaghmaeil, M. Mirjalili, Antibacterial properties of nano silver finish cellulose fabric, J. Nanostruct. Chem. 3 (2013) 43.

[28] A.I. Wasif, S.K. Laga, Use of nano silver as an antimicrobial agent for cotton, AUTEX Res. J. 9 (2009) 5-13.

[29] IAEA: Elemental analysis of biological materials. International Atomic Energy Agency (IAEA), Veinna, Technical Reports Series No. 197 (1980) 379.

[30] Clinical and Laboratory Standards Institute, Performance Standards for Antimicrobial Disk Susceptibility Tests; Approved Standard-Ninth Edition, Clinical and Laboratory Standards Institute document M2-A9 (ISBN 1-56238-586-0), 940 West Valley Road, Suite 1400, Wayne, Pennsylvania 19087-1898 USA, 2006.

[31] I. Perelshtein, G. Applerot, N. Perkas, Sonochemical coating of silver nanoparticles on textile fabrics (nylon, polyester and cotton) and their antibacterial activity and their antibacterial activity, Nanotechnology 19 (2008) 245705.

[32] B. Liu, W.Z. Chen, S.W. Jin, Synthesis, structural characterization, and luminescence of new silver aggregates containing short Ag-Ag contacts stabilized by functionalized bis (N-heterocyclic carbene) ligands, Organometallics 26 (2007) 3660-3667.

[33] Q. Lu, X. Hu, X.Q. Wang, J.A. Kluge, S.Z. Lu, P. Cebe, D.L. Kaplan, Water-insoluble silk films with silk I structure, Acta Biomater., 6 (2010) 1380-1387.

[34] X. Zou, E. Ying, S. Dong, Preparation of novel silver gold bimetallic nanostructures by seeding with silver nanoplates and application in surface enhanced Raman scattering, J. Colloid Interf. Sci. 306 (2007) 307-315.

[35] X.X. Feng, L.L. Zhang, J.Y. Chen, Y.H. Guo, H.P. Zhang, C.I. Jia, Preparation and characterization of novel nanocomposite films formed from silk fibroin and nano-$TiO_2$, Int. J. Biol. Macromol. 40 (2007) 105-111.

[36] L. Piao, K.H. Lee, B.K. Min, W. Kim, Y.R. Do, S. Yoon, A facile synthetic method of silver nanoparticles with a continuous size range from sub-10 nm to 40 nm, Bull. Korean Chem. Soc. 32 (2011) 117-121.

[37] F. Chen, Y. Liu, R.E. Wasylishen, Z.H. Kuznicki, Solid-state NMR and TGA studies of silver reduction in chabazite, J. Nanosci. Nanotechno. 12 (2012) 1988-1993.

[38] M.A.M. Khan, S. Kumar, M. Ahamed, S.A. Alrokayan, M.S. AlSalhi, Structural and thermal studies of silver nanoparticles and electrical transport study of their thin film, Nanoscale Res. Lett. 6 (2011) 434.

[39] S.A. Khan, A. Ahmad, M.I. Khan, M. Yusuf, M. Shahid, N. Manzoor, F. Mohammad, Antimicrobial activity of wool yarn dyed with *Rheum emodi* L. (Indian Rhubarb), Dyes Pigments 95 (2012) 206-214.

# 4

# Drying of calcium carbonate in a batch spouted bed dryer: Optimization and kinetics modeling

Sadegh Beigi[1], Mohammad Amin Sobati[*,1], Amir Charkhi[2]

[1] School of Chemical Engineering, Iran University of Science and Technology (IUST), Tehran, Iran

[2] Material and Nuclear Fuel research school, Nuclear Science and Technology Research Institute, Tehran, Iran

## HIGHLIGHTS

- A new batch spouted bed dryer was investigated for drying calcium carbonate.

- A new criterion has been introduced to measure the effective efficiency of the drying process.

- The drying kinetics have been modelled using semi-theoretical approaches.

## GRAPHICAL ABSTRACT

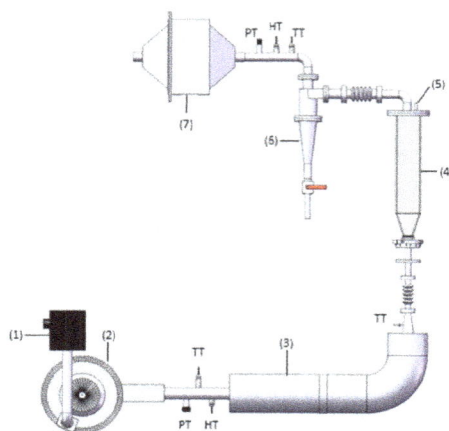

## ARTICLE INFO

*Keywords:*
Spouted bed dryer
Drying kinetics
Taguchi method
Drying effective efficiency
Modeling

## ABSTRACT

In the present work, the drying of calcium carbonate in a batch spouted bed dryer with inert particles has been investigated experimentally. The effect of several operating parameters including air temperature (90, 100, and 110 ºC), air velocity ($U_{ms}$, 1.2 $U_{ms}$, and 1.5 $U_{ms}$), and dry solid mass (5, 10, 20 g) has been studied. The Taguchi method has been applied to determine the optimal parameters and also to reduce the number of required experimental runs. It has been found that the dryer performance was affected by all parameters. It has also been found that drying with 5 g dry solid at a temperature of 100 ºC and a velocity of 1.2 $U_{ms}$ leads to maximum drying efficiency. Additionally, the effect of air inlet velocity and temperature on the drying kinetics of calcium carbonate has been investigated. Several semi-theoretical models with temperature and velocity dependent parameters have been selected to estimate the drying kinetics. The performance of all fitted models was acceptable but the logarithmic model was the best model in terms of the statistical analysis.

* Corresponding author:  ; E-mail address: sobati@iust.ac.ir

# 1. Introduction

Calcium carbonate is applied in different industries such as chemical [1], pharmaceutical [2,3], food [4], plastics [5,6], paper [7], paints [8], ceramic materials [9,10], etc. Hence, the demand for calcium carbonate has been growing quickly in recent years.

Calcium carbonate is produced in powders, granules and slurries form by two general methods: (1) ground calcium carbonate, commonly referred to as GCC, which is produced by intensive milling or grinding of natural calcium carbonate and (2) precipitated calcium carbonate (PCC) which is produced by the reaction of aqueous calcium hydroxide suspension with carbon dioxide. Generally, the product should be conventionally dewatered and dried [11-13]. For instance, PCC has four processing steps including (1) calcination, (2) lime slaking, (3) carbonation, and (4) drying [14]. Accordingly, drying is required in order to prepare the final calcium carbonate for market. Also, nanoparticles of calcium carbonate are required in some of the above-mentioned applications [6,15].

Different dryers, such as spray dryers, fluidized bed dryers, and spouted bed dryers, have been proposed for drying slurries and pastes. Spouted bed dryers with inert particles are a relatively new technology with a number of advantages such as high drying rates, uniform temperature distribution, uniform size particle production, and low drying cost for solution, slurries, and pasty materials [16-20].

Gishler and Mathur applied a spouted bed dryer in 1954 for drying wheat in Canada [21]. The spouted bed is a gas-solid contactor in which the fluid jet is introduced through a nozzle located at the center of the conical bottom of the bed. Intensive interaction between the solid particles and gas occurs in this dryer. Three regions are distinguished for gas-solids flow in a spouted bed: (1) a spout zone located in the center of the bed (2) an annulus zone located between the spout and the column wall and (3) a fountain zone located above the spout [22]. The particle flow with respect to the gas flow is co-current and counter-current in the spout and annulus zones, respectively.

Regardless of the hydrodynamic configuration of the dryer, the principle behind this technology is based on the drying of a thin layer of slurry that coats the surface of the inert particles. Depending on the type of dryer, these particles can be vibrated, fluidized, or spouted either by a hot air or combination of hot air and a mechanical device installed inside the dryer, such as an agitator or a conveyor screw [23].

In the spouted bed dryer with inert particles, the slurry is directly sprayed onto the bed and is deposited as a thin-film layer on the surface of the inert particles. The bed of particles is spouted with the hot gas causing the film on the particles' surface to dry and become fragile due to the large surface area of inert particles and intensive spouting. Then, particle-particle and particle-wall collisions lead to the conversion of the dried film to a dried powder. In the final step, the dried powders are entrained by the exit stream and are recovered using suitable solid separation equipment. It should be noted that the required energy for water evaporation is supplied via (1) a direct method through the hot air and (2) an indirect method through the inert particles heated by the hot spouting air [24]. The bed temperature is approximately uniform due to rapid circulation. The main operational parameters influencing the drying rate are the inlet air flow rate and temperature which in turn control the heat transfer and solids circulation rates [25].

Kudra and Mujumdar [26] have reviewed the application of the inert particles in different dryers. Schneider and Bridgwater [27] have investigated the drying of inorganic suspensions in a spouted bed using different inert particles. The effect of liquid injection on the spouting velocity, stability of the spouting regime, fountain height, and bed pressure drop was examined in their studies. The drying of a micro-particle slurry and a salt-water solution using a powder-particle spouted bed have been investigated by Guo et al. [28]. According to their results, the agglomeration of particles was observed when the drying efficiency was above 60%. Passos et al. [25] have investigated the drying performance for pastes in a conical spouted bed. Their study was focused on the investigation of the effect of column dimensions, fluid flow characteristics, and paste properties on the drying performance. Based on their study, a criterion is provided for the design of conical spouted bed dryers used for suspensions drying.

Nakazato et al. [29] have applied a powder-particle spouted bed to the production of fine calcium carbonate with a mean particle size around 1 μm. They have investigated the effect of operating conditions, such as superficial gas velocity and static bed height, on the particle size of the product (i.e., $CaCO_3$ powder). Benali and Amazouz [30] have proposed a jet spouted bed dryer to substantially reduce the stickiness of meat-rendering slurry (MRS). They examined the effects of calcium carbonate as drying-aid agents on the reduction of slurry stickiness. Arsenijevic et al. [31,32] have studied the drying of calcium carbonate and calcium stearate in a draft tube spouted bed and investigated the effects of the operating conditions on the dryer throughput and product quality. Moreover, they proposed and verified a model for predicting the particle circulation rate. Almeida et al. [33] have experimentally analyzed the fluid dynamic, thermal and mass transfer behavior of pasty materials, such as calcium carbonate, sewage sludge, and skimmed milk, during transient drying in spouted beds.

Calcium carbonate drying is influenced by parameters such as hot inlet air velocity, inlet drying temperature, dry solid mass, and so on. These parameters do not have the same effect on the drying process and their simultaneous impacts are very complex and debatable. Therefore, the Taguchi method has been used to determine the influence of each of these parameters and quantify their influence on the drying efficiency [34,35]. Taguchi techniques have been widely used in engineering design for the optimization and identification of critical parameters [36,37].

In the present research, the objectives are to investigate the drying kinetics of calcium carbonate slurry in a spouted bed dryer and to optimize the drying using the Taguchi method. The spouted bed dryer was selected and designed among different types of dryers. In this regard, experimental runs were conducted in a laboratory-scale spouted bed dryer with inert particles. Then, the effects of drying conditions, such as air temperature, hot air velocity, and dry solid mass, were discussed and optimized using the Taguchi method. Afterwards, an experimental run was carried out in optimal conditions. Finally, several semi-theoretical models were applied to model the drying kinetics.

## 2. Theory

### 2.1. Taguchi method

Taguchi has introduced a robust design method for ordered categorical response data. This method uses the cumulative frequencies of each category and each parameter setting to analyze data, determine optimal levels and apply the analysis of variance (ANOVA)

[38]. Taguchi methodology proposes a particular method using an orthogonal array (OA) to reduce the number of experimental trials [39].

In the Taguchi method, the signal-to-noise ratio (S/N) is used to express the variability and it is calculated from experimental data by a loss function [40]. In the Taguchi method, the signal-to-noise (S/N) ratio is applied as an objective function for the optimization [36,41]. There are three categories of performance characteristics to analyze the signal-to-noise ratio (SNR) which are given by the following equations:

Larger is better:

$$SNR = -10 \log_{10}\left[\frac{1}{n}\sum \frac{1}{y^2}\right] \qquad (1)$$

Nominal is the best:

$$SNR = -10 \log_{10}\left[\frac{1}{n}\sum \frac{\bar{y}}{s_y^2}\right] \qquad (2)$$

Smaller is better:

$$SNR = -10 \log_{10}\left[\frac{1}{n}\sum y^2\right] \qquad (3)$$

Where n, $y$, $\bar{y}$, and $s_y^2$ are the number of observations, the observed data, the average value of the observed data, and the variance of observed data, respectively.

### 2.2. Mathematical modeling of the falling period of drying rate

There are several models available in the literature to estimate the drying kinetics. The developed models of drying can be categorized into three categories: empirical, semi-theoretical, and theoretical models [42,43]. In the present study, the drying kinetics of calcium carbonate in the spouted bed is evaluated using some of the semi-theoretical correlations. These methods are proposed based on Fick's second law using a number of additional exponent terms [44]. These models are summarized in Table 1.

**Table 1.** Semi-theoretical models for drying kinetics.

| No. | Model | Equation | | Ref. |
|---|---|---|---|---|
| 1 | Lewis | $MR = exp(-kt)$ | (4) | [45] |
| 2 | Page | $MR = exp(-kt^n)$ | (5) | [46] |
| 3 | Henderosn and Pabis | $MR = a.exp(-kt)$ | (6) | [47] |
| 4 | Logarithmic | $MR = a.exp(-kt)+b$ | (7) | [48] |
| 5 | Balbay and Sahin | $MR = (1-a).exp(-kt^n)+b$ | (8) | [49] |

$MR$, dimensionless solid moisture, is calculated as follows:

$$MR = \frac{X - X_{eq}}{X_0 - X_{eq}} \qquad (9)$$

It should be noted that $X_0, X, X_{eq}$ are the initial moisture content, the moisture content, and the equilibrium moisture content on the dry basis, respectively. It should be noted that $X_{eq}$ was assumed to be zero because the values of equilibrium moisture content, $X_{eq}$, are small in comparison with $X$ and $X_0$ [50,51].

The $k$ constant in the above-mentioned correlations can be assumed as a function of temperature, and inlet air velocity [52]:

$$k = k_0 \, U^m exp\left(\frac{-E}{RT}\right) \qquad (10)$$

where, $U, m, E, k_0$, and $R$ are the gas superficial velocity, power constant, the activation energy of process, the pre-exponential factor, and the universal gas constant, respectively.

It should be noted that a derivative-free method, such as the simplex search method of Lagarias et al., was used to determine the model parameters [53].

## 2.3. Statistical analysis

The estimation capability of the fitted models was evaluated in terms of some statistical criteria such as root mean square error ($E_{RMS}$), the coefficient of determination ($R^2$), and Chi-square [54]:

$$R^2 = 1 - \frac{\sum_{i=0}^{N}\left(M_{exp,i} - M_{pre,i}\right)^2}{\sum_{i=0}^{N}\left(M_{exp,i} - \bar{M}_{exp,i}\right)^2} \qquad (11)$$

$$E_{RMS} = \left[\frac{1}{N}\sum_{i=0}^{N}\left(M_{exp,i} - M_{pre,i}\right)^2\right]^{1/2} \qquad (12)$$

$$\chi^2 = \frac{\sum_{i=0}^{N}\left(M_{exp,i} - M_{pre,i}\right)^2}{N-Z} \qquad (13)$$

It should be noted that $M_{exp,i}$, $M_{pre,i}$, $\bar{M}_{exp,i}$, N and Z are the experimental data, the estimated data, the mean value of the experimental data, the number of experimental data, and the number of constants in the drying correlation. The best fitted correlation is the one with the lowest values of $E_{RMS}$ and $\chi^2$ close to zero, and the highest value of $R^2$ close to 1.

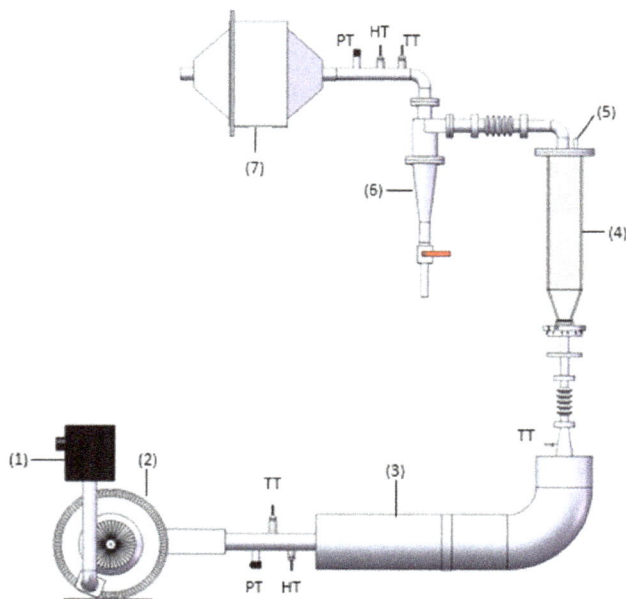

**Fig. 1.** Experimental setup, (1) pre-filter, (2) side-channel blower, (3) heater, (4) spouted bed dryer, (5) the feed port, (6) cyclone , and (7) HEPA filter; TT: temperature transmitter; HT: humidity transmitter; PT: pressure transmitter.

## 3. Experimental

### 3.1. Experimental Setup

A schematic view of the experimental setup is shown in Figure 1. The main drying equipment was made of an upright cylindrical column (inside diameter: 9 cm, outside diameter: 10 cm) and a base conical section (cone angle of 45°). A 1.5 cm nozzle was also located at the center of the lower end of the conical section. The height of the cylindrical and the conical sections were 40 cm and 9.5 cm, respectively. Glass beads (diameter: 1 mm, density: 1602 kg/m³) were used as inert particles. The air inlet port was equipped with a stainless steel screen in order to avoid the glass particles discharging into the air duct.

The spouting air was supplied through a side channel blower (2RB 420-7HH46, GREEN-CO). The maximum air flow rate was 140 m³/h at a pressure of 50 mbar. Some auxiliary accessories including an inlet vacuum filter, relief valve, and silencer were also applied along with the side channel blower. A 3 kW electrical heater was used for heating the air before it was introduced into the drying chamber. A cyclone and HEPA filter (Camfil Ireland Co.) were used to separate the dried powder from the outlet air stream.

The rotating speed of the blower was changed to

adjust the inlet air flow rate. A digital thermo-anemometer mounted at the entry point of the blower was used to measure the air inlet velocity. Three K-type thermocouples (accuracy of ±0.1°C) were used to measure the air temperature at the inlet, middle, and the outlet of the drying chamber. In addition, two temperature-moisture transmitters (HTemp-wire, HW groups s.r.o) and two pressure transmitters (KM11-ASHCROFT, Japan) were also installed after the blower and before the cyclone to detect the inlet and outlet air moisture and pressure with an accuracy of 1% R.H. and 1 mbar, respectively. A human-machine interface (HMI: the model of MT-6070IE, Weintek) was applied to collect and monitor the measured data.

### 3.2. Experimental design

The main function in this study is to improve the drying of calcium carbonate in a spouted bed dryer and the side effect is the variation in the effective efficiency. The considered factors affecting the drying process are inlet air temperature, inlet air velocity, and dry solid mass. It should be noted that the initial moisture content, ambient temperature, and operator skill have been considered as noise factors in this study.

The quality characteristic (response) in this study is the effective efficiency because energy consumption is generally considered the most important parameter in all industrial processes. The effective efficiency is defined as the ratio of the power consumption to the total input power.

The whole input power during the process includes the blower and the heater powers:

$$P_{in,tot} = \dot{m}C_p(T_{out} - T_{in}) + \frac{\rho g Q \Delta P}{\eta} \qquad (14)$$

The first term on the right side is the heater power which is consumed to increase the air temperature from the ambient temperature to the dryer inlet temperature, and the second term is the blower power. $\dot{m}$, $C_p$, $\rho$, Q, $\Delta P$, and $\eta$ are the mass flow rate of air (kg/s), specific heat capacity of air (kJ/kg.K), air density (kg/m³), volumetric flow rate (m³/s), dryer pressure drop (pa), and blower efficiency, respectively.

It should be noted that the specific heat capacity and density of the air are considered as dry air because of the low inlet humidity of the air (lower than 5%) and the use of the same inlet air in all experiments.

The amount of energy consumed to evaporate the moisture contained in the wet feed, considering a negligible final moisture content of the solid product, can be calculated as follows:

$$E_{ev} = m_s X_0 \lambda \qquad (15)$$

where $m_s$, $X_0$ and $\lambda$ are the mass of dry solid (g), initial moisture content on the dry basis (i.e., g water/g dry solid), and latent heat of vaporization of water at the dryer inlet temperature, respectively.

The amount of power consumed to evaporate moisture is obtained by dividing the evaporation energy to the drying time. Consequently, the effective efficiency can be expressed as follows:

$$\eta_{eff} = \frac{m_s X_0 \lambda / t}{\dot{m}C_p(T_{out} - T_{in}) + \rho g Q \Delta P / \eta} \qquad (16)$$

Considering this definition for the effective efficiency, the quality characteristic is established as the maximum as the best. Therefore, the objective function is calculated by Eq. (1).

In order to determine levels, three or more levels should be selected if a factor has a nonlinear or dynamic relationship to the response variable [55]. The examined factors in the drying experiments were the inlet air temperature, inlet air velocity, and dry solid mass mentioned earlier. Before designing the experiments, three levels for each factor, low, medium, and high, are determined for conducting the experiments. The factors and their levels are summarized in Table 2.

**Table 2.** Factors and their levels for the drying experiments.

| Factors | Levels | | |
|---|---|---|---|
| | 1 | 2 | 3 |
| Inlet air temperature (°C) | 90 | 100 | 110 |
| Inlet air velocity (m/s) | 1 $U_{ms}$ | 1.2 $U_{ms}$ | 1.5 $U_{ms}$ |
| Dry solid mass (gr) | 5 | 10 | 20 |

It should be mentioned that $U_{ms}$ is the minimum spouting velocity which is determined by the Mathur and Epstein method (1974) [21] and its value is 6.06 m/s.

In the Taguchi method, the combination of factors is selected from the orthogonal array (OA). In this study, a $L_9$ ($3^3$) orthogonal array is selected which has nine rows and three "3 level" columns. Table 3 shows the required 9 experimental runs.

**Table 3.** Design of experiments using L$_9$ array.

| Exp. No. | Parameters | | |
| --- | --- | --- | --- |
| | Inlet air temperature (°C) | Inlet air velocity (m/s) | Dry solid mass (g) |
| 1 | 90 | 6.06 | 5 |
| 2 | 90 | 7.20 | 10 |
| 3 | 90 | 9.00 | 20 |
| 4 | 100 | 6.06 | 10 |
| 5 | 100 | 7.20 | 20 |
| 6 | 100 | 9.00 | 5 |
| 7 | 110 | 6.06 | 20 |
| 8 | 110 | 7.20 | 5 |
| 9 | 110 | 9.00 | 10 |

### 3.2.1. Implementation of the experiments and data collection

Calcium carbonate particles were provided by the Local Corporation (Zagros Powder Co., Iran). Nine drying experimental runs were designed and carried out in a conventional spouted bed (Figure 1) at the selected factor levels according to the L$_9$ orthogonal array (as seen in Table 3).

In each experimental run, first the conical base of the spouted bed was filled with the glass beads to an initial static bed height of 115 mm. Afterward, the bed was preheated by the hot air at the specified temperature and flow rate for about 30 min. The inlet air temperature was maintained constant with an accuracy of ±1°C. After achievement of steady state conditions, the feed slurry (i.e., 30% w/w calcium carbonate solution) was fed at the top of the bed. During the drying experiments, the temperature and relative humidity of inlet and outlet air were measured and monitored every 0.5 s. The experimental run was terminated when the relative humidity of the outlet air becomes almost the same as the inlet air.

At the end of each experimental run, the glass bead particles were discharged from the bed in order to be washed with water and dried for the next run.

## 4. Results and discussion

In order to obtain the solid moisture content, the overall

mass balance in the bed is used as follows:

$$m_s \frac{dX}{dt} = G_{in} \left( Y_{out} - Y_{in} \right) \qquad (17)$$

where $X$, $G_{in}$, $Y_{out}$, and $Y_{in}$ are the solid moisture content, mass flow rate of air, outlet and inlet air humidity, respectively.

So the moisture content at different times is calculated after integration of Eq. (17) as follows:

$$X = X_0 + \int_0^t \frac{G_{in}}{m_s} \left( Y_{out} - Y_{in} \right) dt \qquad (18)$$

Relevant data such as the moisture content of air (Y) [56] and saturation vapor pressure of the air ($P^{sat}$) [57] were described using Eqs. (19) and (20), respectively.

$$Y = 0.62198 \frac{P^{sat}.RH}{P_a - P^{sat}.RH} \qquad (19)$$

$$ln \left( \frac{P^{sat}}{P_c} \right) = \frac{T_c}{T} \left( C_1 \vartheta + C_2 \vartheta^{1.5} + C_3 \vartheta^3 \right. \qquad (20)$$
$$\left. + C_4 \vartheta^{3.5} + C_5 \vartheta^4 + C_6 \vartheta^{7.5} \right)$$

$$\vartheta = 1 - \frac{T}{T_c} \qquad (21)$$

where $RH$, $P_a$, $P_c$, $T_c$, and $C_i$ are the relative humidity of air, ambient pressure (mbar), critical pressure (220640 mbar), critical temperature (647.096 K), and coefficients, respectively.

The variation of the moisture content of calcium carbonate versus time in the spouted bed dryer during the 9 different drying experimental runs is shown in Figure 2.

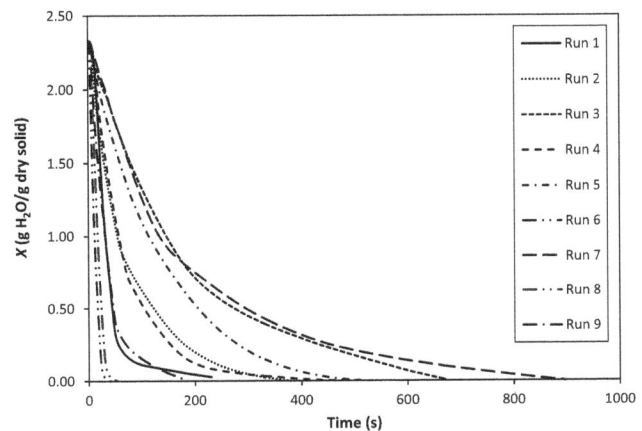

**Fig. 2.** Variation of moisture content versus drying time using Taguchi experiments

The observed values for drying time, the pressure drop of the dryer, the outlet temperature, and the measured value of effective efficiency are summarized in Table 4.

**Table 4.** Experimental results for $L_9$ orthogonal array.

| Run | Drying time (s) | Pressure drop (mbar) | Outlet temperature (°C) | Effective efficiency (%) |
|-----|-----------------|----------------------|-------------------------|--------------------------|
| 1 | 230 | 0.073 | 41.1 | 8.670 |
| 2 | 380 | 0.101 | 43.1 | 7.778 |
| 3 | 667 | 0.148 | 44.8 | 5.818 |
| 4 | 518 | 0.072 | 45.6 | 6.979 |
| 5 | 511 | 0.099 | 45.7 | 10.630 |
| 6 | 40 | 0.151 | 54.5 | 22.504 |
| 7 | 896 | 0.074 | 46.6 | 7.269 |
| 8 | 53 | 0.106 | 53.2 | 22.874 |
| 9 | 186 | 0.152 | 57.6 | 8.946 |

The results of the analysis of variance (ANOVA) are summarized in Table 5.

**Table 5.** The ANOVA results.

| Factor | DF | Adj SS | Adj MS | F-value |
|--------|-----|--------|--------|---------|
| Inlet air temperature | 2 | 66.95 | 33.475 | 4.15 |
| Inlet air velocity | 2 | 62.15 | 31.075 | 3.85 |
| Dry solid mass | 2 | 204.55 | 102.273 | 12.67 |
| Error | 2 | 16.14 | 8.071 | |
| Total | 8 | 349.79 | | |

As can be observed, the dry solid mass is the main parameter affecting the effective efficiency in the dryer.

### 4.1. Data analysis and the optimal point determination

The target of the experimental design is to get the optimum parameters of drying by maximizing the effective efficiency using the Taguchi method; therefore, the larger effective efficiency the better type analysis was used in calculation of the S/N ratio. The average S/N ratios for the effective efficiency is shown in Figure 3. According to Figure 3, the highest S/N ratio values were computed when the drying conditions were at the inlet air temperature of 110 °C, inlet air velocity of 1.2 $U_{ms}$, and dry solid mass of 5 g. When the S/N ratio has the highest value, the corresponding factor level is close to the optimum. The optimum value for each factor was clearly detected from the linear plot. These optimum

values for the calcium carbonate drying process are as follows: 100 °C for the inlet air temperature (level 2), 1.2 $U_{ms}$ for the inlet air velocity (level 2) and 5 g for the dry solid mass (level 1).

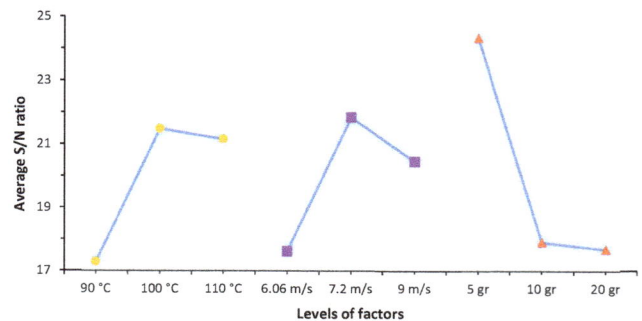

**Fig. 3.** Plots of factor effects.

### 4.2. Validation of the experiments

An experiment was carried out at the optimum values of the factors. The results of this experiment is shown in Table 6. As can be observed, the effective efficiency at the optimal point suggested by the Taguchi method is larger in comparison with other experimental runs (Table 4).

**Table 6.** Experimental results for the optimal point.

| Drying time (s) | Pressure drop (mbar) | Outlet temperature (°C) | Effective efficiency (%) |
|-----------------|----------------------|-------------------------|--------------------------|
| 54 | 0.101 | 48.4 | 24.949 |

The estimation of the response at the optimum condition based on the Taguchi method was 22.599 %. It should be noted that the prediction error is equal to 9.42 %.

### 4.3. Drying kinetics of calcium carbonate in the spouted bed dryer

The effect of air inlet velocity and temperature on the drying kinetics of calcium carbonate has been investigated. Drying experiments have been conducted at three different temperatures (90, 100 and 110 °C), and three different air inlet velocities (1, 1.2 and 1.5 times the minimum spouting velocity).

The trends of moisture content of calcium carbonate versus drying time at the different temperatures and different inlet air velocities are shown in Figures 4 and 5, respectively. As can be observed, an increase in the

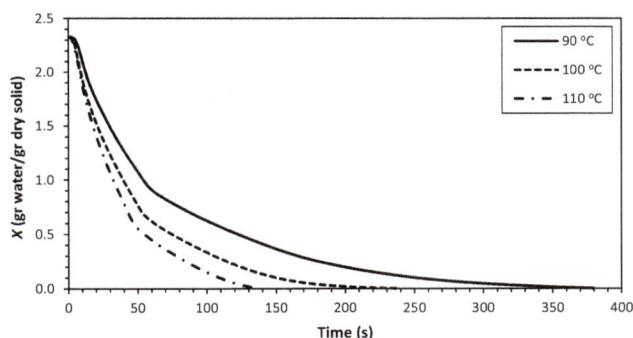

**Fig. 4.** Variation of moisture content versus drying time at different temperatures (gas velocity = 1.2 $U_{ms}$).

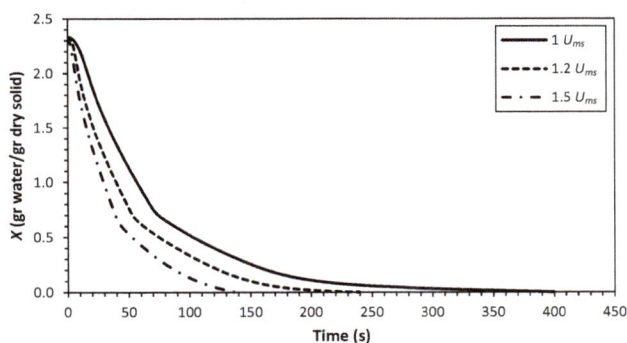

**Fig. 5.** Variation of moisture content versus drying time at different gas velocities (gas temperature = 100 °C).

shows the best fitting capability. Therefore, this model is selected as the best model of the drying kinetics of calcium carbonate in the spouted bed dryer.

The trends of estimated moisture ratio using the Logarithmic model versus drying time are shown in Figures 6 and 7.

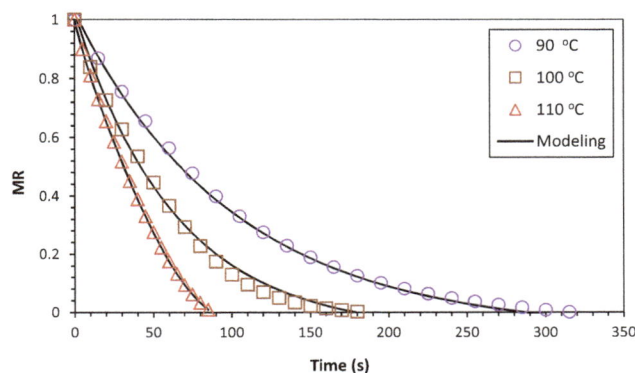

**Fig. 6.** Estimated moisture ratio by Logarithmic model versus drying time at different temperatures (gas velocity =1.2 $U_{ms}$).

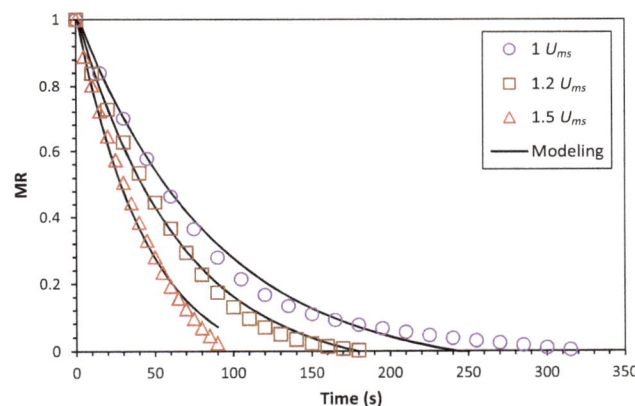

**Fig. 7.** Estimated moisture ratio by Logarithmic model versus drying time at different gas velocities (gas temperature =100 °C).

### 4.4. Performance of the fitted models

An attempt was made to fit the drying correlations introduced in Table 1 to the experimental data. Regression analysis was carried out to determine the values of the correlation parameters. Table 7 shows the values of the fitted parameters and the results of the statistical analysis. It should be mentioned that the experimental data was normalized and applied for the modeling.

As can be observed in Table 7, the logarithmic model with the lowest values of $E_{RMS}$ and Chi-square ($\chi^2$), and the highest value of determination of coefficient ($R^2$),

inlet air velocity and temperature results in a decrease in the drying time.

### 5. Conclusion

In the present work, the optimization of calcium carbonate drying in a conventional spouted bed with inert particles was implemented using experiments designed based on the Taguchi method. Inlet air

**Table 7.** Statistical results and parameters of the examined drying models for calcium carbonate drying in the spouted bed dryer.

| Model | $k_0$ (s$^{-1}$) | $m$ | $E$ (J.mole$^{-1}$) | $n$ | $a$ | $b$ | $R^2$ | $E_{RMS}$ | $\chi^2$ |
|---|---|---|---|---|---|---|---|---|---|
| Lewis | 9367.1 | 1.6278 | 50854.2 | — | — | — | 0.9839 | 0.0795 | 0.0014 |
| Page | 80997.0 | 2.0231 | 64990.2 | 1.2126 | — | — | 0.9943 | 0.0483 | 0.0005 |
| Henderson and Pabis | 10041.0 | 1.6378 | 50994.7 | — | 1.0459 | — | 0.9865 | 0.0727 | 0.0012 |
| Logarithmic | 12958.0 | 1.7454 | 52954.4 | — | 1.0917 | -0.0669 | 0.9966 | 0.0114 | 0.0002 |
| Bablay and Sahin | 99196.0 | 2.0686 | 63996.2 | 1.2286 | 0.0098 | -0.0103 | 0.9948 | 0.0460 | 0.0005 |

temperature, inlet air velocity, and dry solid mass were selected as the control factors. Effective efficiency was considered as the criterion to optimize the drying process. Effective efficiency was defined as the ratio of the total energy consumption for water evaporation to the total input energy.

According to the orthogonal array, 9 experiments were carried out. Signal to noise ratio analysis was used to find the influence of the control factors on the drying effective efficiency. Optimization of drying was performed using the "larger is the better" criterion. It was found that when the S/N ratio has the highest value the corresponding factor level is optimum. These optimum values for the calcium carbonate drying process were as follows: 100 °C for inlet air temperature, 1.2 $U_{ms}$ for inlet air velocity, and 5 g of dry solid mass. An experiment was performed under the optimal conditions and the effective efficiency value was found to be 24.95%. It should be noted that this value was larger in comparison with the calculated efficiency of all previous experiments.

The effect of air temperature in the range of 90-110 °C and air velocity in the range of 1-1.5 times of the minimum spouting velocity (i.e., 6.06 m/s) were determined. According to the results, the moisture ratio of the calcium carbonate decreases exponentially with time. Moreover, a shorter drying time was obtained using a drying air temperature of 110 °C and an air velocity of 1.2 $U_{ms}$.

Several available drying models with temperature and velocity dependent constant were examined to estimate the drying kinetics of calcium carbonate in the falling rate period. The fitted models were compared on the basis of the coefficient of determination ($R^2$), root mean square error ($E_{RMS}$), and Chi-square. The coefficients of the models for each experimental run were calculated. Among the proposed models, the Logarithmic model was found to be the best one to describe the drying behavior of calcium carbonate.

## References

[1] G. Zhu, H. Li, S. Li, X. Hou, D. Xu, R. Lin, Q. Tang, Crystallization behavior and kinetics of calcium carbonate in highly alkaline and supersaturated system, J. Cryst. Growth, 428 (2015) 16-23.

[2] T. Stirnimann, S. Atria, J. Schoelkopf, P.A. Gane, R. Alles, J. Huwyler, M. Puchkov, Compaction of functionalized calcium carbonate, a porous and crystalline microparticulate material with a lamellar surface, Int. J. Pharm. 466 (2014) 266-275.

[3] C. Bacher, P. Olsen, P. Bertelsen, J. Kristensen, J. Sonnergaard, Improving the compaction properties of roller compacted calcium carbonate, Int. J. Pharm. 342 (2007) 115-123.

[4] T. Paseephol, D.M. Small, F. Sherkat, Lactulose production from milk concentration permeate using calcium carbonate-based catalysts, Food Chem. 111 (2008) 283-290.

[5] H. Zhang, J. Chen, H. Zhou, G. Wang, J. Yun, Preparation of nano-sized precipitated calcium carbonate for PVC plastisol rheology modification, J. Mater. Sci. 21 (2002) 1305-1306.

[6] M. Di Lorenzo, M. Errico, M. Avella, Thermal and morphological characterization of poly (ethylene terephthalate)/calcium carbonate nanocomposites, J. Mater. Sci. 37 (2002) 2351-2358.

[7] J. Gullichsen, C-J. Fogelholm, Papermaking science and technology book 6A: chemical pulping, Finish Paper Engineers Association and TAPPI, Finland, 1999.

[8] C. Petersen, C. Heldmann, D. Johannsmann, Internal stresses during film formation of polymer latices, Langmuir, 15 (1999) 7745-7751.

[9] F. He, J. Zhang, F. Yang, J. Zhu, X. Tian, X. Chen, In vitro degradation and cell response of calcium carbonate composite ceramic in comparison with other synthetic bone substitute materials, Mater. Sci. Eng. C. 50 (2015) 257-265.

[10] L. Simão, R. Caldato, M. Innocentini, O. Montedo, Permeability of porous ceramic based on calcium carbonate as pore generating agent, Ceram. Int. 41 (2015) 4782-4788.

[11] C. Nover, H. Dillenburg, US Patent No. 0276897A1, (issued Dec. 15, 2005).

[12] J.B. Foster, EP Patent No. 1790616A1, (issued May 30, 2007).

[13] J.B. Foster, EP Patent No. 1790616B1, (issued Mar. 9, 2011).

[14] S. Teir, S. Eloneva, R. Zevenhoven, Production of precipitated calcium carbonate from calcium silicates and carbon dioxide, Energ. Convers. Manage. 46 (2005) 2954-2979.

[15] T. Vehmas, U. Kanerva, E. Holt, Spray-dry agglomerated nanoparticles in ordinary portland cement matrix, Mater. Sci. Appl. 5 (2014) 837-844.

[16] M. Markowski, I. Białobrzewski, A. Modrzewska, Kinetics of spouted-bed drying of barley: Diffusivities for sphere and ellipsoid, J. Food Eng. 96 (2010) 380-387.

[17] N. Epstein, J.R. Grace, Spouted and spout-fluid beds: fundamentals and applications, Cambridge University Press, 2010.

[18] A.D.A. Araújo, R.M. Coelho, C.P.M. Fontes, A.R.A. Silva, J.M.C. da Costa, S. Rodrigues, Production and spouted bed drying of acerola juice containing oligosaccharides, Food Bioprod. Process. 94 (2015) 565-571.

[19] Z.L. Arsenijević, Z.B. Grbavcić, R.V. Garić-Grulović, Drying of suspensions in the draft tube spouted bed, Can. J. Chem. Eng. 82 (2004) 450-464.

[20] S. Tia, C. Tangsatitkulchai, P. Dumronglaohapun, Continuous drying of slurry in a jet spouted bed, Drying Technol. 13 (1995) 1825-1840.

[21] K. Mathur, N. Epstein, Spouted Beds, Academic Press, New York, 1974.

[22] F.G. Cunha, K.G. Santos, C.H. Ataíde, N. Epstein, M.A. Barrozo, Annatto powder production in a spouted bed: an experimental and CFD study, Ind. Eng. Chem. Res. 48 (2008) 976-982.

[23] Z.B. Grbavcic, Z.L. Arsenijevic, R.V. Garic-Grulovic, Drying of slurries in fluidized bed of inert particles, Drying Technol. 22 (2004) 1793-1812.

[24] M. Passos, A. Mujumdar, Effect of cohesive forces on fluidized and spouted beds of wet particles, Powder Technol. 110 (2000) 222-238.

[25] M. Passos, G. Massarani, J. Freire, and A. Mujumdar, Drying of pastes in spouted beds of inert particles: Design criteria and modeling, Drying Technol. 15 (1997) 605-624.

[26] T. Kudra, A.S. Mujumdar, Advanced drying technologies, Second Ed. CRC press, 2009.

[27] T. Schneider, J. Bridgwater, The stability of wet spouted beds, Drying Technol. 11 (1993) 277-301.

[28] Q. Guo, S. Hikida, Y. Takahashi, N. Nakagawa, K. Kato, Drying of microparticle slurry and salt-water solution by a powder-particle spouted bed, J. Chem. Eng. Jpn. 29 (1996) 152-158.

[29] T. Nakazato, Y. Liu, K. Sato, K. Kato, Semi-dry process for production of very fine calcium carbonate powder by a powder-particle spouted bed, J. Chem. Eng. Jpn. 35 (2002) 409-414.

[30] M. Benali M. Amazouz, Effect of drying aid agents on processing of sticky materials, Dev. Chem. Eng.

Min. Process. 10 (2002) 401-414.

[31] Z.L. Arsenijević, Ž.B. Grbavčić, R.V. Garić-Grulović, Prediction of the particle circulation rate in a draft tube spouted bed suspension dryer, J. Serb. Chem. Soc. 71 (2006) 401-412.

[32] Z.L. Arsenijević, Ž.B. Grbavčić, R.V. Garić-Grulović, Drying of solutions and suspensions in the modified spouted bed with draft tube, J. Therm. Sci. 6 (2002) 47-70.

[33] A. Almeida, F. Freire, J. Freire, Transient analysis of pasty material drying in a spouted bed of inert particles, Dry. Technol. 28 (2010) 330-340.

[34] S.M. Tasirin, S.K. Kamarudin, J.A. Ghani, K. Lee, Optimization of drying parameters of bird's eye chilli in a fluidized bed dryer, J. Food Eng. 80 (2007) 695-700.

[35] R. Moreno, G. Antolín, A. Reyes, Thermal behaviour of forest biomass drying in a mechanically agitated fluidized bed, Lat. Am. Appl. Res. 37 (2007) 105-113.

[36] K. Uday, J. Prathyusha, D. Singh, P. Apte, Application of the Taguchi method in establishing criticality of parameters that influence cracking characteristics of fine-grained soils, Dry. Technol. 33 (2015) 1138-1149.

[37] S.K. Karna, R. Sahai, An overview on Taguchi method, Int. J. Eng. Math. Sci. 1 (2012) 1-7.

[38] S.M. Tasirin, I. Puspasari, L.J. Xing, Z. Yaakob, J.A. Ghani, Energy optimization of fluidized bed drying of orange peel using Taguchi method, World Appl. Sci. J. 26 (2013) 1602-1609.

[39] S. Athreya, Y. Venkatesh, Application of Taguchi method for optimization of process parameters in improving the surface roughness of lathe facing operation, Int. Ref. J. Eng. Sci. 1 (2012) 13-19.

[40] H.-H. Chen, C.-C. Chung, H.-Y. Wang, T.-C. Huang, Application of Taguchi method to optimize extracted ginger oil in different drying conditions, IPCBEE May, 9 (2011) 310-316.

[41] J. López-Cacho, P.L. González-R, B. Talero, A. Rabasco, M. González-Rodríguez, Robust optimization of alginate-carbopol 940 bead formulations, Sci. World J. (2012) 1-15, Article ID 605610.

[42] M. Perea-Flores, V. Garibay-Febles, J.J. Chanona-Perez, G. Calderon-Dominguez, J.V. Mendez-Mendez, E. Palacios-González, G.F. Gutierrez-Lopez, Mathematical modelling of castor oil seeds (*Ricinus communis*) drying kinetics in fluidized bed at high

temperatures, Ind. Crops Prod. 38 (2012) 64-71.

[43] E.K. Akpinar, Determination of suitable thin layer drying curve model for some vegetables and fruits, J. Food Eng. 73 (2006) 75-84.

[44] S. Azzouz, A. Guizani, W. Jomaa, A. Belghith, Moisture diffusivity and drying kinetic equation of convective drying of grapes, J. Food Eng. 55 (2002) 323-330.

[45] W.K. Lewis, The rate of drying of solid materials, Ind. Eng. Chem. 13 (1921) 427-432.

[46] G.E. Page, Factors influencing the maximum rates of air drying shelled corn in thin layers, M.Sc. Thesis, Purdue University, West Lafayette, 1949.

[47] S. Hendreson, S. Pabis, Grain drying theory. I. Temperature effect on drying coefficients, J. Agr. Eng. Res. 6 (1961) 169-174.

[48] A. Yagcioglu, Drying characteristic of laurel leaves under different conditions, In: A. Bascetincelik (ED.), Proceedings of the 7th International Congress on Agricultural Mechanization and Energy, Adana, Turkey, (1999) 565-569.

[49] A. Balbay, Ö. Şahin, Microwave drying kinetics of a thin-layer liquorice root, Dry. Technol. 30 (2012) 859-864.

[50] G. Dadalı, D. Kılıç Apar, B. Özbek, Microwave drying kinetics of okra, Dry. Technol. 25 (2007) 917-924.

[51] O. Yaldýz, C. Ertekýn, Thin layer solar drying of some vegetables, Dry. Technol. 19 (2001) 583-597.

[52] A. Magalhães, C. Pinho, Spouted bed drying of cork stoppers, Chem. Eng. Process. Process Intensif., 47 (2008) 2395-2401.

[53] J.C. Lagarias, J.A. Reeds, M.H. Wright, P.E. Wright, Convergence properties of the Nelder-Mead simplex method in low dimensions, SIAM J. Optim. 9 (1998) 112-147.

[54] J. Stoer, R. Bulirsch, Introduction to numerical analysis, Second Ed., Springer-Verlag New York, 2013.

[55] M. Satter, Optimization of copra drying factors by Taguchi method, 4th International Conference on Mechanical Engineering, Dhaka, Bangladesh (ICME2001) (2001) III 23-27.

[56] A.S. Mujumdar, Principles, classification, and selection of dryers, Handbook of Industrial Drying, Fourth Ed, CRC Press, 2014.

[57] W. Wagner, A. Pruß, The IAPWS formulation 1995 for the thermodynamic properties of ordinary water substance for general and scientific use, J. Phys. Chem. Ref. Data. 31 (2002) 387-535.

# 5

# Application of response surface methodology for thorium(IV) removal using Amberlite IR-120 and IRA-400: Ion exchange equilibrium and kinetics

Ehsan Zamani Souderjani[1], Ali Reza Keshtkar [2,*], Mohammad Ali Mousavian[1]

[1] Department of Chemical Engineering, Collage of Engineering, University of Tehran, Tehran, Iran

[2] Nuclear Fuel Cycle School, Nuclear Science and Technology Research Institute, Tehran, Iran

## HIGHLIGHTS

- A novel PVA/TiO$_2$/ZnO/TMPTMS nanofiber adsorbent was fabricated by the electrospinning method.

- The effects of pH, initial Th(IV) concentration and the amount of adsorbent were investigated.

- The properties of the prepared novel adsorbent were determined by FTIR analysis.

- The adsorption capacity of Th(IV) in a single system was reported.

- The mechanism of Th(IV) adsorption was recognized.

## GRAPHICAL ABSTRACT

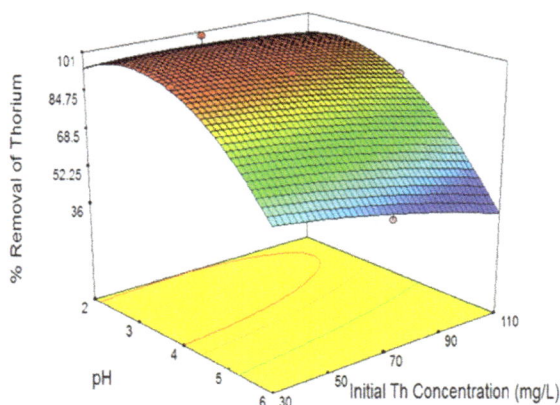

## ARTICLE INFO

Keywords:
Th(IV) removal
Response surface methodology (RSM)
Central composite design
Ion exchange resin

## ABSTRACT

In this work, thorium(IV) removal from aqueous solutions was investigated in batch systems of cationic and anionic resins of Amberlite IR-120 and IRA-400. In this way, the effects of pH, initial Th(IV) concentration and the amount of adsorbent were investigated. A Central composite design (CCD) under response surface methodology (RSM) was employed to determine the optimized condition. The results showed that the maximum removal efficiency of Th(IV) onto IR-120 and IRA-400 either discretely or in combination, albeit with equal mass fraction, was determined as follows: 98.09%, 65.70% and 72.19% at pH=3.23, 6 and 4.07, initial Th(IV) concentration of 78.2, 30 and 55.4 mg.L$^{-1}$ and 2.08, 2.5 and 2.2 g.L$^{-1}$ of resin, respectively. The kinetic and equilibrium data were accurately described by the pseudo-second order and Langmuir models. The results showed that IR-120 is a suitable adsorbent for thorium removal from aqueous solutions.

* Corresponding author:  ; E-mail address: akeshtkar@aeoi.org.ir

## 1. Introduction

During the last century, thorium has been widely used in a variety of industries such as electronic equipment [1]. Thorium is a naturally occurring radioactive element which is distributed over the earth's crust. The toxic nature of this radionuclide has been a public health dilemma for many years, even at trace levels [2]. It must be mentioned that Thorium can affect human health by causing such diseases as lung and liver cancers [3,4]. Thorium nitrate mainly localizes in the liver, spleen and marrow and it may precipitates in a hydroxide form [5]. Thorium, uranium and other actinides also lead to irrecoverable damage to the environment [6]. Due to the above facts, the removal of Thorium from aqueous solutions is necessary and several methods have been proposed [7]. One of the easiest and consequently environmental friendly way to remove this element is by adsorption. Sorption using simple and cheap adsorbents is gaining a lot of importance and there is ongoing research for the development of newer, cheap and easily available sorbents to avoid regeneration [8,9].

Different types of materials have been used as adsorbents for thorium adsorption, such as activated carbons, zeolites [10], alumina, silica [11], crystalline tin oxide nanoparticles [12] and polyvinylalcohol/titanium oxide nanofiber [13], but to our knowlege there is no data on adsorption of thorium on Amberlite IR-120 and IRA-400.

Ion exchange is one of the most popular methods for the removal of metal ions from aqueous solutions [14]. The ion exchange process has been developed as a major option for separation of metals from aqueous solutions such as purification of wastewater. The hydrogen ions released from the cationic ion exchange resin neutralize the hydroxide ions. The influence of complex formation on ion exchange adsorption equilibrium and on the distribution of metal ions between the liquid and resin phase has been extensively studied [15-17]. Ion exchange resins are usable at different pH values and high temperatures. Also, they are insoluble in most organic and aqueous solutions. These resins contain a covalent bonding between the charged functional groups and the cross linked polymer matrix [18]. The practical application of ion exchange resins is in a continuous electro deionization (CEDI) system. In this process, the dilute compartment is occupied with ion exchange resins that enhance the transport of cations or anions under the driving force of a direct current. In CEDI operation, hydrogen and hydroxyl ions were produced which regenerate the resins electrochemically without using any chemicals for regeneration [19]. The main advantages of this method are a higher recovery ratio and less pollution. This process has been adapted for the removal of sulfate, chloride and metal ions such as potassium, calcium and magnesium by using Amberlite IR-120 and IRA-400 [20-23]. Many studies have been reported on adsorption of metal ions using ion exchange resins such as Amberlite XAD-4 [24,25], IR-120 [26], CG-400 [27], and IRN-77 [28]. Recently, Singh et.al [29] studied the extraction of Thorium using ionic liquid based solvent systems.

In the recent studies the effect of individual parameters has been reported in order to maintain the other process parameters at unspecified levels. This approach can not consider the combined effect of all process parameters. Therefore, it requires a number of experiments to determine optimum levels, which may still be unreliable. Response surface methodology (RSM) can eliminate these limitations of a classical method by optimizing all the process parameters collectively by utilizing a statistical experimental design [30].

The aim of the present study is optimizing the removal of Th(IV) ions from aqueous solutions using resins of Amberlite IR-120 and IRA-400 in a batch mode. Central composite design (CCD) was used to determine the quantitative relationship between the response and the levels of the experimental factors. In the this work the effects of pH, initial Th(IV) concentration and adsorbent dosage on the adsorption process were investigated using this statistical design, and subsequently those levels were optimized.

## 2. Materials and methods

### 2.1. Materials

A strong basic anion exchange resin (Amberlite IRA-400 Cl) and a strong acidic cation exchange resin (Amberlite IR-120 Na) were purchased from BDH Chemical Co. Their properties are listed in Table 1. The solution of 100 mg.L$^{-1}$ of thorium was prepared from Th(NO$_3$)$_4$.5H$_2$O (98.5% purity, Merck Co.) by dissolving the salt in double distilled water. More diluted solutions were prepared daily as required. The pH of the solutions was adjusted using HNO$_3$ and NaOH. All the experiments were carried out at 20±2 °C.

## 2.2. Preparation of adsorbent

The cationic and anionic resins were initially immersed in HCl (10%) and stirred for 30 minutes, separately. After being rinsed twice with deionized water, the anionic resin was regenerated with NaOH solution (4%). The resins were again rinsed three times with deionized water, after that the cake was separated from water via filtration and finally dried completely at 60 °C.

## 2.3. Batch adsorption studies

The behavior of Th(IV) adsorption towards the cationic and anionic resins were examined in flasks containing 100 mL Th(IV) solution by shaking the flasks at 150 rpm at a period contact time of 15 h. Batch adsorption experiments were performed at room temperature (20±2 °C) to study the effects of pH, initial Th(IV) ion concentration and the dosage of adsorbent on optimization of Th(IV) removal by CCD under RSM. The concentration of Th(IV) ion before and after equilibrium adsorption was measured utilizing an inductivity coupled plasma atomic emission spectrophotometer (ICP-AES, Thermo Jarrel Ash, Model Trace Scan). The adsorption capacity and metal removal were defined as follows [31]:

$$q_e = \frac{(C_0 - C_e)V}{m} \tag{1}$$

$$R = \frac{C_0 - C_e}{C_0} \times 100 \tag{2}$$

where $q_e$ (mg.g$^{-1}$) is the equilibrium capacity and $R$ is the metal adsorbed percentage by ion exchange resins; $C_0$ and $C_e$ (mg.L$^{-1}$) are the initial and equilibrium metal ion concentrations, respectively; and $V$ and m are the liquid volume (L) and the weight of dried used adsorbent (g), respectively.

## 2.4. Experimental design

Optimum conditions for the adsorption of Th(IV) by IR-120 and IRA-400 either discretely or in combination, albeit with equal mass fraction, were determined by means of CCD under RSM. The RSM consists of a group of empirical techniques to evaluate a relationship between a cluster of controlled experimental factors and measured responses due to one or more selected criteria. In this study the optimization was carried out by focusing on the effect of three variables including cationic and anionic resins dosages, initial Th(IV) concentration and pH. The independent variables used in this study were coded according to Eq. (3).

$$x_i = \frac{X_i - X_0}{\Delta X} \tag{3}$$

where $x_i$ is the dimensionless coded value of the ith independent variable, $X_i$ is the real value of the independent variable, $X_0$ is the value of $X_i$ at the center point and $\Delta X$ is the step change value. The behavior of the system is defined by the following empirical second-order polynomial model [32]:

$$y = \beta_0 + \sum_{i=1}^{z} \beta_i X_i + \sum_{i=1}^{z-1} \sum_{j=1}^{z} \beta_{ij} X_i X_j + \sum_{i=1}^{z} \beta_{ii} X_i^2 + \epsilon \tag{4}$$

where $y$ is the predicted response, $\beta_0$ is the intercept term, $\beta_i$ is the linear effect, $\beta_{ii}$ is the quadratic effect, $\beta_{ij}$ is the interaction effect, $X_i$ is the input variable affecting the response of $y$, $X_i^2$ is the square effect, $X_iX_j$ is the interaction effect and $\varepsilon$ is a random error.

DESIGN EXPERT 7.0 (Stat-Ease, Inc, Minneapolis, MN, USA) software was utilized for graphical analysis and regression of the obtained data. In this way, a design of 20 experiments was formulated and then all variables were coded at five levels: $-\alpha$, $-1$, $0$, $+1$ and $+\alpha$. Finally, the optimum values of the selected variables were found by solving the regression equation. The range and the level of these variables in coded values from RSM studies are given in Table 2.

**Table 1.** Properties of the cationic and anionic resins.*

| Name | Matrix | Functional group | Total exchange capacity (eq.kg$^{-1}$) | Harmonic mean size (mm) |
|---|---|---|---|---|
| Amberlite IR-120 Na | Styrene DVB copolymer | $R\text{-}SO_3^-$ | $5 \geq$ of dry mass | 0.62-0.83 |
| Amberlite IRA-400 Cl | Polystyrene DVB | $-N^+R_3$ | 2.6-3 of dry mass | 0.3-0.9 |

* Obtained from the manufacturer

**Table 2.** Experimental ranges and levels of the independent variables.

| Independent variables | Range and level | | | | |
|---|---|---|---|---|---|
| | $-\alpha$ | $-1$ | 0 | $+1$ | $+\alpha$ |
| pH (A) | 2 | 3 | 4 | 5 | 6 |
| Initial thorium concentration, mg.L$^{-1}$ (B) | 30 | 50 | 70 | 90 | 110 |
| Adsorbent dosage, g.L$^{-1}$ (C) | 0.5 | 1 | 1.5 | 2 | 2.5 |

## 3. Results and discussion

### 3.1. Fitting the process models

The results are obtained with an experimental design aimed at identifying the best levels of the selected variables. The full factorial central composite design matrix of orthogonal and real values of adsorbent dosage (0.5–2.5 g.L$^{-1}$), pH (2–6) and initial metal ions concentration (30–110 mg.L$^{-1}$) along with the observed responses for removal of thorium(IV) have been shown in Table 3.

The second-order polynomial equation was used to find out the relationship between the variables and response. In this way, the regression equation coefficients were calculated and the data were fitted to a second-order polynomial equation. In order to check the acceptability of the model, the analysis of variance (ANOVA) was used for a removal study of Th(IV) ion with IR-120 and IRA-400 either discretely or in combination. The test for the significance of the regression model and the results of ANOVA are reported in Table 4. It must be mentioned that a Prob > F less than 0.05 indicate the significance of the model terms.

Three verification experiments were conducted for adsorption of Th(IV) ions onto a resin of IR-120 and IRA-400 and a mixture of both. To verify the accuracy of the data the corresponding correlation coefficient values ($R^2$) and adjusted $R^2$ values were obtained. The mentioned values in the case of using IR-120 are 0.9871 and 0.9755, in the case of IRA-400 are 0.9921 and 0.9850 and for a composite of both are 0.9943 and 0.9891, respectively. Due to the proximity of these

**Table 3.** Full factorial central composite design matrix of orthogonal and real values along with the observed responses for removal of thorium(IV).

| Run order | Real (coded) values | | | Removal efficiency (%) | | | | | |
|---|---|---|---|---|---|---|---|---|---|
| | A | B | C | IR-120 | | IRA-400 | | IR-120+IRA-400 | |
| | | | | Experimental | Predicted | Experimental | Predicted | Experimental | Predicted |
| 1 | 3(-1) | 50(-1) | 1(-1) | 89.82 | 92.85 | 7.87 | 7.05 | 67.19 | 66.45 |
| 2 | 5(+1) | 50(-1) | 1(-1) | 66.13 | 65.13 | 16.02 | 16.53 | 38.63 | 40.13 |
| 3 | 3(-1) | 90(+1) | 1(-1) | 86.36 | 77.37 | 3.10 | 3.09 | 56.61 | 55.33 |
| 4 | 5(+1) | 90(+1) | 1(-1) | 55.52 | 54.43 | 8.24 | 8.81 | 26.74 | 26.73 |
| 5 | 3(-1) | 50(-1) | 2(+1) | 99.46 | 99.61 | 16.60 | 15.19 | 70.76 | 71.25 |
| 6 | 5(+1) | 50(-1) | 2(+1) | 88.91 | 89.19 | 32.47 | 31.63 | 66.02 | 67.81 |
| 7 | 3(-1) | 90(+1) | 2(+1) | 97.28 | 98.17 | 14.28 | 12.95 | 68.85 | 67.85 |
| 8 | 5(+1) | 90(+1) | 2(+1) | 84.04 | 82.45 | 25.64 | 25.63 | 60.89 | 62.13 |
| 9 | 2(-α) | 70(0) | 1.5(0) | 99.13 | 95.84 | 7.37 | 8.83 | 65.91 | 67.43 |
| 10 | 6(+α) | 70(0) | 1.5(0) | 47.37 | 49.96 | 31.40 | 30.89 | 37.37 | 35.39 |
| 11 | 4(0) | 30(-α) | 1.5(0) | 98.23 | 96.74 | 20.75 | 21.63 | 70.31 | 69.05 |
| 12 | 4(0) | 110(+α) | 1.5(0) | 84.65 | 84.66 | 11.67 | 11.67 | 51.49 | 52.25 |
| 13 | 4(0) | 70(0) | 0.5(-α) | 61.63 | 61.48 | 2.23 | 1.69 | 31.36 | 31.87 |
| 14 | 4(0) | 70(0) | 2.5(+α) | 99.29 | 98.64 | 25.29 | 26.65 | 72.11 | 71.07 |
| 15 | 4(0) | 70(0) | 1.5(0) | 93.16 | 92.98 | 9.12 | 9.81 | 64.10 | 64.81 |
| 16 | 4(0) | 70(0) | 1.5(0) | 93.06 | 92.98 | 9.73 | 9.81 | 64.94 | 64.81 |
| 17 | 4(0) | 70(0) | 1.5(0) | 93.13 | 92.98 | 9.68 | 9.81 | 64.91 | 64.81 |
| 18 | 4(0) | 70(0) | 1.5(0) | 93.33 | 92.98 | 10.11 | 9.81 | 65.35 | 64.81 |
| 19 | 4(0) | 70(0) | 1.5(0) | 92.98 | 92.98 | 9.44 | 9.81 | 65.22 | 64.81 |
| 20 | 4(0) | 70(0) | 1.5(0) | 93.71 | 92.98 | 9.92 | 9.81 | 64.84 | 64.81 |

**Table 4.** Analysis of variance (ANOVA) for Th(IV) removal by ion exchange resins.

| Source | Sum of squares (coded) | | | DF | Prob>F | | |
|---|---|---|---|---|---|---|---|
| | IR-120 | IRA-400 | R-120+IR-A400 | | IR-120 | IRA-400 | I-R120+IR-A400 |
| Model | 4541.04 | 1442.58 | 3666.32 | 9 | <0.0001 | <0.0001 | <0.0001 |
| A | 2103.14 | 490.40 | 1027.36 | 1 | <0.0001 | <0.0001 | <0.0001 |
| B | 145.68 | 99.30 | 281.82 | 1 | 0.0006 | <0.0001 | <0.0001 |
| C | 1379.75 | 623.50 | 1617.05 | 1 | <0.0001 | <0.0001 | <0.0001 |
| AB | 12.10 | 7.07 | 2.57 | 1 | 0.1837 | 0.0325 | 0.2965 |
| AC | 118.12 | 24.29 | 261.40 | 1 | 0.0012 | 0.0010 | <0.0001 |
| BC | 6.16 | 1.45 | 29.76 | 1 | 0.3323 | 0.2882 | 0.0038 |
| $A^2$ | 632.75 | 157.11 | 282.97 | 1 | <0.0001 | <0.0001 | <0.0001 |
| $B^2$ | 8.14 | 73.18 | 27.18 | 1 | 0.2686 | <0.0001 | 0.0050 |
| $C^2$ | 261.76 | 30.07 | 258.43 | 1 | <0.0001 | 0.0005 | <0.0001 |
| Residual | 59.34 | 11.48 | 21.15 | 10 | | | |
| Lack-of-fit | 58.99 | 10.87 | 20.20 | 5 | <0.0001 | 0.0034 | 0.0022 |
| Pure error | 0.35 | 0.62 | 0.95 | 5 | | | |

values to 1.0, it can be understood that a good agreement has been achieved between the correlated data and the experimental data . The final responses for the removal of Th(IV) ions using IR-120, IRA-400 and a composite of both of them (IR-120 and IRA-400) in terms of coded factors are reported in Eqs. (5-7):

$$\% \, Removal = +92.98 - 11.47 \times A - 3.02 \times B + 9.29 \times C - 1.23 \times A \times B + 3.84 \times A \times C + 0.88 \times B \times C - 5.02 \times A^2 - 0.57 \times B^2 - 3.23 \times C^2 \qquad (5)$$

$$\% \, Removal = +9.81 + 5.54 \times A - 2.49 \times B + 6.24 \times C - 0.94 \times A \times B + 1.74 \times A \times C + 0.43 \times B \times C + 2.5 \times A^2 + 1.71 \times B^2 + 1.09 \times C^2 \qquad (6)$$

$$\% \, Removal = +64.81 - 8.01 \times A - 4.20 \times B + 10.05 \times C - 0.57 \times A \times B + 5.72 \times A \times C + 1.93 \times B \times C - 3.35 \times A^2 - 1.04 \times B^2 - 3.21 \times C^2 \qquad (7)$$

where in these equations A, B and C are the coded terms for pH, initial Th(IV) concentration and adsorbent dosage, respectively.

### 3.2. Interaction effects of two variables

It is obvious that the sensitivity of the response to the two interacting variables can include three dimensional graphs by holding the other variable at the central values. On the basis of quadratic polynomial Eqs. (5–7) of the response surface methodology, the effects of interacting

variables were analyzed. The positive linear coefficient indicates that the Th(IV) removal percentage increased as the variable increased. Figure 1(a-c) shows the simultaneous effect of pH and initial metal ion concentration on Th(VI) removal efficiency in an aqueous solution with the resin of IR-120 (Figure 1a), IRA-400 (Figure 1b) and in combination (Figure 1c).

The 3-dimentional surface plot shows that the removal efficiency of Th(IV) for the three adsorbents was subject to changes in pH. According to Figures 1a and 1c, the removal efficiency of Th(IV) increased with pH ranging from 2.0 to 3.5, this phenomenon can be explained by the reduction of active sites of the adsorbent at low pH because of protonation of the functional groups. Moreover, at low pH the competition between $Th^{+4}$ ions and hydrogen ions increases resulting in a decrease in removal efficiency. However, the removal efficiency of Th(IV) decreases when the pH is above 4.0, which can be due to the formation of different thorium species, such as $[Th_2(OH)_2]^{6+}$, $[Th_3(OH)_5]^{7+}$ and $[Th_4(OH)_8]^{8+}$, which have a lower adsorption tendency [33,34]. The response plots show that the thorium removal efficiency firstly increases with an increase in the metal ions concentration and then decreases. The decrease in the percentage removal of metals ions can be attributed to the fact that all the adsorbents have a limited number of active sites, which are saturated above a certain initial metal ions concentration. Figure 1b shows that the removal efficiency of Th(IV) increases with an increase in pH from 2 to 6. It shows the ability of anionic resin to

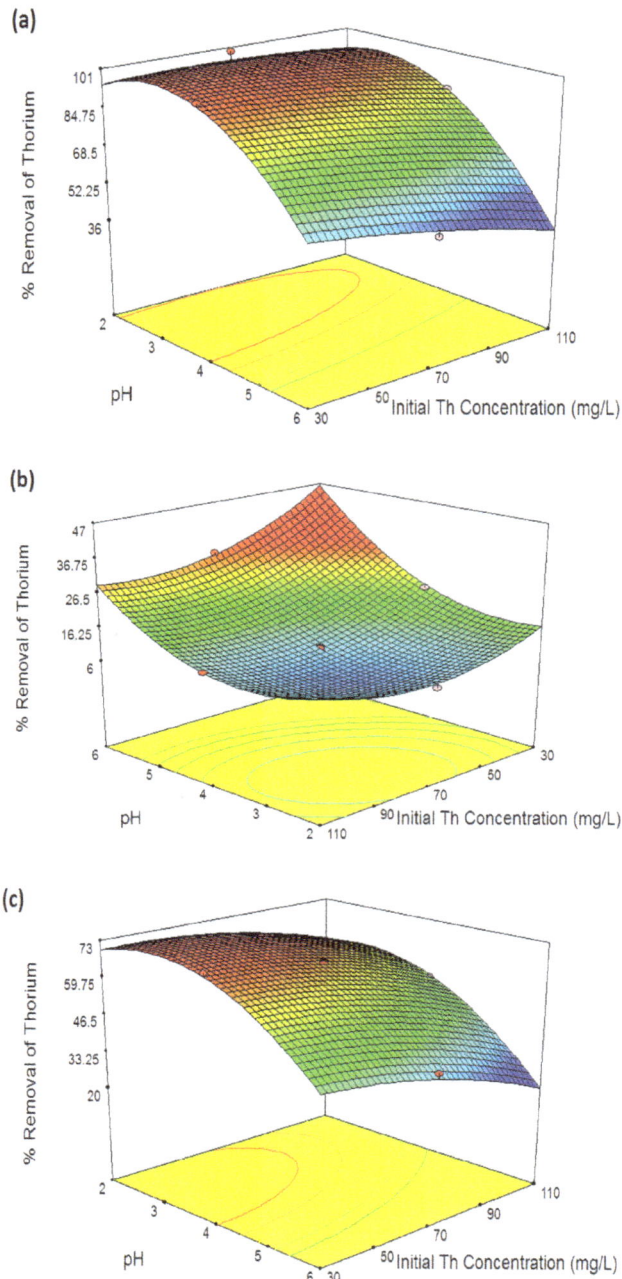

**Fig. 1.** Three-dimensional plots showing effect of initial Th(IV) ion concentration [mg.L$^{-1}$] and pH on the removal efficiency of Th(IV) using IR-120 (a), IRA-400 (b) and in combination (c), keeping adsorbent dosage 1.5 g.L$^{-1}$.

**Fig. 2.** Three-dimensional plots showing effect of pH and adsorbent dosage [mg.L$^{-1}$] on the removal efficiency of Th(IV) using IR-120 (a), IRA-400 (b) and in combination (c), keeping initial Th(IV) ion concentration 70 mg.L$^{-1}$.

to remove thorium hydroxide is more than Th$^{+4}$ ions. The three-dimensional response plots show the interactive effect of the pH and the adsorbent dosage (Figure 2).

In the case of using IR-120, it was proven that at an initial Th(IV) ion concentration of 70 mg.L$^{-1}$ and pH=4, the removal efficiency of Th(IV) increased from 61.63 to 99.29, and in the case of IRA-400 the same increased from 2.23 to 25.29. It was also proven that as the adsorbent dosage increased, the number of the

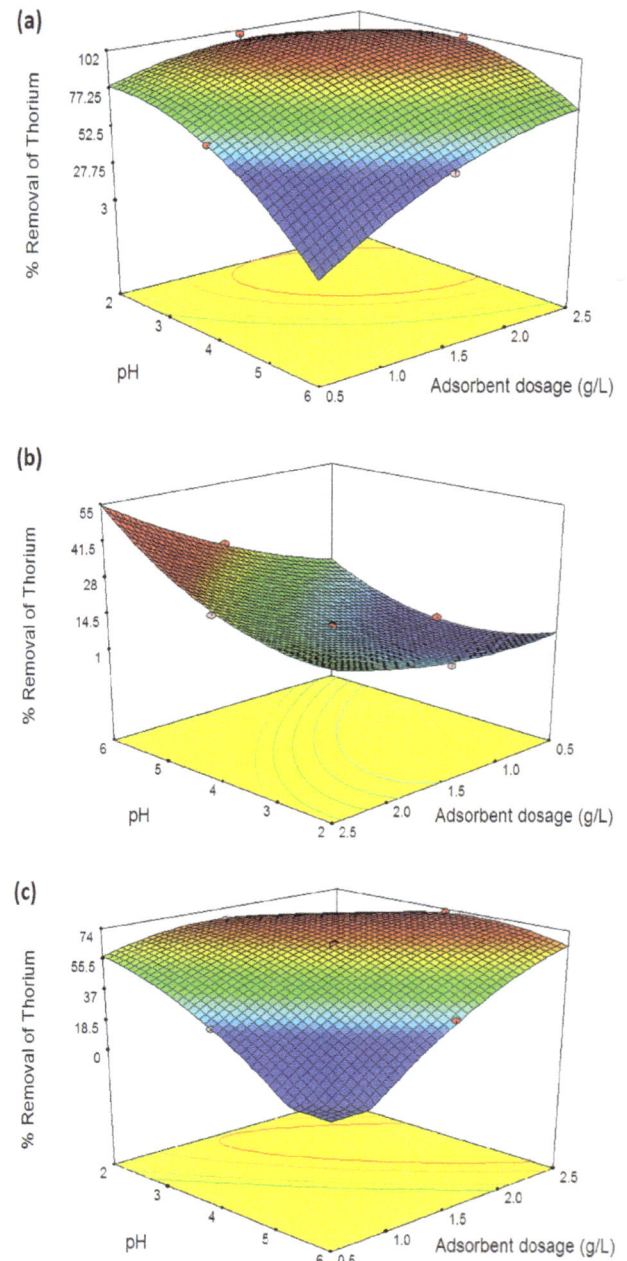

unoccupied effective sites and the surface area also increased, improving the percentage of adsorption [35]. The interactive effects of the adsorbent dosage and initial Th(IV) ion concentration can be inferred from the response plot Figure 3(a-c), holding the pH equal to 4.

When the initial Th(IV) ion concentration increases from 30 to 110 mg.L$^{-1}$, the removal efficiency of Th(IV) decreases because of increasing driving force of the mass transfer and limited number of active sites on the

**(a)**

**(b)**

**(c)**

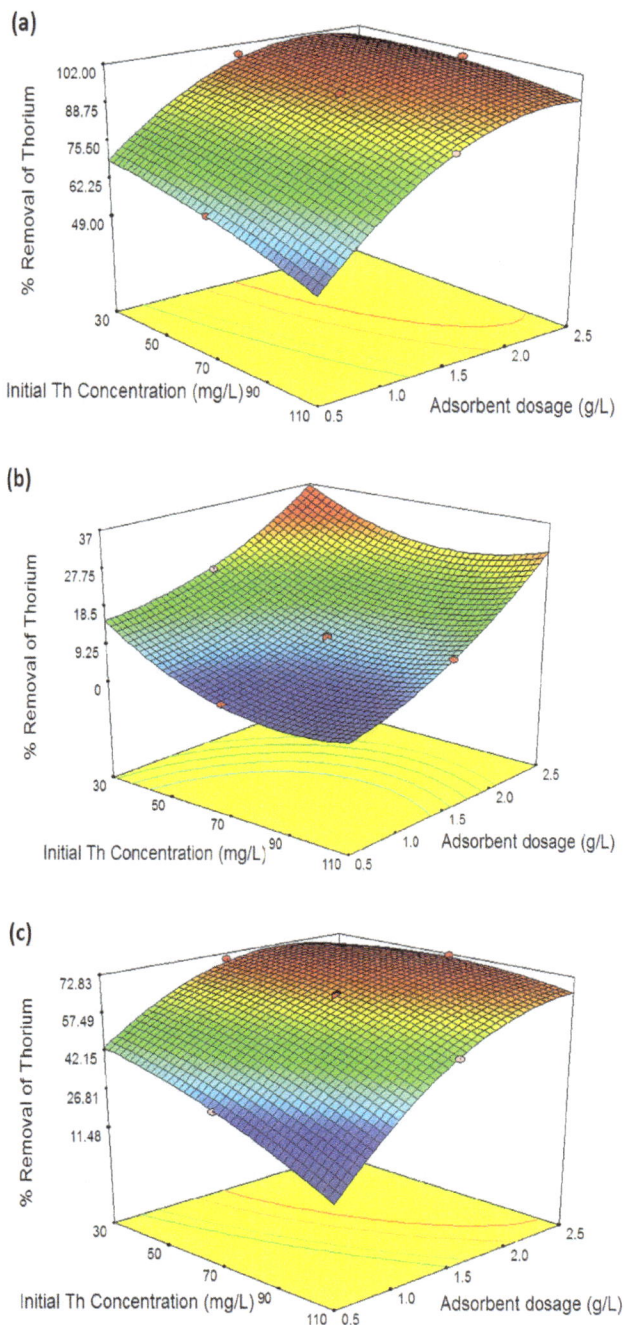

**Fig. 3.** Three-dimensional plots showing effect of initial Th(IV) ion concentration [mg.L$^{-1}$] and adsorbent dosage [g.L$^{-1}$] on the removal efficiency of Th(IV) using IR-120 (a), IRA-400 (b) and in combination (c), keeping pH 4.

and limited number of active sites on the adsorbent [36]. At the adsorbent dosage of 1.5 g.L$^{-1}$, this is evidenced by greater removal efficiency of Th(IV) (98.23, 20.75 and 70.31) at initial Th(IV) ion concentrations of 30 mg L$^{-1}$ in comparison to 84.65, 11.67 and 51.49 at initial concentration of 110 mg.L$^{-1}$ with IR-120, IRA-400 and a mix of the adsorbents, respectively. According to the P-value (>0.05) reported in Table 4, it can be seen

that the interaction between the resin dosage and the metal ion concentration for IR-120 and IRA-400 is not significant.

### 3.3 Optimization of Th(IV) removal

Optimization of the independent variables to maximize the removal efficiency of Th(IV) was performed using the quadratic model within the studied experimental range. The model optimization suggested the optimum values of the selected three independent process variables as; pH 3.23, 6 and 4.07; initial Th(IV) ion concentration of 78.2, 30.0 and 55.4 mg.L$^{-1}$; and the adsorbent dose of 2.08, 2.5 and 2.2 g.L$^{-1}$, to achieve the maximum reduction (99.82% , 66.31% and 73.15%) of the Th(IV) by resin of IR-120, IRA-400 and a mix of both as an adsorbent, respectively.

The removal efficiencies in specified conditions were obtained and then compared with the predicted values. As can be seen from Table 5, the percentage error between the removal efficiency of the experimental and predicted values was in the range of 0.70–4.39%. The experimental removal efficiency of Th(IV) onto resin of IR-120, IRA-400 and a mix of both, under optimum conditions was determined to be 98.09% , 63.52% and 72.19%, respectively. The reported errors showed that there is a good agreement between the experimental data and the predicted data using RSM.

### 3.4. Kinetic study

In this work, the adsorption of Th (IV) on Amberlite ion exchange resins is examined at different time intervals. Figure 4 shows the effect of time on the adsorption of Th (IV) at pH = 4, initial Th(IV) ion concentration of 70 mg L$^{-1}$ and adsorbent dosage of 1.5 g.L$^{-1}$.

In this study, different kinetic models were tested to check the mechanism involved during the adsorption process, i.e. the pseudo-first order and the pseudo-second-order. The applicability of these kinetic models was determined by measuring the correlation coefficients ($R^2$) as well as the closeness of the experimental and calculated adsorption capacity values.

Pseudo first-order:    $q_t = q_e \left(1 - exp(-k_1 t)\right)$    (8)

Pseudo second-order:    $q_t = \dfrac{k_2 q_e^2 t}{1 + k_2 q_e t}$    (9)

**Table 5.** Verification experiments for adsorption of Th(IV) ions onto resin of IR-120, IRA-400 and the mix of them.

| Adsorbent | No. | Condition | | | Removal efficiency of Th(IV) | | Error (%) |
|---|---|---|---|---|---|---|---|
| | | pH | Initial concentration (mg.g$^{-1}$) | Adsorbent dose (g.L$^{-1}$) | Observed values | Predicted values | |
| IR-120 | 1 | 2 | 70 | 2 | 95.64 | 94.22 | 1.48 |
| | 2 | 3 | 50 | 2.5 | 97.70 | 96.87 | 0.85 |
| | 3 | 4 | 30 | 2.5 | 97.92 | 98.88 | 0.98 |
| | Opt | 3.23 | 78.2 | 2.08 | 98.09 | 99.82 | 1.76 |
| IRA-400 | 1 | 5 | 50 | 2.5 | 40.89 | 42.45 | 3.81 |
| | 2 | 5 | 30 | 2 | 43.56 | 44.71 | 2.64 |
| | 3 | 6 | 50 | 2.5 | 60.33 | 59.91 | 0.70 |
| | Opt | 6 | 30 | 2.5 | 63.52 | 66.31 | 4.39 |
| IR-120 + IRA-400 | 1 | 3 | 50 | 2.5 | 66.44 | 64.02 | 3.64 |
| | 2 | 4 | 50 | 2.5 | 70.79 | 71.37 | 0.82 |
| | 3 | 5 | 70 | 2 | 64.95 | 66.01 | 1.63 |
| | Opt | 4.07 | 55.4 | 2.2 | 72.19 | 73.15 | 1.33 |

where $k_1$ (min$^{-1}$) is the pseudo-first-order rate constant and $k_2$ (g.mg$^{-1}$.min$^{-1}$) is the pseudo second-order rate constant. $q_t$ and $q_e$ (mg.g$^{-1}$) are the amounts of Th(IV) adsorbed on the adsorbent at time $t$ and in equilibrium, respectively. In Table 6 the parameters of the kinetic models and the correlation coefficient of the models are reported.

As can be found from Table 6, in the case of the pseudo-second order reaction the $R^2$ values were better than those of the pseudo-first order. Also the experimental qe values were close to the qe values obtained from the slope of the linear plot of $t/q_t$ versus t suggesting that the experimental data fits well into the pseudo-second-order model system (Table 6). In this way, the adsorption procedure was supportive of the pseudo-second order equation, which showed that the adsorption involved in the chemical reaction incorporated the physical adsorption, in like manner [37].

### 3.5. Adsorption isotherms

The initial concentration plays an important role to overcome all mass transfer resistances of metal ions in adsorption from an aqueous phase to a solid phase. The Langmuir adsorption isotherm suggests monolayer sorption on a homogeneous surface without any interaction between the adsorbed molecules; this model can be written as follows [38]:

$$q_e = \frac{q_m K_L C_e}{1 + K_L C_e} \tag{10}$$

where $q_e$ (mg.g$^{-1}$) is the equilibrium metal ion concentration on the sorbent, $C_e$ (mg.L$^{-1}$) is the equilibrium metal ion concentration in the solution, $q_m$ (mg.g$^{-1}$) is the monolayer sorption capacity of the sorbent (mg g$^{-1}$), and $K_L$ is the Langmuir sorption constant (L.mg$^{-1}$) relating the free energy of sorption. A multilayer sorption with a heterogeneous energetic distribution of active sites can be accompanied by interactions between the adsorbed molecules which is proposed by the Freundlich isotherm model [39]:

$$q_e = K_F C_e^{1/n} \tag{11}$$

**Table 6.** Kinetic parameters for the adsorption of Th(IV) onto ion exchange resins.

| Experimental $q_e$ (mg.g$^{-1}$)* | Pseudo-first order | | | Pseudo-second order | | |
|---|---|---|---|---|---|---|
| | $k_1$ (min$^{-1}$)×10$^{-3}$ | $q_e$ (mg.g$^{-1}$) | $R^2$ | $k_2$ (g.mg$^{-1}$.min$^{-1}$)×10$^{-4}$ | $q_e$ (mg.g$^{-1}$) | $R^2$ |
| (a) 43.39 | 9.21 | 46.18 | 0.786 | 4.55 | 44.84 | 0.996 |
| (b) 4.58 | 7.14 | 4.36 | 0.906 | 25.48 | 4.99 | 0.998 |
| (c) 30.24 | 9.67 | 39.59 | 0.884 | 4.94 | 30.00 | 0.998 |

* Kind of adsorbent: (a) IR-120 , (b) IRA-400 and (c) Mix of them

Where $K_F$ is a constant relating the sorption capacity and $1/n$ is an empirical parameter relating the sorption intensity, which varies with the heterogeneity of the material. Table 7 shows the parameters of these isotherm models. The linearized Langmuir sorption isotherm is plotted in Figure 5. It was found that the metal ion removal mechanism is related to the initial metal ion concentration. In other words, adsorption of metal ion takes place at definite sites only when the metal ion concentration is low, but as the concentration is increased, the sites are saturated and the exchange sites are filled [40]. Based on the correlation coefficients (Table 7), the Langmuir equation gives a better fit for the experimental data than the Freundlich one for the adsorption of Th(IV). Langmuir isotherm suggests monolayer coverage of the thorium species on the surface of the resins. The $n>1$ obtained in the Freundlich model depicts a favorable adsorption.

**Table 7.** Langmuir and Freundlich constants for the adsorption of Th(IV) onto ion exchange resins.

| Isotherm model | Adsorbent | | |
|---|---|---|---|
| | IR-120 | IRA-400 | IR-120 + IRA-400 |
| **Langmuir** | | | |
| qm (mg.g⁻¹) | 24.15 | 2.43 | 13.11 |
| $K_L$ (L.mg⁻¹) | 0.151 | 0.026 | 0.045 |
| $R^2$ | 0.9903 | 0.9854 | 0.9916 |
| **Freundlich** | | | |
| $K_F$ (mg.g⁻¹) | 4.921 | 0.209 | 1.094 |
| $n$ | 2.654 | 2.228 | 1.964 |
| $R^2$ | 0.9486 | 0.9227 | 0.8892 |

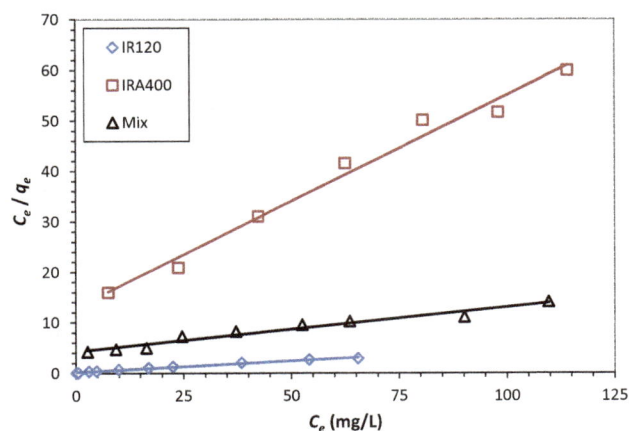

**Fig 5.** The Langmuir adsorption isotherm models of Th(IV) adsorption onto resins of IR-120, IRA-400 and the mix of them at T= 25 °C , pH= 4 , adsorbent dose = 1.5 g.L⁻¹.

In Table 8 the maximum sorption capacity of Th(IV) ion onto the Amberlite IR-120Na⁺ is compared with that of the other adsorbents reported in the literature. As can be seen in Table 8, the maximum sorption capacity of Th(IV) ion onto the resin of IR-120 is in the same range of maximum sorption capacity of Th(IV) ion obtained by other researchers.

**Table 8.** Comparison of the various sorbents used for Th(IV) uptake.

| Adsorbent | $q_{max, Th}$ (mg.g⁻¹) | Ref. |
|---|---|---|
| XAD-4-o-phenylene dioxydiacetic acid | 26.22 | [39] |
| XAD-4-octacarboxymethyl-C-methyl calix resorcinaren | 62.65 | [40] |
| Merrifield chloromethylated resin-4-ethoxy-4-ethyl-N,N-bis-2-ethyl hexyl butanamide | 46.41 | [41] |
| PAN/zeolite | 9.28 | [42] |
| Crystalline tin oxide nanoparticles | 62.5 | [11] |
| Carboxylate-functionalised graft copolymer derived from titanium dioxide -densified cellulose | 92.2 | [43] |
| Amberlite IR-120Na⁺ | 24.15 | This work |

*3.6. FTIR analysis*

In order to determine the main functional groups of the Amberlite IR-120 in Th(IV) adsorption, the infrared spectroscopy characteristics were compared before and after adsorption (Figure 6).

In Figure 6 the band at 2924 cm⁻¹ was attributed to C−H stretching vibrations. The intensive peak at 1633 cm⁻¹was produced by −C=O stretching vibration. The bands observed at 1413 cm⁻¹ assigned the stretching of the O−S−O group and at 836 cm⁻¹ presented the aromatic out of plane C−H band [41]. The comparison of FTIR spectra before and after adsorption of Th(IV) onto IR-120 shows that the intensities of the peak series in 1126 and 1035 cm⁻¹ regions were changed because of the metal complex formation between thorium ions and sulfonate groups of the resin [42]. The sulfonate ($-SO_3^-$) groups band participating in ion exchange are located under 1035-1153 cm⁻¹ regions for the cationic resin.

The infrared spectroscopy of Amberlite IRA-400 is shown in Figure 7. The broad and strong band ranging from 3300 to 3600 cm⁻¹ might be due to −OH groups in both Figures 6 and 7.

The −C=O stretching vibration is shown at 1629 and 1634 cm⁻¹ before and after adsorption, respectively. A primary amine band can be seen at the range of 1440 to 1560 cm$^{-1}$ and a secondary amine band is presented at the range of 1000 to 1350 cm$^{-1}$; therefore, it can be concluded that there are two types of amines in the anionic resin. Furthermore, a peak at 1383 cm$^{-1}$ is related to the $NO_3^-$ functional group after adsorption of Th(IV) onto IRA-400.

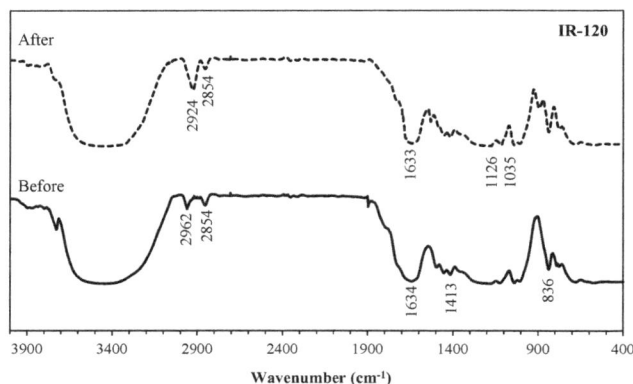

**Fig. 6.** FTIR spectra of the Amberlite IR-120 adsorbent before and after Th(IV) adsorption.

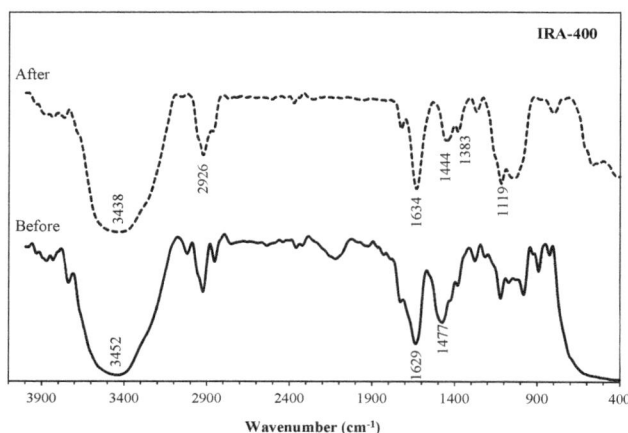

**Fig. 7.** FTIR spectra of the Amberlite IRA-400 adsorbent before and after Th(IV) adsorption.

## 4. Conclusion

The objective of this study was to explore the optimum process conditions via the response surface methodological approach, as required while using Amberlite IR-120, IRA-400 and a mix of both to remove thorium(IV) from the aqueous solutions. On the basis of the RSM approach using the central composite model for the experimental design and fitness of polynomial equation, maximum removal efficiencies of Th(IV) onto IR-120, IRA-400 and a mix of both with equal mass fraction were determined as 98.09%, 65.70% and 72.19% at pH of 3.23, 6 and 4.07, initial Th(IV)

concentration of 78.2, 30 and 55.4 mg.L$^{-1}$ and the resin amount of 2.08, 2.5 and 2.2 g.L$^{-1}$, respectively.

The equilibrium adsorption data were correlated with Langmuir and Freundlich isotherm equations. The statistical parameters indicate that the Langmuir equation was the best fit and there was a good agreement between model and experimental data. The maximum monolayer adsorption capacities of Th(IV) onto IR-120, IRA-400 and a mix of both were found to be 24.15, 2.43 and 13.11 mg.g$^{-1}$, respectively. The kinetic of thorium adsorption was very well described by the pseudo-second-order kinetic model with $R^2$ values exceeding 0.996, 0.998 and 0.998 related to IR-120, IRA-400 and a mix of both, respectively. FTIR results showed that sulfonate groups were mainly involved in Th(IV) adsorption. Finally, Amberlite IR-120Na$^+$ could be utilized successfully in the removal of thorium ion from aqueous solutions.

## References

[1] D.I. Ryabchikov, E.K. Gol'braikh, The Analytical Chemistry of Thorium, Pergamon Press, Oxford, 1963.

[2] G.R. Choppin, Actinide speciation in the environment, Radiochim. Acta, 91 (2003) 645-650.

[3] J.D. Van Horn, H. Huang, Uranium(VI) bio-coordination chemistry from biochemical, solution and protein structural data, Coord. Chem. Rev. 250 (2006) 765-775.

[4] A.R. Keshtkar, M. Irani, M.A. Moosavian, Removal of uranium(VI) from aqueous solutions by adsorption using a novel electrospun PVA/TEOS/APTES hybrid nanofiber membrane: comparison with casting PVA/TEOS/APTES hybrid Membrane, J. Radioanal. Nucl. Ch. 295 (2013) 563-571.

[5] M. Metaxas, V. Kasselouri-Rigopoulou, P. Galiatsatou, C. Konstantopoulou, D. Oikonomou, Thorium removal by different adsorbents, J. Hazard. Mater. 97 (2003) 71-82.

[6] Z. Hongxia, D. Zheng, T. Zuyi, Sorption of thorium (IV) ions on gibbsite: effects of contact time, pH, ionic strength, concentration, phosphate and fulvic acid, Colloid. Surface. A, 278 (2006) 46-52.

[7] A. Hamta, M.R. Dehghani, Application of polyethylene glycol based aqueous two-phase systems for extraction of heavy metals, J. Mol. Liq. 231 (2017) 20-24.

[8] S. Chandramouleeswaran, J. Ramkumar, V. Sudarsan, A.V.R. Reddy, Boroaluminosilicate glasses: novel sorbents for separation of Th and U, J. Hazard. Mater. 198 (2011) 159-164.

[9] J. Ramkumar, S. Chandramouleeswaran, V. Sudarsan, R.K. Vatsa, S. Shobha, V.K. Shrikhande, G.P. Kothiyal, T. Mukherjee, Boroaluminosilicate glasses as ion exchange materials, J. Non-Cryst. Solids, 356 (2010) 2813-2819.

[10] A. Dyer, L.C. Jozefowicz, The removal of thorium from aqueous solutions using zeolites, J. Radioanal. Nucl. Ch. 159 (1992) 47-62.

[11] L. Weijuan, T. Zuyi, Comparative study on Th(IV) sorption on alumina and silica from aqueous solutions, J. Radioanal. Nucl. Ch. 254 (2002) 187-192.

[12] A. Nilchi, T. Shariati Dehaghan, S. Rasouli Garmarodi, Kinetics, isotherm and thermodynamics for uranium and thorium ions adsorption from aqueous solutions by crystalline tin oxide nanoparticles, Desalination, 321 (2013) 67-71.

[13] S. Abbasizadeh, A.R. Keshtkar, M.A. Mousavian, Preparation of a novel electrospun polyvinyl alcohol/titanium oxide nanofiber adsorbent modified with mercapto groups for uranium(VI) and thorium(IV) removal from aqueous solution, Chem. Eng J. 220 (2013) 161-171.

[14] J.S. Kentish, G.W. Stevens, Innovations in separation technology for the recycling and re-use of liquid waste streams, Chem. Eng. J. 84 (2001) 149-159.

[15] A.A. Khan and R.P. Singh, Adsorption thermodynamics of carbofuran on Sn (IV) arsenosilicate in $H^+$, $Na^+$ and $Ca^{2+}$ forms, Colloid. Surface. A, 24 (1987) 33-42.

[16] C.H. Lee, J.S. Kim, M.Y. Suh, W. Lee, A chelating resin containing 4-(2- thiazolylazo) resorcinol as the functional group synthesis and sorption behaviours for trace metal ions, Anal. Chim. Acta, 339 (1997) 303-312.

[17] J.P. Rawat, K.P.S. Muktawat, Thermodynamics of ion-exchange on ferric antimonite, J. Inorg. Nucl. Chem. 43 (1981) 2121-2128.

[18] D.C. Sherrington, Preparation, structure and morphology of polymer supports, Chem. Commun. 21 (1998) 2275-2286.

[19] J.H. Song, K.H. Yeon, S.H. Moon, Effect of current density on ionic transport and water dissociation phenomena in a continuous electrodeionization (CEDI), J. Membrane Sci. 291 (2007) 165-171.

[20] J.S. Park, J.H. Song, K.H. Yeon, S.H. Moon, Removal of hardness ion from tap water using electromembrane processes, Desalination, 202 (2007) 1-8.

[21] H.J. Lee, M.K. Hong, S.H. Moon, A feasibility study on water softening by electrodeionization with the periodic polarity change, Desalination, 284 (2012) 221-227.

[22] T. Ho, A. Kurup, T. Davis, J. Hestekin, Wafer chemistry and properties for ion removal by wafer enhanced electrodeionization, Sep. Sci. Technol. 45 (2010) 433-446.

[23] K. Dermentzis, Continuous electrodeionization through electrostatic shielding, Electrochim. Acta, 53 (2008) 2953-2962.

[24] B.N. Singh, B. Maiti, Separation and pre-concentration of U(VI) on XAD-4 modified with 8-hydroxy quinoline, Talanta, 69 (2006) 393-396.

[25] S. Chandramouleeswaran, Jayshree Ramkumar, n-Benzoyl-n-phenylhydroxylamine impregnated Amberlite XAD-4 beads for selective removal of thorium, J. Hazard. Mater. 280 (2014) 514-523.

[26] A. Demirbas, E. Pehlivan, F. Gode, T. Altun, G. Arslan, Adsorption of Cu(II), Zn(II), Ni(II), Pb(II) and Cd(II) from aqueous solution on Amberlite IR-120 synthetic resin, J. Colloid Interf. Sci. 282 (2005) 20-25.

[27] F. Semnani, Z. Asadi, M. Samadfam, H. Sepehrian, Uranium(VI) sorption behavior onto amberlite CG-400 anion exchange resin:Effects of pH, contact time, temperature and presence of phosphate, Ann. Nucl. Energy, 48 (2012) 21-24.

[28] S. Rengaraj, K.H. Yeon, S.Y. Kang, J.U. Lee, K.W. Kim, S.H. Moon, Studies on adsorptive removal of Co(II), Cr(III) and Ni(II) by IRN77 cation-exchange resin, J. Hazard. Mater. 92 (2002) 185-198.

[29] M. Singh, A. Sengupta, Sk. Jayabun, T. Ippili. Understanding the extraction mechanism, radiolytic stability and stripping behavior of thorium by ionic liquid based solvent systems: evidence of ion-exchange and solvation mechanism, J. Radioanal. Nucl. Ch. 311 (2017) 195-208.

[30] M. Elibol, D. Ozer, Response surface analysis of lipase production by freely suspended Rhizopus arrhizus, Process Biochem. 38 (2002) 367-372.

[31] M. Gavrilescu, Removal of heavy metals from the environment by biosorption, Eng. Life Sci. 4 (2004) 219-232.

[32] F. Ghorbani, H. Younesi, S.M. Ghasempouri, A.A. Zinatizadeh, M. Amini, A. Daneshi, Application of response surface methodology for optimization of cadmium biosorption in an aqueous solution by saccharomyces serevisiae, Chem. Eng. J. 145 (2008) 267-275.

[33] V.K. Gupta, B. Gupta, A. Rastogi, S. Agarwal, A. Nayak, A comparative investigation on adsorption performances of mesoporous activated carbon prepared from waste rubber tire and activated carbon for a hazardous azo dye-Acid Blue 113, J. Hazard. Mater. 186 (2011) 891-901.

[34] V.K. Gupta, A. Mittal, A. Malviya, and J. Mittal, Adsorption of carmoisine a from waste materials-bottom ash and deoiled soya, J. Colloid. Interface Sci., 355 (2009) 24–33.

[35] T.S. Anirudhan, S. Jalajamony, Ethylthio-semicarbazide intercalated organophilic calcined hydrotalcite as a potential sorbent for the removal of uranium(VI) and thorium(IV) ions from aqueous solutions, J. Env. Sci. 25 (2013) 717-725.

[36] G.H. Mirzabe, A.R. Keshtkar, Application of response surface methodology for thorium adsorption on PVA/Fe3O4/SiO2/APTES nanohybrid adsorbent, J. Ind. Eng. Chem. 26 (2015) 277-285.

[37] T.S. Anirudhan, S. Rijith, A.R. Tharun, Adsorptive removal of thorium(IV) from aqueous solution using poly (methacrylic acid)-grafted chitosan/bentonite composite matrix: Process design and equilibrium studies, Colloid. Surface. A, 368 (2010) 13-22.

[38] I. Langmuir, The constitution and fundamental properties of solids and liquids. part I. solids, J. Am. Chem. Soc. 38 (1916) 2221-2295.

[39] Y.S. Ho, G. Mckay, Pseudo-second order model for sorption processes, Process Biochem. 34 (1999) 451-465.

# 6

# Investigation of mass transfer coefficients in irregular packed liquid-liquid extraction columns in the presence of various nanoparticles

Ali Vesal, Ahmad Rahbar-Kelishami*, Toraj Mohammadi

*1* *Department of Chemical Engineering, Iran University of Science and Technology (IUST), Tehran, Iran*

## HIGHLIGHTS

- In this study, the effect of various nanoparticles on the mass transfer coefficient was investigated.

- Maximum enhancements in mass transfer coefficient of 35%, 245% and 207% were achieved in the presence of $SiO_2$, $TiO_2$ and $ZrO_2$, respectively.

- A new conceptual model was proposed for prediction of the effective diffusivity as a function of nanoparticle concentration, drop size and drop Reynolds number with a high accuracy.

## GRAPHICAL ABSTRACT

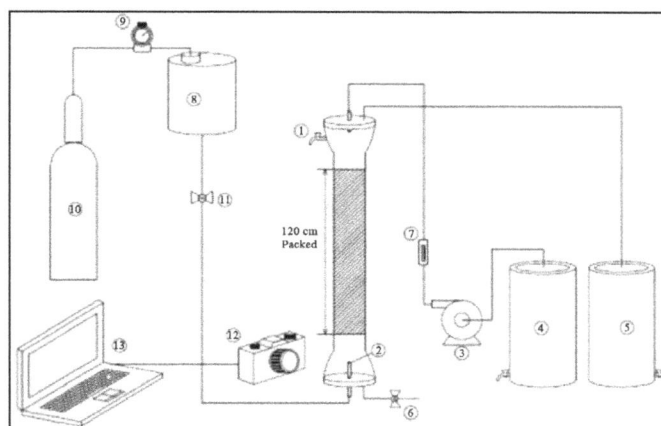

## ARTICLE INFO

*Keywords:*
Liquid-liquid extraction
Nanoparticles
Mass transfer coefficient
Hydrophobic
Hydrophilic

## ABSTRACT

In the present study, the effect of various nanofluids on mass transfer coefficients in an irregular packed liquid-liquid extraction column was investigated. The chemical system of toluene-acetic acid-water was used. 10 nm $SiO_2$, $TiO_2$ and $ZrO_2$ nanoparticles with various concentrations were dispersed in toluene-acid acetic to provide nanofluids. The influence of concentration and hydrophobicity/hydrophilicity of nanoparticle on mass transfer coefficient was discussed. The experimental results show that the mass transfer coefficient enhancement depends on the kind and the concentration of nanoparticles. The Maximum enhancement of 35%, 245% and 207% was achieved for 0.05 vol% of $SiO_2$, $TiO_2$ and $ZrO_2$ nanofluids, respectively. A new conceptual model was proposed for prediction of the effective diffusivity as a function of nanoparticle concentration, drop size and drop Reynolds number.

\* *Corresponding author:   ;  E-mail address: ahmadrahbar@iust.ac.ir*

## 1. Introduction

Nanofluids include various nanoparticles, such as metallic or nonmetallic, with a size less than 100 nm and have varied applications such as heat exchanger, medical, nuclear reactor, fuel cell, cooling of electronics, cameras, and displays [1].

Many researchers have worked on improving mass transfer with nanoparticles [2-11]. Recently, some researchers investigated the enhancement of mass transfer coefficients by nanoparticles in a liquid-liquid extraction process. Bahmanyar et al. evaluated mass transfer coefficients in a pulsed liquid-liquid extraction column using $SiO_2$ / kerosene nanofluids. They found that the mass transfer coefficient increased by 4-60% [12,13]. Saien et al. investigated mass transfer from nanofluids single drop in liquid-liquid extraction using two different nanofluids ($\gamma$-$Al_2O_3$/toluene and $Fe_3O_4$ / toluene) with and without a magnetic field [14,15]. They reported a maximum enhancement of 157% for the mass transfer rate. Rahbar et al. investigated the effect of type and concentration of nanoparticles on mass transfer coefficients [16]. Most researchers introduced the Brownian motion of the nanoparticles as the basic propellant for mass transfer enhancement.

In this work, the effect of various nanoparticles with different hydrophobic or hydrophilic properties and the same average size (10 nm) on the mass transfer coefficient in irregular packed liquid-liquid extraction columns has been investigated.

## 2. Experimental

### 2.1. Materials

Acetic acid (Merck, 99.9% w/w), toluene (Merck, 99% w/w) and deionized water were used. Deionized water and toluene with 0.05 vol% of acetic acid were used as the continuous phase and dispersed phase, respectively. Spherical $SiO_2$, $TiO_2$, and $ZrO_2$ nanoparticles with purities of more than 99.9% and the same average size (10 nm) were purchased from the TECNAN Company. The properties of the chemical materials and nanoparticles are given in Tables 1 and 2, respectively.

### 2.2. Experimental set-up

The experimental setup (Figure 1) consists of a Pyrex

**Table 1.** Physical properties of chemical materials at 20 °C [17].

| Property | Dispersed phase | Continuous phase |
|---|---|---|
| $\rho$ (kg/m³) | 882.7 | 1009.7 |
| $\mu$ (mPa.s) | 0.611 | 1.016 |
| $\gamma$ (mN/m) | 27.5 - 30.1 | |
| $D_d$ (m²/s) | 2.92 ×10⁻⁹ | |

**Table 2.** Properties of nanoparticles.

| Nanoparticle | Density (g/mL) | Specific surface (m²/g) | Purity (%) |
|---|---|---|---|
| $SiO_2$ | 2.2 | 180-270 | +99.9 |
| $TiO_2$ | 3.84 | 100-150 | +99.9 |
| $ZrO_2$ | 5.68 | 70-105 | +99.9 |

glass column (5.2 cm diameter and 1.6 m height) as the contactor. The stainless steel Raschig Ring random packings (0.9 porosity and 10 mm diameter) were used to fill the contactor. The column contains a perforated stainless steel tray to hold the packing. The packing height is 1.2 m in the column.

Three containers were used for the continuous phase, dispersed phase and the extract. The dispersed phase container was installed at a height of 2.5 meters from the ground level to supply sufficient pressure for the push dispersed phase into the continuous phase. The air pressure was applied on the dispersed phase. The dispersed phase enters the column bottom by a steel nozzle (17 cm length, 9.5 mm external diameter and 4.5 mm inner diameter). A solenoid valve was used to regulate the flow rate of the dispersed phase. The water enters through a peripheral pump at the top of the column. A rotameter was applied to have a constant water flow rate of 50 ml/min. Sampling was performed using the valve at the top of the column.

### 2.3. Preparation of nanofluids

The nanofluids were prepared by dispersing various concentrations of nanoparticles (0, 0.01, 0.05 and 0.1 vol%) into the dispersed phase. For this purpose a Hielscher ultrasonic vibrator was used for about one hour duration. The stability of nanofluids was evaluated by the sedimentation method.

### 2.4. Operation procedure

Before each experiment, both the continuous and

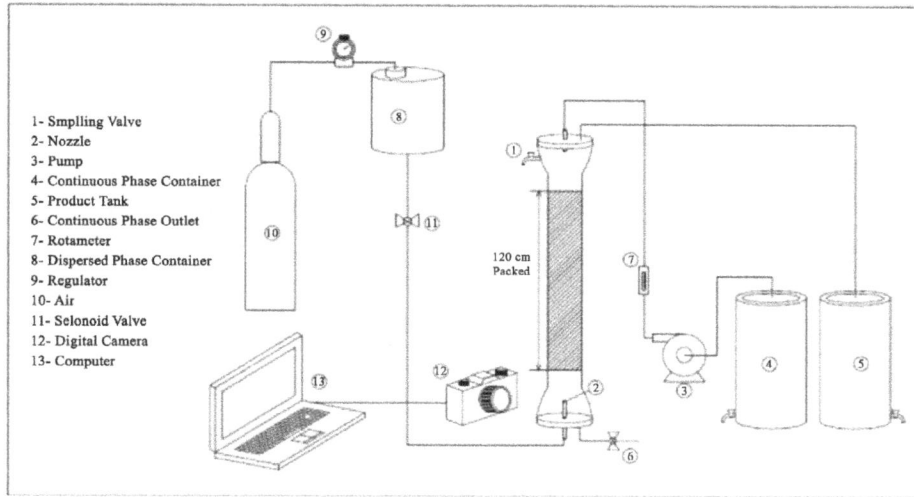

**Fig. 1.** Schematic diagram of the experimental setup.

the dispersed phases were saturated by each other. To beginning, the continuous phase entered from the top of the column, which was filled to the desired height. Afterward, the discharge valve of the continuous phase was opened such its level remains constant. The dispersed phase was then injected into the continuous phase from the bottom. The flow rate of the dispersed phase was set by the solenoid valve. If the pressure of the dispersed phase was not enough for dispersion into the continuous phase, the pressure was set by the pressure valve.

The diameter of the dispersed phase droplets was determined by taking digital photos. The mean diameter of the droplets was calculated by:

$$d = \frac{\sum n_i d_i^3}{\sum n_i d_i^2} \tag{1}$$

where $n_i$ is the number of droplets and $d_i$ is the measured droplet diameter.

In a steady state condition, sampling was carried out from the dispersed phase by the sampling valve and then was separated from the continuous phase by a decanter. The acetic acid concentration in the sample was determined by titration with a 0.1 N NaOH and in the presence of phenolphthalein indicator. 5 ml of sample was used for each titration.

Hold-up was determined by the shutdown method. At the end of each test run, the inlet and outlet valves of dispersed phase were closed simultaneously, and the droplets were allowed to coalesce at the interface. The hold-up was then calculated by:

$$\varphi = \frac{V_D}{V_D + V_C} \tag{2}$$

where, $V_D$ and $V_C$ are the collected volumes of the dispersed and continuous phases, respectively.

Mass transfer direction was from the dispersed phase to continuous phase. All experiments were performed at 25 °C.

*2.5. Determination of the mass transfer coefficient*

The mass transfer coefficient in the extractor is one of the most important parameters in industrial design. Considering the mass transfer during measured contact time, the mass balance relation is:

$$\dot{m}_1 - \dot{m}_2 = \frac{dm}{dt} \tag{3}$$

$$-N_{Ar}.S_r.M_A = \frac{d(n_A M_A)}{dt} = V_A \frac{dC_A}{dt} \tag{4}$$

$$-K_d(C_A - C_A^*).4\pi r^2 = \frac{4}{3}\pi r^3.\frac{dC_A}{dt} \tag{5}$$

The above mass balance is valid only if the droplet diameter and mass transfer coefficient remain constant while the drop rises through the column. By Integration of the equation, assuming the continuous phase to be completely mixed, the mass transfer coefficient can be obtained:

$$K_d = -\frac{d}{6t}\ln(1-E) \tag{6}$$

where

$$E = \frac{C_{A0} - C_A}{C_{A0} - C_A^*} \tag{7}$$

$C_{A0}$ is the initial solute concentration, $C_A$ is the final solute concentration in a specific position, and $C_A^*$ is the equilibrium solute concentration. $C_A$ was measured by titration and $C_A^*$ can be assumed zero because the solute concentration in the continuous phase is negligible, $d$ is the mean droplet diameter and $t$ is the contact time which was obtained by [18]:

$$t = \frac{LS\,\varepsilon\varphi}{Q_d} \tag{8}$$

where $L$, $S$ and $\varepsilon$ are height, cross-sectional area, and voidage of the column, respectively. $Q_d$ and $\varphi$ are dispersed phase volume flow rate and hold-up.

## 3. Results and discussion

### 3.1. Stability of nanofuids

The nanofluids stability is shown in Figure 2. It was observed that the stability of $SiO_2$ was more than $TiO_2$ and the stability of $TiO_2$ was also more than $ZrO_2$. It was concluded that the nanoparticles with lower density and hydrophobic property have better distribution stability.

### 3.2. Mass transfer coefficient

Experimental data for mass transfer coefficient s in the absence of nanoparticles are presented in Table 3. Tables 4, 5 and 6 provide all experimental data for mass transfer coefficients of $SiO_2$, $TiO_2$ and $ZrO_2$, respectively. It should be noted that each experiment was repeated two times and the average value of these two experiments was reported in the paper.

Figure 3 shows the effect of $SiO_2$ nanoparticles on the mass transfer coefficient. It was found that the mass transfer coefficient improves in the presence of $SiO_2$ nanoparticles and a maximum enhancement of 35% in a concentration of 0.05 vol% was achieved. This enhancement may be due to microconvection caused by Brownian motion of nanoparticles. It can be observed from Figure 3 that in higher concentrations of

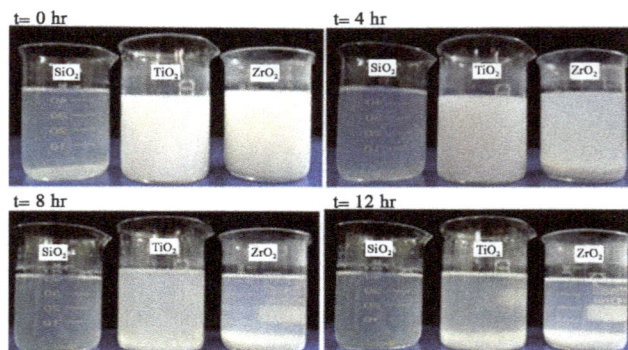

**Fig. 2.** Comparison of distribution stability for different nanoparticles in 0.1 vol%.

$SiO_2$ (usually more than 0.05 vol%) the mass transfer coefficient reduced due to aggregation of nanoparticles and a consequent reduction in the Brownian motion velocity.

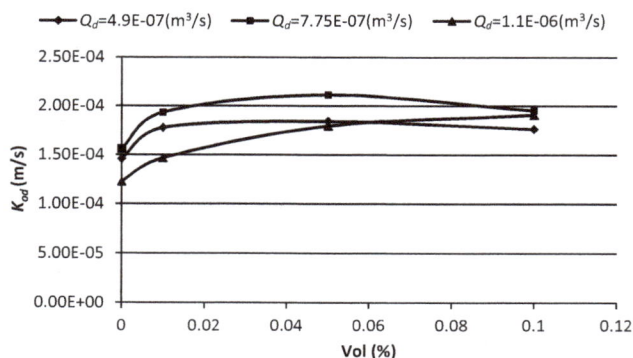

**Fig. 3.** Variation of mass transfer coefficient with concentration of $SiO_2$.

Figures 4-6 show mass transfer coefficient enhancement in the presence of $TiO_2$ and $ZrO_2$ nanoparticles. As observed, maximum enhancements of mass transfer coefficient were 245% and 207% for $TiO_2$ and $ZrO_2$ (in 0.05 vol%), respectively, which was very significant. This high enhancement is not only caused by Brownian motion of nanoparticles. As shown in Figure 7, hydrophilic nanoparticles ($TiO_2$ and $ZrO_2$) tend to transfer from the organic phase to the aqueous phase which created turbulence at the interfacial surface, and consequently the mass transfer coefficient was significantly enhanced.

**Table 3.** Experimental data for mass transfer coefficients at the absence of nanoparticles.

| $Q_d$ (m³/s) | $d$ (mm) | Dynamic Hold up | $E \times 100$ | $t$ (s) | $K_d \times 10^4$ (m/s) | $D_{eff}$ (m²/s) |
|---|---|---|---|---|---|---|
| $4.88 \times 10^{-7}$ | 9.11 | 0.006 | 93.56 | 28.6 | 1.46 | $9.96 \times 10^{-8}$ |
| $7.75 \times 10^{-7}$ | 8.71 | 0.008 | 92.41 | 24.0 | 1.56 | $1.03 \times 10^{-7}$ |
| $1.09 \times 10^{-6}$ | 8.30 | 0.012 | 89.66 | 25.6 | 1.23 | $7.73 \times 10^{-8}$ |

**Table 4.** Experimental data for mass transfer coefficients of $SiO_2$ particles.

| Concentration (vol%) | $Q_d$ (m³/s) | $d$ (mm) | Dynamic Hold-up | $E \times 100$ | $t$ (s) | $K_d \times 10^4$ (m/s) | $D_{eff}$ (m²/s) |
|---|---|---|---|---|---|---|---|
| 0.01 | 4.88×10⁻⁷ | 9.03 | 0.006 | 95.86 | 27.0 | 1.78 | 1.26×10⁻⁷ |
|  | 7.75×10⁻⁷ | 9.63 | 0.008 | 93.10 | 22.2 | 1.93 | 1.32×10⁻⁷ |
|  | 1.09×10⁻⁶ | 9.71 | 0.013 | 90.80 | 26.3 | 1.47 | 9.57×10⁻⁸ |
| 0.05 | 4.88×10⁻⁷ | 11.06 | 0.008 | 97.70 | 37.8 | 1.84 | 1.13×10⁻⁷ |
|  | 7.75×10⁻⁷ | 11.75 | 0.009 | 94.48 | 26.9 | 2.11 | 1.91×10⁻⁷ |
|  | 1.09×10⁻⁶ | 12.15 | 0.014 | 93.10 | 30.2 | 1.79 | 1.52×10⁻⁷ |
| 0.10 | 4.88×10⁻⁷ | 11.87 | 0.009 | 97.47 | 41.3 | 1.76 | 1.43×10⁻⁷ |
|  | 7.75×10⁻⁷ | 12.65 | 0.010 | 93.79 | 30.0 | 1.95 | 1.49×10⁻⁷ |
|  | 1.09×10⁻⁶ | 14.00 | 0.015 | 92.18 | 31.2 | 1.91 | 2.28×10⁻⁷ |

**Table 5.** Experimental data for mass transfer coefficients of $TiO_2$ particles.

| Concentration (vol%) | $Q_d$ (m³/s) | $d$ (mm) | Dynamic Hold-up | $E \times 100$ | $t$ (s) | $K_d \times 10^4$ (m/s) | $D_{eff}$ (m²/s) |
|---|---|---|---|---|---|---|---|
| 0.01 | 4.88×10⁻⁷ | 9.2 | 0.003 | 98.85 | 14.1 | 4.86 | 3.75×10⁻⁷ |
|  | 7.75×10⁻⁷ | 10.2 | 0.006 | 97.24 | 16.9 | 3.61 | 2.60×10⁻⁷ |
|  | 1.09×10⁻⁶ | 10.7 | 0.009 | 96.78 | 18.1 | 3.39 | 2.44×10⁻⁷ |
| 0.05 | 4.88×10⁻⁷ | 10.5 | 0.004 | 99.20 | 16.8 | 5.03 | 3.38×10⁻⁷ |
|  | 7.75×10⁻⁷ | 10.6 | 0.007 | 98.16 | 19.5 | 3.63 | 3.45×10⁻⁷ |
|  | 1.09×10⁻⁶ | 11 | 0.011 | 97.70 | 22.6 | 3.06 | 2.78×10⁻⁷ |
| 0.10 | 4.88×10⁻⁷ | 10.6 | 0.004 | 99.35 | 18.1 | 4.91 | 4.42×10⁻⁷ |
|  | 7.75×10⁻⁷ | 10.8 | 0.006 | 97.24 | 18.2 | 3.56 | 2.94×10⁻⁷ |
|  | 1.09×10⁻⁶ | 12 | 0.010 | 96.09 | 21.1 | 3.08 | 3.90×10⁻⁷ |

**Table 6.** Experimental data for mass transfer coefficients of $ZrO_2$ particles.

| Concentration (vol%) | $Q_d$ (m³/s) | $d$ (mm) | Dynamic Hold-up | $E \times 100$ | $t$ (s) | $K_d \times 10^4$ (m/s) | $D_{eff}$ (m²/s) |
|---|---|---|---|---|---|---|---|
| 0.01 | 4.88×10⁻⁷ | 9.5 | 0.003 | 97.70 | 14.1 | 4.24 | 3.25×10⁻⁷ |
|  | 7.75×10⁻⁷ | 10.8 | 0.007 | 96.78 | 19.5 | 3.18 | 2.27×10⁻⁷ |
|  | 1.09×10⁻⁶ | 11.5 | 0.009 | 96.09 | 18.7 | 3.33 | 2.39×10⁻⁷ |
| 0.05 | 4.88×10⁻⁷ | 10.1 | 0.004 | 98.85 | 16.8 | 4.48 | 2.99×10⁻⁷ |
|  | 7.75×10⁻⁷ | 11.0 | 0.006 | 97.47 | 19.0 | 3.55 | 3.37×10⁻⁷ |
|  | 1.09×10⁻⁶ | 12.2 | 0.009 | 96.55 | 19.6 | 3.50 | 3.21×10⁻⁷ |
| 0.10 | 4.88×10⁻⁷ | 10.2 | 0.004 | 98.53 | 16.8 | 4.26 | 3.80×10⁻⁷ |
|  | 7.75×10⁻⁷ | 11.2 | 0.006 | 95.86 | 16.9 | 3.50 | 2.88×10⁻⁷ |
|  | 1.09×10⁻⁶ | 12.5 | 0.008 | 94.25 | 17.2 | 3.47 | 4.44×10⁻⁷ |

It can also be observed from Figures 4 and 5 that increasing the concentration of $TiO_2$ and $ZrO_2$ more than 0.05 vol% has no significant effect on the mass transfer coefficient. This is due to aggregation of nanoparticles. Aggregation of nanoparticles causes clusters of a hard solid media in the liquid phase. So it acts as an obstacle. As shown in Figure 6, mass transfer coefficient enhancement in the presence of $TiO_2$ and $ZrO_2$ was more significant than for $SiO_2$ because the mass transfer mechanism was affected by the transfer of hydrophilic nanoparticles ($TiO_2$ and $ZrO_2$) from the dispersed phase to the continuous phase, while $SiO_2$ was probably

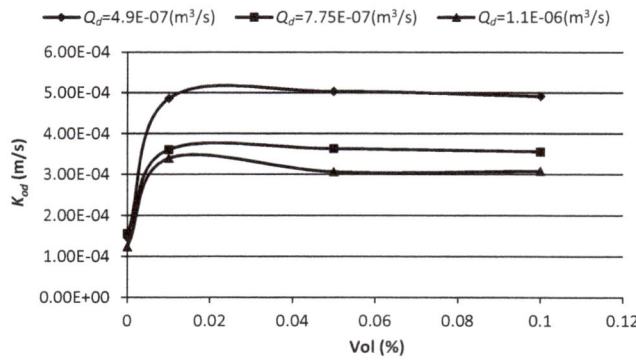

**Fig. 4.** Variation of mass transfer coefficient with concentration of TiO$_2$.

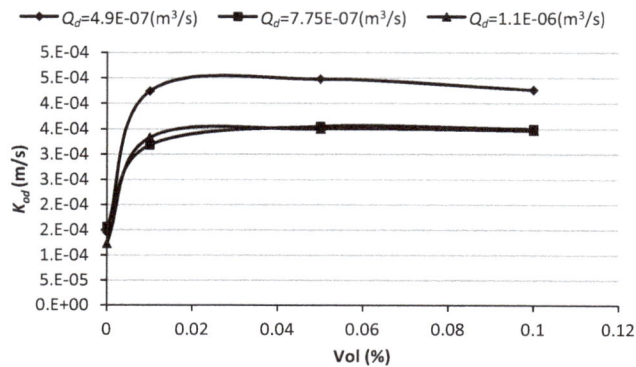

**Fig. 5.** Variation of mass transfer coefficient with concentration of ZrO$_2$.

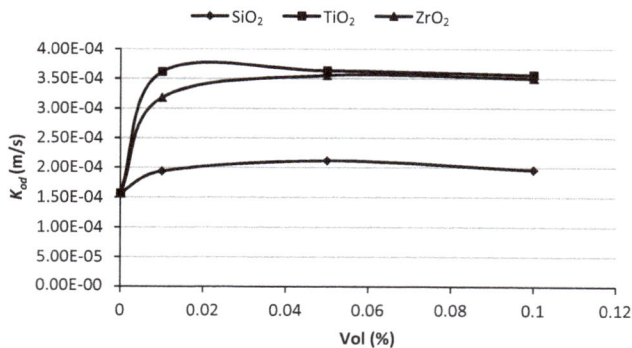

**Fig. 6.** Variation of mass transfer coefficient with concentration for various nanoparticles in $Q_d$ = 7.75E-07 (m$^3$/s).

**Fig. 7.** Hydrophilic and hydrophobic properties of nanoparticles.

affected by the Brownian motion of the nanoparticles.

It can also be concluded from Figure 6 that the mass transfer coefficient in the presence of TiO$_2$ was more than ZrO$_2$. This is due to the super hydrophilicity and lower density of TiO$_2$ with respect to ZrO$_2$.

### 3.3. Predictive correlation for the effective diffusivity

The effective diffusivity was calculated from the experimental values of the mass transfer coefficients. To this purpose, molecular diffusivity ($D_d$) in the Newman equation (6) was replaced with effective diffusivity ($D_{eff}$) [19].

$$K_{od} = -\frac{d}{6t} \ln\left[ \frac{6}{\pi^2} \sum_{n=1}^{\infty} \frac{1}{n^2} \exp(-\frac{4n^2\pi^2 D_d t}{d^2}) \right] \qquad (9)$$

The calculated effective diffusivity for all drop size, nanoparticles concentrations (for $n_c > 0$), and drop Reynolds number was fit to determine a predictive correlation. The predictive correlation is:

$$D_{eff} = a_1 \, \mathrm{Re}^{a_2} (d + a_3 n_c) - a_4 \qquad (10)$$

The coefficients of the predictive model are reported in Table 7.

The average absolute relative error (*AARE*) for the effective diffusivity calculated with this predictive model compared with the experimental results is 9%. The % *AARE* is calculated by:

$$\%AARE = \frac{1}{N} \sum_{i=1}^{N} \left| \frac{Model - Experiment}{Experiment} \right| \times 100 \qquad (11)$$

where, $N$ is the number of data.

A comparison of the experimental effective diffusivities with those calculated by the proposed model is shown in Figure 8. This figure indicates that the suggested correlation can estimate the effective diffusivities with high accuracy. The predictive model shows that the effective diffusivity of hydrophilic nanoparticles (TiO$_2$ and ZrO$_2$) is also more dependant on nanoparticle concentration.

**Table 7.** The coefficients of the Predictive model.

| Nanoparticle | a$_1$ | a$_2$ | a$_3$ | a$_4$ |
|---|---|---|---|---|
| SiO$_2$ | $1.27 \times 10^{-6}$ | 0.443 | $-2.65 \times 10^{-3}$ | $1.95 \times 10^{-8}$ |
| TiO$_2$ & ZrO$_2$ | $2.75 \times 10^{-6}$ | 0.443 | $1.31 \times 10^{-2}$ | $3.35 \times 10^{-8}$ |

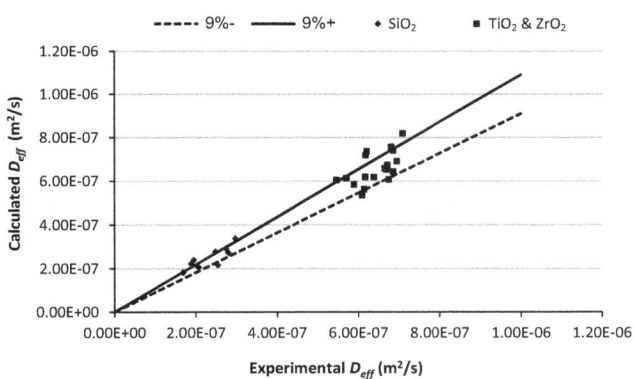

**Fig. 8.** Comparison of experimental and calculated $D_{eff}$.

## 4. Conclusion

In this study the effect of various nanoparticles with the same average size (10 nm) and different hydrophobic/hydrophilic property on the mass transfer coefficient was investigated. It was found through experimental data that the mass transfer coefficient was enhanced in the presence of nanoparticles. Maximum enhancements in the mass transfer coefficient of 35%, 245% and 207% were achieved in the presence of $SiO_2$, $TiO_2$ and $ZrO_2$, respectively, at a concentration of 0.05 vol%. A moderate enhancement in mass transfer coefficient (0-35%) may be due to Brownian motion of nanoparticles. But, the high enhancement in the mass transfer coefficient (more than 100%) in the presence of hydrophilic nanoparticles ($TiO_2$ and $ZrO_2$) was due to transfer of these nanoparticles from the organic to aqueous phase, which created a high turbulence on the interfacial surface. Unfortunately, the suspension of $TiO_2$ and $ZrO_2$ was too unstable and $TiO_2$ is too expensive. So, to achieve better mass transfer coefficient enhancement, the $TiO_2$ nanoparticle is suggested for this chemical system. The proposed model for prediction of effective diffusivity agreed well with the experimental data.

## References

[1] R. Saidura, K.Y. Leongb, H.A. Mohammad, A review on applications and challenges of nanofluids, Renew. Sust. Energ. Rev. 15 (2011) 1646-1668.

[2] S. Krishnamurthy, P. Bhattacharya, P.E. Phelan, R.S. Prasher, Enhanced mass transport in nanofluids, Nano Lett. 6 (2006) 419-42.

[3] B. Olle, S. Bucak, T.C. Holmes, L. Bromberg, T.A. Hatton, D.I.C. Wang, Enhancement of oxygen mass transfer using functionalized magnetic nanoparticles, Ind. Eng. Chem. Res., 45 (2006) 4355-4363.

[4] H. Zhu, B.H. Shanks, T.J. Heindel, Enhancing CO-water mass transfer by functionalized MCM-41 nanoparticles, Ind. Eng. Chem. Res. 47 (2008) 7881-7887.

[5] X. Fang, Y. Xuan, Q. Li, Experimental investigation on enhanced mass transfer in nanofluids, Appl. Phys. Lett. 95 (2009) 203108.

[6] J. Veilleux, S. Coulombe, A total internal reflection fluorescence microscopy study of mass diffusion enhancement in water-based alumina nanofluids, Appl. Phys. 108 (2010) 104316-104318.

[7] J.K. Lee, J. Koo, H. Hong, Y.T. Kang, The effects of nanoparticles on absorption heat and mass transfer performance in $NH_3/H_2O$ binary nanofluids, Int. J. Refrig. 33 (2010) 269-275.

[8] C. Pang, W. Wub, W. Sheng, H. Zhang, Y.T. Kang, Mass transfer enhancement by binary nanofluids ($NH_3/H_2O$ + Ag nanoparticles) for bubble absorption process, In. J. Refrig. 35 (2012) 2240-2247.

[9] I.T. Pineda, J.W. Lee, I. Jung, Y.T. Kang, $CO_2$ absorption enhancement by methanol-based $Al_2O_3$ and $SiO_2$ nanofluids in a tray column absorber, Int. J. Refrig. 35 (2012) 1402-1409.

[10] H. Beiki, M. Nasr Esfahany, N. Etesami, Laminar forced convective mass transfer of g-$Al_2O_3$/ electrolyte nanofluid in a circular tube, Int. J. Therm. Sci. 64 (2013) 251-256.

[11] H. Beiki, M. Nasr Esfahany, N. Etesami, Turbulent mass transfer of $Al_2O_3$ and $TiO_2$ electrolyte nanofluids in circular tube, Microfluid. Nanofluid. 15 (2013) 501-508.

[12] A. Bahmanyar, N. Khoobi, M.R. Mozdianfard, H. Bahmanyar, The influence of nanoparticles on hydrodynamic characteristics and mass transfer performance in a pulsed liquid-liquid extraction column, Chem. Eng. Process. 50 (2011) 1198-1206.

[13] A. Bahmanyar; N. Khoobi, M.M.A. Moharrer, H. Bahmanyar, Mass transfer from nanofluid drops in a pulsed liquid-liquid extraction column, Chem. Eng. Res. Des. 92 (2014) 2313-2323.

[14] J. Saien, H. Bamdadi, Mass transfer from nanofluid single drops inliquid-liquid extraction process, Ind. Eng. Chem. Res. 51 (2012) 5157-5166.

[15] J. Saien, H. Bamdadi, Sh. Daliri, Liquid-liquid extraction intensification with magnetite nanofluid single drops under oscillating magnetic field, Ind.

Eng. Chem. 21 (2014) 1152-1159.

[16] A. Rahbar-Kelishami, S.N. Ashrafizadeh, M. Rahnamaee, The effect of type and concentration of nano-particles on the mass transfer coefficients: experimental and Sherwood number correlating, Sep. Sci. Technol. 50 (2015) 1776-1784.

[17] A. Rahbar, Z. Azizi, H. Bahmanyar, M.A. Moosavian, Prediction of enhancement factor for mass transfer coefficient in regular packed liquid-liquid extraction columns, Can. J. Chem. Eng. 89 (2011) 508-519.

[18] R.E. Treybal, Mass transfer operations, 3rd Ed., McGraw Hill, Japan, 1990.

[19] A.B. Newman, The drying of porous solids: Diffusions and surface emission equations, T. Am. Inst. Chem. Eng. 27 (1931) 203-220.

# 7

# CFD simulation of pervaporation of organic aqueous mixture through silicalite nanopore zeolite membrane

**Mansoor Kazemimoghadam[1,\*], Zahra Amiri-Rigi[2]**

[1] *Department of Chemical Engineering, Malek-Ashtar University of Technology, Tehran, Iran*

[2] *Department of Chemical Engineering, South Tehran Branch, Islamic Azad University, Tehran, Iran*

## HIGHLIGHTS

- Silicalite nanopore zeolite membranes were synthesized by in-situ liquid phase hydrothermal method and studied by XRD and SEM techniques.

- Pervaporation tests were carried out for evaluation of the performance of the membranes in the separation of water-UDMH mixtures.

- A comprehensive steady state model was developed for CFD simulation of pervaporation using the finite element method.

## GRAPHICAL ABSTRACT

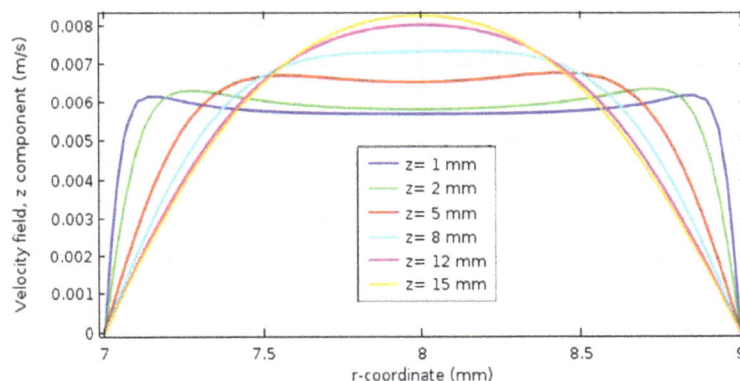

## ARTICLE INFO

*Keywords:*
Silicalite
Zeolite membrane
Pervaporation
Water–UDMH separation
CFD simulation

## ABSTRACT

Nanopore silicalite type membranes were prepared on the outer surface of a porous-mullite tube by in situ liquid phase hydrothermal synthesis. The hydrothermal crystallization was carried out under an autogenously pressure, at a static condition and temperature of 180 °C with tetrapropylammonium bromide (TPABr) as a template agent. The molar composition of the starting gel of silicalite zeolite membrane was: $Na_2O/SiO_2$=0.287-0.450, $H_2O/SiO_2$ = 8-15, $TPABr/SiO_2$ = 0.01-0.04. The zeolites calcinations were carried out in air at 530°C, to burn off the template (TPABr) within the zeolites. X-ray diffraction (XRD) patterns of the membranes consisted of peaks corresponding to the support and zeolite. The crystal species were characterized by XRD, and morphology of the supports subjected to crystallization was characterized by scanning electron microscopy (SEM). Performance of silicalite nanoporous membranes was studied for separation of water-unsymmetrical dimethylhydrazine (UDMH) mixtures using pervaporation (PV). Finally, a comprehensive steady state model was developed for the pervaporation of a water-UDMH mixture by COMSOL Multiphysics software version 5.2. The developed model was strongly capable of predicting the effect of various dimensional factors on concentration and velocity distribution within the membrane module. The best silicalite zeolite membranes had a water flux of 3.34 kg/m².h at 27 °C. The best PV selectivity for silicalite membranes obtained was 53.

\* *Corresponding author: ; E-mail address: mzkazemi@gmail.com*

## 1. Introduction

Unsymmetrical dimethylhydrazine (UDMH) is an organic derivative of hydrazine which is usually used as a propellant. This hazardous material has many other new applications and is widely applied as an oxygen scavenger for boiler-feed water, a starting material for drug and dye intermediates, a catalyst for polymerization reactions, etc. UDMH is very corrosive and its vapor is extremely toxic and carcinogenic [1-3]. Removal of highly hazardous UDMH from water is important for the recovery of valuable organic products, for the recycling of process water and for the treatment of waste water [4]. Generally, traditional azeotropic distillation or extraction can be used to separate organic compounds from their aqueous solutions. However, for low organic concentrations or thermally sensitive organic compounds, distillation is very expensive and also extremely dangerous due to the explosive nature of UDMH.

There has been a growing interest in membrane-based PV technology due to the extreme effectiveness as well as mild operating conditions of this separation technique. Since only a fraction of the solution to be removed is vaporized, the energy consumption will be decreased significantly. Moreover, compared to conventional distillation, PV is a simple technique because only a pump is required to maintain the driving force.

At present, dehydration of organic solvents is the major market of PV. High separation factors and water permeate fluxes are reported in previous studies on pervaporation dehydration of isopropanol, ethanol, n-butanol, n-butyl-acetate, ethylene glycol and acetic acid aqueous solution [5-8]. Uragami et al. investigated the effect of immersion time in $CaCl_2$ or $MgCl_2$ methanol solutions on the permeation flux and separation factor of pervaporation dehydration of ethanol aqueous solution using Alg-DNA/$Mg^{2+}$ membrane [9]. Their results showed that after immersing the membrane in methanol solutions, the separation factor increased remarkably for the first 12 hours, after which it started to fall [9].

Zeolite membranes are usually used in pervaporation processes due to their strong potential. These membranes are synthetized using various methods such as hydrothermal in-situ crystallization, chemical vapor phase technique and spray seed coating. Zeolite NaA membranes were reported to be excellent materials for solvent dehydration by PV [10]. But under slightly severe conditions and under hydrothermal stresses, zeolite NaA membranes behaved unsuitably due to hydrolysis. There have been only a few attempts to develop hydrophilic highly siliceous zeolite membranes of different Si/Al ratios with improved hydrothermal stabilities.

In PV, the feed mixture is contacted with a nonporous perm selective membrane. In general, separation is explained by the steps of sorption into, diffusion through and desorption from the membrane. The latter is usually considered to be fast and taking place at equilibrium, while diffusion is kinetically controlled and the slowest step of the process. Permeation is dependent on sorption and diffusion steps. The driving force for the filtration is created by maintaining a pressure lower than the saturation pressure on the permeate side of the membrane. The mechanism of separation is usually explained in terms of sorption-diffusion processes [11-13].

Many studies have been conducted to model concentration distribution within the membrane module in order to commercialize PV separation systems. There are two major approaches to PV simulations: Molecular Dynamic (MD) simulation and Computational Fluid Dynamic (CFD) simulation. Based on MD, Huang et al. developed a model to explain free-volume form and the flexibility and stiffness of polymer chain. Their results, obtained from MD simulations, were in good harmony with the chemical structure of the polyelectrolyte complex membranes (PECMs) [14]. Jain et al. (2017) proposed a mathematical model for the purification of n-heptane/thiophene model gasoline using a tubular pervaporation membrane module [15]. Their findings showed that the dimensional factors had positive effects on separation performance of pervaporation membranes.

Based on CFD simulation, Moulik et al. developed a steady state model to predict concentration distribution within the membrane module in pervaporation of acetic acid solution [16]. Their results were in good agreement with experimental data, but their model was not perfect, since they didn't model the concentration distribution within the feed section, which significantly affects the concentration profile on the membrane side. Prasad et al. also developed a 2D steady state model using the CFD technique [17]. They also modeled the membrane section only and assumed the conditions to be steady state.

As understood, a comprehensive model is required, capable of predicting concentration distribution within both membrane and feed sub-domains. In this paper, preparation methods of the nanopore silicalite-1 zeolite membrane on mullite support are reported. Performances of the membranes prepared by hydrothermal in situ crystallization were studied in separation of the water-UDMH by PV. Finally, a comprehensive steady state 2D model was proposed based on solving Navier-Stokes equations of mass and momentum transfer, simultaneously. The conservation equations were solved using COMSOL Multiphysics software version 5.2. COMSOL applies the Finite Element Method (FEM) to solve the equations numerically. Effect of various membrane dimensions and feed flow rates was investigated to find the optimum operating conditions. The model obtained here was masterfully capable of predicting concentration distribution of water through both membrane and feed sides of the separation module. The results indicated that the effect of dimensional factors related to the membrane module geometry on concentration distribution is very important and cannot be neglected.

## 2. Experimental

### 2.1. Support preparation

In ceramic membranes, thin dense layers are usually deposited over porous supports. The porous supports provide mechanical strength for the thin selective layers. Porous supports can be made from alumina, cordierite, mullite, silica, spinel, zirconia, other refractory oxides and various oxide mixtures, carbon, sintered metals and silicon carbide.

In this research, mullite supports were prepared from kaolin clay. Kaolin is thermally converted to mullite via high temperature calcinations. The reaction takes place when kaolin is utilized as the sole source of silica and alumina. The reaction can be represented by the following equation:

$$3(Al_2O_3.2SiO_2) \longrightarrow 3Al_2O_3.2SiO_2 + 4SiO_2 \qquad (1)$$

Free silica ($4SiO_2$) is generated as a result of this conversion. Free silica was leached out and then porous mullite bodies were prepared. Mullite has several distinct advantages over other materials. Since kaolin is heated to high temperatures to achieve the mullite conversion reaction, strong inter-crystalline bonds

between mullite crystals are formed and this will result in excellent strength and attrition. Leaching time depends on several factors including:
1) the quantity of free silica to be removed,
2) the porosity of body prior to leaching,
3) the concentration of leaching solution, and
4) temperature.

Kaolin (SL-KAD grade) was supplied by WBB cooperation, England. Analysis of the kaolin is listed in Table 1. Cylindrical shaped (tubular) bodies were conveniently made by extruding a mixture of about 75-67% kaolin and 25-33% distilled water. Suitable calcination temperatures and periods are those at which kaolin converts to mullite and free silica. Good results were achieved by calcining for about 3 h at temperatures of about 1250 °C [18].

**Table 1.** Analysis of kaolin clay.

| Component | Percent (%) | Phases | Percent (%) |
|---|---|---|---|
| $SiO_2$ | 51.9 | Kaolinite | 79 |
| $TiO_2$ | 0.1 | Illite | 8 |
| $Al_2O_3$ | 34.1 | Quartz | 10 |
| $Fe_2O_3$ | 1.4 | Feldspar | 3 |
| $K_2O$ | 0.8 | Total | 100 |
| $Na_2O$ | 0.1 | | |
| L.O.I | 11.6 | | |
| Total | 100 | | |

Free silica was removed from the calcined bodies after leaching by strong alkali solutions. Removal of the silica causes mesoporous tubular supports with very high porosity to be made. Free silica removal was carried out using aqueous solutions containing 20% by weight NaOH at a temperature of 80 °C for 5 h. Supports were rinsed using a lot of hot distilled water for a long time in order to remove all remaining NaOH. Porosity of the supports before leaching was 24.3%, while after treatment it increased to 49%. Flux of the supports before and after free silica removal at 1 bar and 20 °C were 6 and 10 kg/m²h, respectively. Porosity of the supports was measured by the water absorption method. Inner and outer diameters and length of the support were 10, 14 and 100 mm, respectively.

### 2.2. Silicalite zeolite membrane synthesis

Zeolite membranes were synthesized on the outer

surface of the porous mullite tubes. The molar gel compositions of the silicalite membranes were: 0.287-0.450 $Na_2O$ : 1.0 $SiO_2$ : 0.01-0.04 TPABr : 8-15 $H_2O$, where TPABr was used as a template [18-23]. Sodium silicate and sodium aluminate were used as the Si and Al sources, respectively. For silicalite-1 preparation, two solutions were prepared; solution A: sodium silicate and solution B: TPABr + $H_2O$ + NaOH. Solution A was added to solution B with stirring. To obtain a homogeneous gel, the mixtures were stirred for 2 h at room temperature.

For membrane preparation, two ends of the supports were closed with rubber caps to avoid any precipitation of the zeolite crystals on the inner surface of the supports during membrane synthesis. The seeded supports were placed vertically in a Teflon autoclave. The solution was carefully poured into the autoclave and then the autoclave was sealed. Crystallization was carried out in an oven at a temperature 180 °C for 24 h. Then, the samples were taken and the synthesized membranes were washed several times with distilled water. The samples were then dried in air at room temperature for 12 h and then dried in the oven at 100 °C for 15 h to remove water occluded in the zeolite crystals and then calcinations were carried out in air at 530 °C for 8 h at a heating rate of 1 °C/min [21, 10, 23-28].

Phase identification was performed by XRD (Philips PW1710, Philips Co., Netherlands) with CuKα radiation. Also, morphological studies were performed using SEM (JEM-1200 or JEM-5600LV equipped with an Oxford ISIS-300 X-ray disperse spectroscopy, EDS).

### 2.3. Pervaporation tests

While the PV system was at steady state after 20 min, weight of permeate was measured at the 30 min period and then flux was calculated (surface area of the zeolite membrane was 44 $cm^2$).

The zeolite membranes were used for long-term dehydration of UDMH aqueous solutions. Experiments were carried out at a temperature of 30 °C and a pressure of 1.5 mbar at the permeate side, within a period of 30-60 min.

The pervaporation setup is presented in Figure 1. Any change of feed concentration due to the permeation is negligible because the amount of permeate is small (max. 2 ml) compared to the total feed volume in the system (0.5 lit). A three stage diaphragm vacuum pump

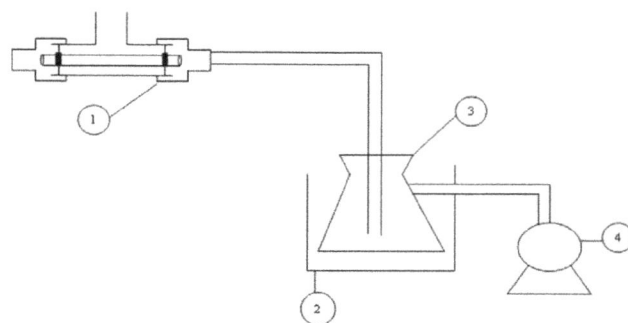

Fig. 1. PV setup; 1: feed container and PV cell, 2: liquid nitrogen trap, 3: permeate container, 4: three stage vacuum pump.

(vacuubrand, GMBH, Germany) was employed to evacuate the permeate side of the membrane to a pressure of approximately 1.5 mbar while the feed side was kept at room pressure. The permeate side was connected to a liquid nitrogen trap via a hose to condense the permeate (vapor). Permeate concentrations were measured using GC (TCD detector, Varian 3400, carrier gas: hydrogen, column: polyethylene glycol, sample size: 5 μm, column and detector temperatures: 120-150 °C, detector flow: 15 ml/min, carrier flow: 5 ml/min, column pressure: 1.6 kPa, GC input pressure: 20 kPa). Performance of PV was evaluated using values of total flux (kg/$m^2$.h) and separation factor (dimensionless). The separation factor of UDMH aqueous solution ($\alpha$) can be calculated from the following equation:

$$\alpha = \frac{\left[y_{H_2O}/y_{UDMH}\right]}{\left[x_{H_2O}/x_{UDMH}\right]} \qquad (2)$$

where $y_{H_2O}$ and $y_{UDMH}$ are the weight fractions of water and UDMH in the permeate, and $x_{H_2O}$ and $x_{UDMH}$ are weight fractions in the feed, respectively [29-31].

## 3. Modeling

Figure 2 represents the schematic diagram of the model domain used in the simulation. Feed solution containing a mixture of 5 wt% UDMH and 95 wt% water flows tangentially through the upper side of the membrane system (z=0) and exits at z=L.

The main assumptions to develop the numerical simulation are as follows:
• Steady state condition is considered,
• Temperature is constant,
• No chemical reaction occurs in the feed stream,
• Feed solution flows only in the z direction,

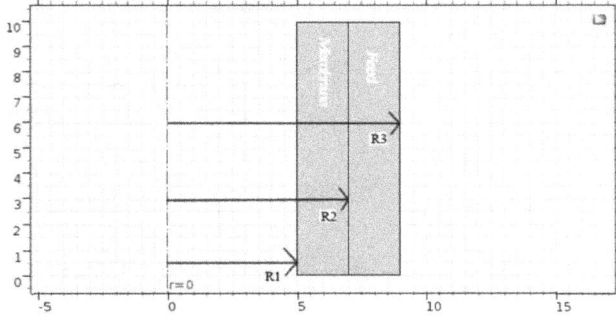

Fig. 2. Vertical diagram of the geometry of the model domain used in simulation

- Feed flow is laminar in the membrane system,
- Thermodynamic equilibrium is considered at the interface of the feed and membrane,
- A small amount of UDMH permeates through the membrane,
- Mass transfer resistance of the support layer is assumed to be negligible,
- Fouling and concentration polarization effects on the PV of UDMH solution are negligible and
- Feed viscosity and density are constant.

Axial and radial diffusions inside the membrane and feed phase are considered in the continuity equations. Moreover, small permeation of UDMH through the membrane is considered in the simulation by applying the selectivity equation (2).

Concentration of UDMH in the permeate side ($y_{UDMH}$) must be determined by the trial and error method. In this method, an initial value for $y_{UDMH}$ is guessed. Then the concentration the permeate side is calculated using model equations. This calculated value will then be compared with the guessed value. If the difference between the old and new values is less than a determined error, the guessed UDMH concentration is considered as the correct concentration. Otherwise, another guess must be made for $y_{UDMH}$.

Mass transport in the membrane system is described using the continuity equation. The following equation presents the differential form of this equation [32]:

$$\frac{\partial C_{H_2O}}{\partial t} + \nabla \cdot \left( -D_{H_2O} \nabla C_{H_2O} + U \cdot C_{H_2O} \right) = R \qquad (3)$$

where $C_{H_2O}$, $D_{H_2O}$, U and R denote the water concentration (mol/m³), water diffusion coefficient (m²/s), velocity vector (m/s) and reaction term (mol/m³.s), respectively. Since no chemical reactions take place in UDMH/water PV, the reaction term is zero. The continuity equation

was defined and solved in COMSOL Multiphysics 5.2 by adding a "transport of diluted species" physic to the whole model.

Velocity distribution was obtained by solving the Navier-Stokes equation for momentum balance, simultaneously with the continuity equation in the feed section. This was done by adding a "laminar flow" physic to the whole model in COMSOL Multiphysics 5.2. The following equation describes the momentum conservation equation [32]:

$$\rho \frac{\partial u}{\partial t} + \rho(u.\nabla)u = \nabla \cdot [-P + \mu(\nabla u + (\nabla u)^T)] + F \qquad (4)$$

$$\nabla \cdot (u) = 0 \qquad (5)$$

where u, $\rho$, P, $\mu$ and F denote the z-component of the velocity vector (m/s), feed density (kg/m³), pressure (Pa), feed viscosity (Pa.s) and body force (N), respectively.

### 3.1. Feed phase simulation

By applying the mentioned assumptions to Eq. (3), the steady state form of the continuity equation for water mass transport in the feed side is obtained:

$$-\frac{1}{r}\frac{\partial}{\partial r}\left(D_{H_2O}r\frac{\partial C_{H_2O-feed}}{\partial r}\right) - \frac{\partial}{\partial z}\left(D_{H_2O}\frac{\partial C_{H_2O-feed}}{\partial z}\right) + u\frac{\partial C_{H_2O-feed}}{\partial z} = 0 \qquad (6)$$

where $C_{H_2O-feed}$ is the water concentration in the feed phase. The simplified form of the momentum transport equations considering the above assumptions will be as follows:

$$\rho\left(u\frac{\partial u}{\partial z}\right) - \frac{1}{r}\frac{\partial}{\partial r}\left(r\mu\frac{\partial u}{\partial r}\right) - \frac{\partial}{\partial z}\left(\mu\frac{\partial u}{\partial z}\right) = -\frac{\partial P}{\partial z} \qquad (7)$$

$$\frac{\partial u}{\partial z} = 0 \qquad (8)$$

where r and z denote the radial and axial coordinates, respectively.

The initial conditions for mass and momentum conservation equations are as follows:

$$C_{H_2O-feed} = C_{0,H_2O} \quad \text{and} \quad u = u_0 \qquad (9)$$

where $C_{0,H_2O}$ is the water initial concentration and $u_0$ is the initial velocity of the feed flow.

The boundary conditions for mass conservation

equations in the feed phase are as follows:

at z = L, Outflow condition                                    (10)

at  z= 0, $C_{H_2O\text{-}feed} = C_{0,H_2O}$                   (11)

at r = $R_3$, No flux condition                                (12)

where $R_3$ is the outer radius of the feed section. At the interface of the membrane-feed, the equilibrium condition is assumed:

at r = $R_2$, $C_{H_2O-feed} = \dfrac{C_{H_2O-membrane}}{p}$    (13)

in which $C_{H_2O\text{-}membrane}$ is the water concentration in the membrane section and p is the partition coefficient obtained from the selectivity equation as follows:

$$p = \frac{y_{UDMH}}{x_{UDMH}} \times \propto = \frac{y_{H_2O}}{x_{H_2O}}$$    (14)

As mentioned earlier, the permeate concentration of UDMH must be determined using the trial and error method, and then is placed in the above equation.

The boundary conditions for momentum transfer equations are as follows:

at z = 0, u = $u_0$, (Inlet velocity)                         (15)

At the outlet, the pressure is atmospheric pressure:

at z = L, P = $P_{atm}$, (Atmospheric pressure)               (16)

at r = $R_2$, u = 0 (No slip condition)                       (17)

at r = $R_3$, u = 0 (No slip condition)                       (18)

### 3.2. Membrane phase simulation

Mass transport of water in the membrane is controlled only by the diffusion mechanism. Therefore, the steady state continuity equation for water can be written as:

$$-\frac{1}{r}\frac{\partial}{\partial r}\left(D_{H_2O-membrane}\, r\, \frac{\partial C_{H_2O-membrane}}{\partial r}\right)$$

$$-\frac{\partial}{\partial z}\left(D_{H_2O-membrane}\, \frac{\partial C_{H_2O-membrane}}{\partial z}\right) = 0$$    (19)

where $D_{H_2O\text{-}membrane}$ is the water diffusion coefficient in the membrane ($m^2/s$).

Membrane phase boundary conditions are given as:

at r = $R_2$, $C_{H_2O\text{-}membrane} = p \times C_{H_2O\text{-}feed}$ (Equilibrium condition)    (20)

at r = $R_1$, $C_{H_2O\text{-}membrane} = 0$ (Dry membrane condition)    (21)

at z = 0 and z = L, $\dfrac{\partial C_{H_2O-membrane}}{\partial z} = 0$ (No flux condition)    (22)

At the permeate-membrane interface, water concentration was assumed to be zero due to the vacuum applied.

### 3.3. Numerical solution of conservation equations

The set of model equations, including mass and momentum transfer equations in the membrane module along with suitable boundary conditions, was solved using COMSOL Multiphysics software version 5.2. Finite element method is used by this software to solve conservation equations numerically. The computational time for solving the equations was about 2 min. "Extra fine mesh" was used for meshing in this simulation. Complete mesh consisted of 30558 domain elements and 975 boundary elements to solve the set of equations. Figure 3 represents the meshes created by COMSOL Multiphysics 5.2 software. Due to the considerable difference between z and r dimensions, a scaling factor equal to 10 was used in the z direction. Therefore, the results were reported in dimensionless length.

## 4. Results and discussion

It is well known that PV performance of a dense polymeric membrane depends on the ability of solvent

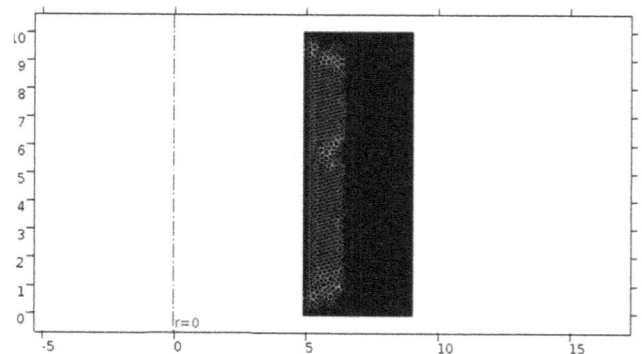

**Fig. 3.** Meshes created by COMSOL Multiphysics 5.2; Complete mesh consisted of 30558 domain elements.

species to be dissolved in the membrane at its interfaces, and their diffusion into the membrane. When a zeolite membrane is used as a separation barrier the solvent species cannot be dissolved in the membrane phase, but they are adsorbed on the zeolite sites of the inorganic materials. Their adsorbed capacities depend on the affinity of the membranes towards the solvents to be removed.

### 4.1. Silicalite-1 performance

The membrane exhibited a high selectivity towards water in water–UDMH mixtures. The permeate water flux reached a value as high as $3.34 \, kg/m^2.h$ for a UDMH concentration of 5 wt%. The fact that the membrane has a high selectivity towards water clearly indicates that the zeolite layer does not have any through-holes, and the transport is diffusive but not convective.

Silicalite-1 membrane showed a water-UDMH ideal selectivity of 10000 at 27°C, indicating its reasonable quality. During PV, water permeates through both zeolite and non-zeolite pores because of its small diameter. The kinetic diameter of UDMH is larger than the diameter of zeolite pores, thus much of the UDMH flux probably passes through the non-zeolite pores. Since the silicalite-1 is a weak hydrophilic membrane, the water flux decreases. The diffusing molecules in these mixtures pass via viscous flow and molecular sieve; whereas, viscous flow requires a pressure gradient across the membrane. If the zeolite is defect-free it has no non-zeolite pores, and thus water can pass only through zeolite pores (Table 2). However, non-zeolite pores usually exist and are larger than the zeolite pores. Non-zeolite pores have a size distribution and may also affect flux and selectivity. Transport through non-zeolite pores has contributions from both surface diffusion and Knudsen diffusion, and possibly from viscous flow.

The Silicalite channel system is shown in Figure 4. The straight elliptical channels running in the b-direction have dimensions of 0.53×0.56 nm, and the sinusoidal

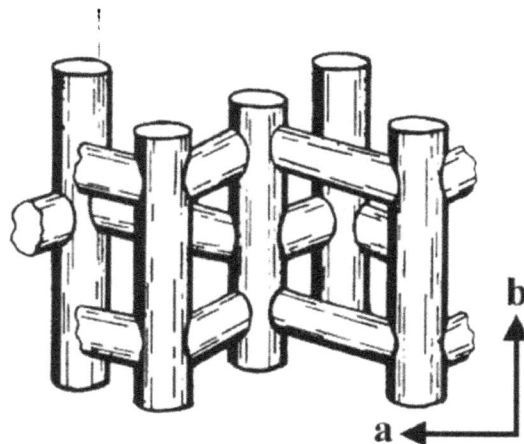

**Fig. 4.** Silicalite channel system.

channels running in the a-direction have dimensions of 0.55×0.51 nm.

Figure 5 shows XRD patterns of the mullite support and the zeolite membranes. Morphology of the support subjected to crystallization was characterized by SEM (Figure 6). Figure 7 shows the morphology of the silicalite-1 membranes (surface and cross section). As seen, most of the crystals lie disorderly on the surface. The SEM photographs of the membranes (cross section) show that the mullite surface is completely covered by a zeolite crystal layer, whose thickness is larger than 40 µm. The crystal layer is composed of two layers; the top layer consists of pure Silicalite crystals and the intermediate one of Silicalite crystals grown into the mullite pores.

As seen in Table 2, the best selectivity for silicalite-1 was 53% and the best water flux was 3.34 $kg/m^2.h$ at 27°C. The best silicalite-1 membranes were prepared using the following gel molar composition: 0.287 $Na_2O$:1.0 $SiO_2$:0.04 TPABr: 15 $H_2O$.

### 4.2. Feed phase simulations

Figure 8 shows the surface concentration distribution of water in half of the feed section at steady state conditions. The UDMH/water solution containing 95 wt% water flows over the outer surface of the membrane

**Table 2.** Flux and separation factor of the silicalite zeolite membranes.

| Sample | $Na_2O/SiO_2$ | $TPABr/SiO_2$ | $H_2O/SiO_2$ | T (°C) | t (h) | UDMH (%) | Flux ($kg/m^2.h$) | Separation factor |
|--------|---------------|---------------|--------------|--------|-------|----------|-------------------|-------------------|
| 1 | 0.450 | 0.01 | 8 | 180 | 24 | 5 | 1.02 | 4 |
| 2 | 0.287 | 0.01 | 8 | 180 | 24 | 5 | 1.67 | 8 |
| 3 | 0.350 | 0.01 | 8 | 180 | 24 | 5 | 1.7 | 23 |
| 4 | 0.350 | 0.04 | 15 | 180 | 24 | 5 | 3.34 | 53 |

(a) Support

(b) Silicalite

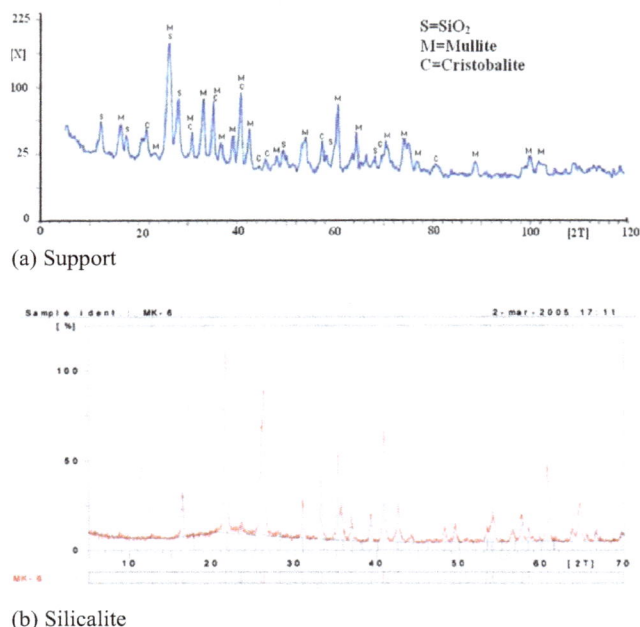

**Fig. 5.** XRD patterns of the (a) support and (b) silicalite zeolite membrane.

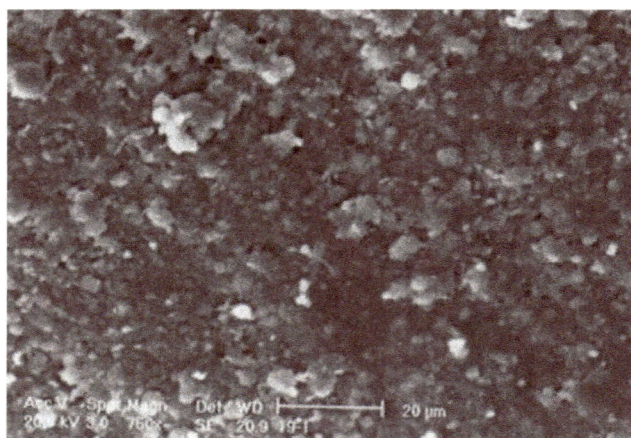

**Fig. 6.** SEM micrograph of the support.

(a) Surface

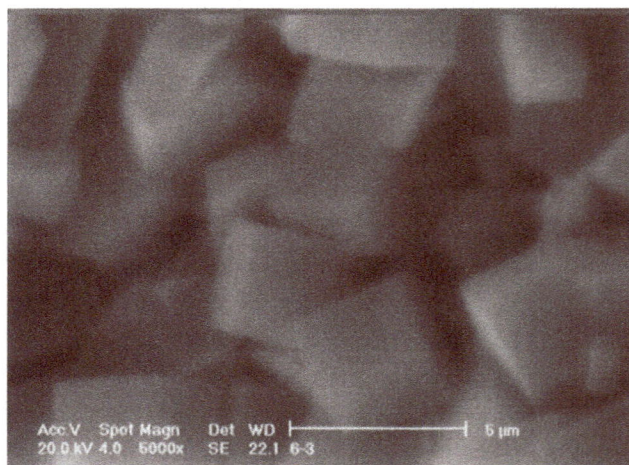

(b) Cross section

**Fig. 7.** SEM of the silicalite zeolite membrane; (a) surface and (b) cross section

module (z=0). As can be seen from the figure 8, a concentration boundary layer is formed on the membrane-feed interface. At z=0, the water concentration is maximum (95 wt%). As the feed solution flows in the feed compartment, water moves towards the membrane surface due to the concentration and pressure differences (driving forces). Therefore, the water concentration on the membrane surface is less than its value at the feed inlet (where water concentration is equal to its initial value, $C_{0, H_2O - feed}$. The water accumulation on the membrane surface was calculated from the membrane selectivity (Eq. 13) and its value in on the membrane side. Since water concentration in the membrane is always less than its value in the feed, the

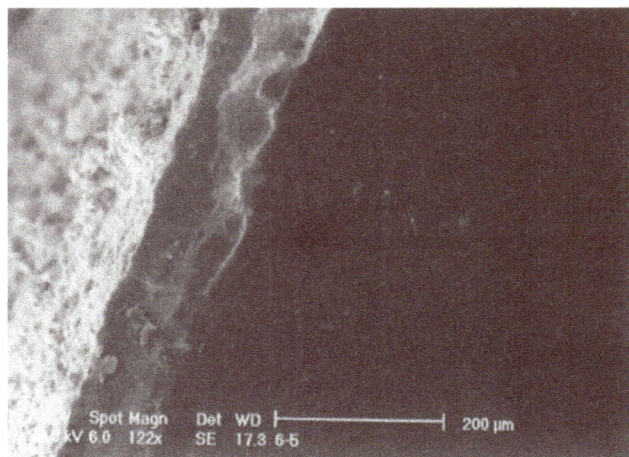

water concentration on the membrane-feed boundary ($r=R_2$) is always less than its value in the feed bulk.

Figure 9 presents the water distribution in the feed phase versus the r-coordinate at different lengths. Water concentration increases along the r direction, as expected. The concentration gradient is great at regions

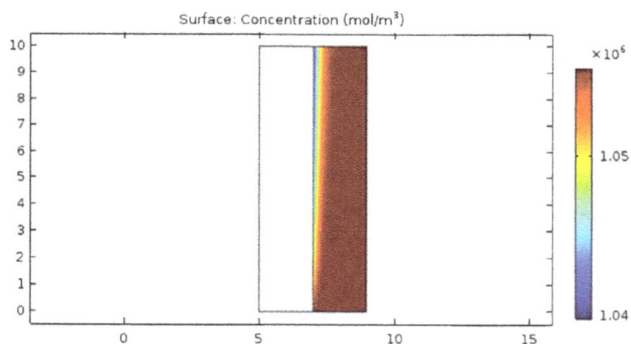

**Fig. 8.** Surface concentration distribution of water in feed phase (1.5 l/min feed flow rate and 30 °C temperature).

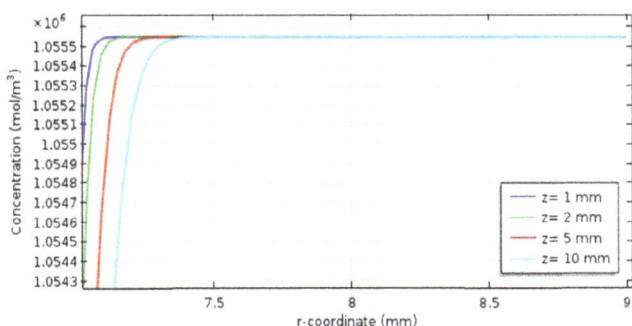

**Fig. 9.** Concentration distribution of water in feed phase vs. radius at various membrane lengths (1.5 l/min feed flow rate and 30 °C temperature).

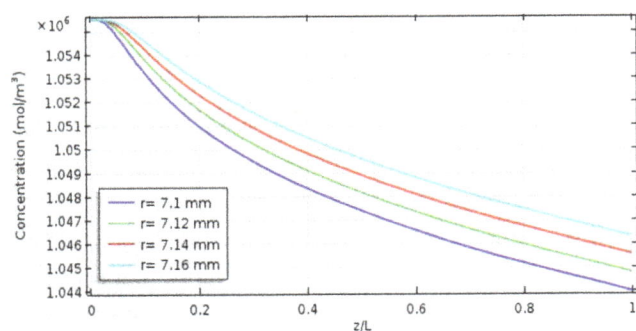

**Fig. 11.** Water concentration in feed section at different feed flow rates (at 30 °C temperature).

near the membrane-feed interface ($r=R_2$) due to the mass transfer towards the membrane at this region.

Figure 10 demonstrates the concentration distribution along the z coordinate at a constant flow rate (1.5 l/min) and different radii. Results indicate that the variation of water concentration along the z coordinate is considerable and cannot be neglected compared to its variation along the r coordinate. The figure also shows that the concentration gradient near the membrane-feed interface (r=7.1 mm) is slightly greater, while it is less at greater radii. This behavior can be attributed to water transfer towards the membrane at this region (greater concentration gradients).

Figure 11 shows the effect of various feed flow rates on water concentration distribution within the feed section. As can be seen, water concentration increases with increasing feed flow rate. This is because higher velocities (or flow rates) would decrease the contact time of the feed flow with the membrane, thus less water has enough time to pass through the membrane. Therefore, much higher concentrations will be obtained at higher feed flow rates.

Figure 12 shows the velocity field in the feed phase of the PV membrane system. The velocity distribution was obtained using numerical solution of momentum balance. This was done by adding a "laminar flow" physic to the whole model in COMSOL. As can be seen from the figure, the velocity profile is fully developed after a short distance. Velocity is zero on the membrane-feed interface and outer boundary of the feed section due to the no slip condition.

Figure 13 shows the velocity profile vs. radius in the feed section. As can be seen, the velocity profile is parabolic and becomes fully developed after a short distance (lengths approximately more than 12 mm). As seen, entrance effects are considered in this simulation, which is one of the advantages of FEM simulation. Figure 14 represents velocity distribution vs. dimensionless length at different feed flow rates. The velocity profile is almost parabolic and reaches its maximum value at the regions close to the feed entrance. Maximum velocity magnitude increases with increasing feed flow rate.

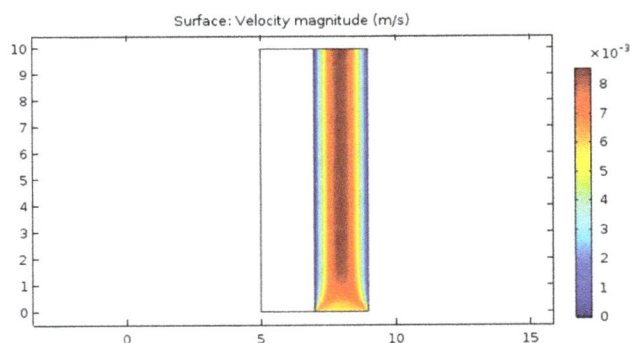

**Fig. 10.** Water concentration distribution in feed phase vs. dimensionless length (1.5 l/min feed flow rate and 30 °C temperature) at various radii.

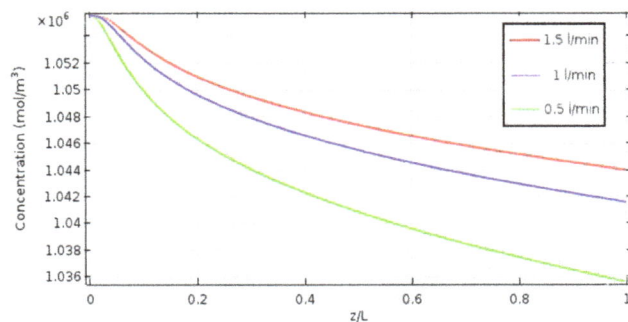

**Fig. 12.** Velocity distribution in the feed phase at 1.5 l/min feed flow rate and 30 °C temperature.

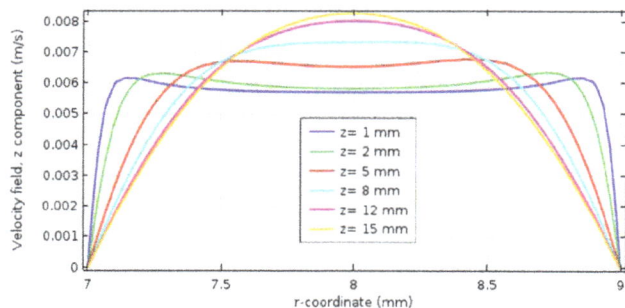

**Fig. 13.** Velocity profile vs. r-coordinate at various membrane lengths (1.5 l/min feed flow rate and 30 °C temperature).

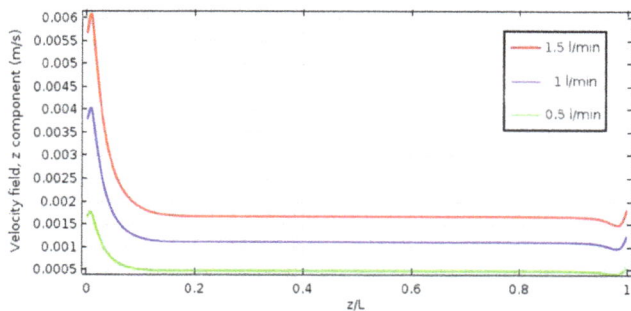

**Fig 14.** Velocity profile vs. dimensionless length at different feed flow rates.

## 4.3. Membrane phase simulation

Figure 15 shows the concentration distribution of water in the membrane phase at steady state conditions. The water transfer mechanism through the membrane was described only by diffusion. Since vacuum condition was assumed at the membrane-permeate interface, the water concentration on this boundary is zero. Water concentration is highest on the membrane-feed interface because it is calculated from its value in the feed section, which is always highest.

Figure 16 demonstrates water concentration distribution within the membrane vs. dimensionless length at a constant flow rate (1.5 l/min) and different membrane radii. Results show that the variation of water concentration along the z coordinate at constant radius is not considerable and can be neglected.

Figure 17 shows the effect of various feed flow rates on water concentration distribution within the membrane. As can be seen from the figure, water concentration increases with increasing feed flow rate. This is because an increase in feed flow rate would result in much higher concentrations in the feed compartment. Since water concentration in the membrane is calculated from its value on the feed side, much more water concentrations in the membrane will be obtained at higher feed flow rates.

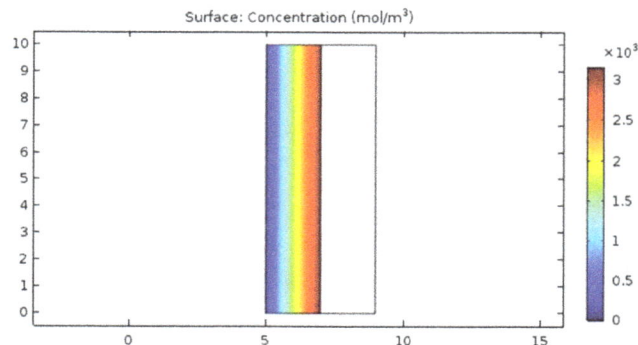

**Fig. 15.** Concentration distribution of water in membrane phase at 1.5 l/min feed flow rate and 30 °C temperature.

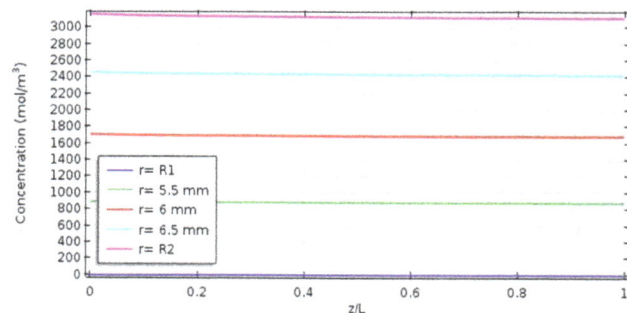

**Fig. 16.** Concentration distribution of water in membrane phase vs. dimensionless length at different radii (1.5 l/min feed flow rate and 30 °C temperature).

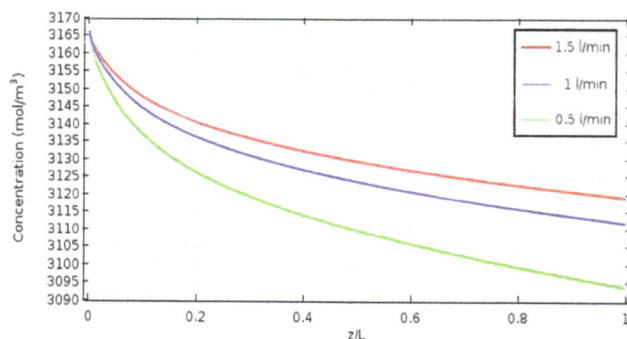

**Fig. 16.** Water concentration profile vs. dimensionless length at different feed flow rates (30 °C temperature).

## 5. Conclusion

Silicalite zeolite membranes were first used for dehydration of water-UDMH mixtures. The membranes were synthesized on the outer surface of porous mullite tubes by the hydrothermal method. The mullite supports were made by extruding kaolin clay. The zeolite membranes showed much higher fluxes and separation factors than commercially available polymeric membranes. The membranes showed good membrane performance for separation of the UDMH-mixtures. It is expected that even significantly higher fluxes, with similar separation factors, can be achieved at higher temperatures. Performance of PV system was modeled using COMSOL Multiphysics software version 5.2. Modeling was conducted by solving mass and momentum equations numerically using the finite element method (FEM). Good modeling results indicated that FEM is a powerful method for simulating membrane separation systems.

Since the silicalite zeolite membranes can withstand high temperatures and harsh environments (pH>12), dehydration of the water-UDMH mixtures can be performed. It was found that PV using the Silicalite zeolite membranes is an effective technique to separate water from the water-UDMH mixtures.

## Nomenclature

| | |
|---|---|
| $C_{0,H_2O}$ | initial water concentration (mol/m$^3$) |
| $C_{H_2O}$ | water concentration (mol/m$^3$) |
| $C_{H_2O\text{-}feed}$ | water concentration in feed phase (mol/m$^3$) |
| $C_{H_2O\text{-}mambrane}$ | water concentration in membrane phase (mol/m$^3$) |
| $D_{H_2O}$ | water diffusion coefficient (m$^2$/s) |
| $D_{H_2O\text{-}mambrane}$ | water diffusion coefficient in membrane (m$^2$/s) |
| F | body force (N) |
| L | membrane length (mm) |
| p | partition coefficient |
| P | pressure (Pa) |
| $P_{atm}$ | atmospheric pressure (Pa) |
| r | radial coordinate |
| $R_1$ | permeate-membrane radius (mm) |
| $R_2$ | membrane-feed radius (mm) |
| $R_3$ | Outer radius of the feed section (mm) |
| R | reaction term (mol/m$^3$.s) |
| α | selectivity |
| t | separation time (s) |
| U | velocity vector (m/s) |
| u | z-component velocity (m/s) |
| $x_{UDMH}$ | UDMH wt% in feed |
| $x_{H_2O}$ | water wt% in feed |
| $y_{UDMH}$ | UDMH wt% in permeate |
| $y_{H_2O}$ | water wt% in permeate |
| z | axial coordinate |
| ρ | density (kg/m$^3$) |
| μ | viscosity (Pa.s) |

## References

[1] R. Ravindra K.R. Krovvidi, A.A. Khan, A.K. Rao, D.S.C. studies of states of water, hydrazine and hydrazine hydrate in ethyl cellulose membrane, Polymer, 40 (1999) 1159-1165.

[2] R. Ravindra, A. Kameswara, A. Khan, A qualitative evaluation of water and monomethyl hydrazine in ethyl cellulose membrane, J. Appl. Polym. Sci. 72 (1999) 689-700.

[3] S. Sridhar, G. Susheela, G.J. Reddy, A.A. Khan, Cross linked chitosan membranes: characterization and study of dimethylhydrazine dehydration by pervaporation, Polym. Int. 50 (2001) 1156-1165.

[4] S. Moulik, K.P. Kumar, S. Bohra, S. Sridhar, Pervaporation performance of PPO membranes in dehydration of highly hazardous MMH and UDMH liquid propellants, J. Hazard. Mater. 288 (2015) 69-79.

[5] Y-L. Liao, C-C. Hu, J-Y. Lai, Y-L. Liu, Cross-linked polybenzoxazine based membrane exhibiting in-situ self-promoted separation performance for pervaporation dehydration on isopropanol aqueous solutions, J. Membrane Sci. 531 (2017) 10-15.

[6] Y.M. Xu, T-S. Chung, High-performance UiO-66/polymide mixed matrix membranes for ethanol, isopropanol and n-butanol dehydration via pervaporation, J. Membrane Sci. 531 (2017) 16-26.

[7] S. Zhang, Y. Zou, T. Wei, C. Mu, X. Liu, Z. Tong, Pervaporation dehydration of binary and ternary mixtures of n-butyl acetate, n-butanol and water using PVA-CS blended membranes, Sep. Purif. Technol.173 (2017) 314-322.

[8] J. Liu, R. Bernstein, High-flux thin-film composite polyelectrolyte hydrogel membranes for ethanol dehydration by pervaporation, J. Membrane Sci. 534 (2017) 83-91.

[9] T. Uragami, M. Banno, T. Miyata, Dehydration of an ethanol/water azeotrope through alginate-DNA membranes cross-linked with metal ions by pervaporation, Carbohyd. Polym. 134 (2015) 38-45.

[10] D.A. Fedosov, A.V. Smirnov, V.V. Shkirskiy, T. Voskoboynikov, I.I. Ivanova, Methanol dehydration in NaA zeolite membrane reactor, J. Membrane Sci. 486 (2015)189-194.

[11] R. Ravindra, S. Sridhar, A.A. Khan, A.K. Rao, Pervaporation of water, hydrazine and monomethyl-hydrazine using ethyl cellulose membranes, Polymer, 41 (2000) 2795-2806.

[12] S. Sridhar, R. Ravindra, A.A. Khan, Recovery of monomethylhydrazine liquid propellant by pervaporation technique, Ind. Eng. Chem. Res. 39 (2001) 2485-2490.

[13] X. Li, I. Kresse, Z.K. Zhou, J. Springer, Effect of temperature and pressure on gas transport in ethyl cellulose membrane, Polymer, 42 (2001) 6801-6810.

[14] Y-H. Huang, Q-F. An, T. Liu, W-S. Hung, C-L. Li, S-H. Huang, C-C. Hu, K-R. Lee, J-Y. Lai, Molecular dynamics simulation and positron annihilation lifetime spectroscopy: Pervaporation dehydration process using polyelectrolyte complex membranes, J. Membrane Sci. 451 (2014) 67-73.

[15] M. Jain, D. Attarde, S.K. Gupta, Removal of thiophenes from FCC gasoline by using a hollow fiber pervaporation module: Modeling, validation and influence of module dimensions and flow directions, Chem. Eng. J. 308 (2017) 632-648.

[16] S. Moulik, S. Nazia, B. Vani, S. Sridhar, Pervaporation separation of acetic acid/water mixtures through sodium alginate/polyaniline polyion complex membrane, Sep. Purif. Technol. 170 (2016) 30-39.

[17] N.S. Prasad, S. Moulik, S. Bohra, K.Y. Rani, S. Sridhar, Solvent resistant chitosan / poly(ether-block-amide) composite membranes for pervaporation of n-methyl-2-pyrrolidone / water mixtures, Carbohyd. Polym. 136 (2016) 1170-1181.

[18] M. Kazemimoghadam, A. Pak, T. Mohammadi, Dehydration of water / 1-1-dimethylhydrazine mixtures by zeolite membranes, Micropor. Mesopor. Mat. 70 (2004) 127-134.

[19] L. Zhoua, T. Wang, Q.T. Nguyenc, J. Li, Y. Long, Z. Ping, Cordierite-supported ZSM-5 membrane: Preparation and pervaporation properties in the dehydration of water-alcohol mixture, Sep. Purif. Technol. 44 (2005) 266-270.

[20] F. Akhtar, E. Sjöberg, D. Korelskiy, M. Rayson, J. Hedlund, L. Bergström, Preparation of graded silicalite-1 substrates for all-zeolite membranes with excellent $CO_2/H_2$ separation performance, J. Membrane Sci. 493 (2015) 206-211.

[21] G. Li, E. Kikuchi, M. Matsukata, A study on the pervaporation of water-acetic acid mixtures through ZSM-5 zeolite membranes, J. Membrane Sci. 218 (2003) 185-194.

[22] J. LP, Q.T. Nguyen, L.Z. Zhou, T. Wang, Y.C. Long, Z.H. Ping, Preparation and properties of ZSM-5 zeolite membrane obtained by low-temperature chemical vapor deposition, Desalination, 147 (2002) 321-326.

[23] T. Masuda, S-H Otani, T. Tsuji, M. Kitamura, S.R. Mukai, Preparation of hydrophilic and acid-proof silicalite-1 zeolite membrane and its application to selective separation of water from water solutions of concentrated acetic acid by pervaporation, Sep. Purif. Technol. 32 (2003) 181-189.

[24] B. Oonkhanond, M.E. Mullins, The preparation and analysis of zeolite ZSM-5 membranes on porous alumina supports, J. Membrane Sci. 194 (2001) 3-13.

[25] M. Nomura, T. Yamaguchi, S-I Nakao, Transport phenomena through intercrystalline and intracrystalline pathways of silicalite zeolite membranes, J. Membrane Sci. 187 (2001) 203-212.

[26] M. A. Baig, F. Patel, K. Alhooshani, O. Muraza, E.N. Wang, T. Laoui, In-situ aging microwave heating synthesis of LTA zeolite layer on mesoporous TiO2 coated porous alumina support, J. Cryst. Growth, 432 (2015) 123-128.

[27] T.C. Bowen, R.D. Noble, J.L. Falconer, Fundamentals and applications of pervaporation through zeolite membranes, J. Membrane Sci. 245 (2004) 1-33.

[28] C. Algieri, P. Bernardo, G. Golemme, G. Barbieri, E. Drioli, Permeation properties of a thin silicalite-1 (MFI) membrane, J. Membrane Sci. 222 (2003) 181-190.

[29] M. Nomura, T. Bin, S-I Nakao, Selective ethanol extraction from fermentation broth using a silicalite membrane, Sep. Purif. Technol. 27 (2002) 59-66.

[30] S. Nai, X. Liu, W. Liu, B. Zhang, Ethanol recovery from its dilute aqueous solution using Fe-ZSM-5 membranes: Effect of defect size and surface hydrophobicity, Micropor. Mesopor. Mat. 215 (2015) 46-50.

[31] A.M. Avila, Z. Yu, S. Fazli, J.A. Sawada, S.M. Kuznicki, Hydrogen-selective natural mordenite in a membrane reactor for ethane dehydrogenation, Micropor. Mesopor. Mat. 190 (2014) 301-308.

[32] R.B. Bird, W.E. Stewart, E.N. Lightfoot, Transport Phenomena, 2nd edition, John Wiley & Sons, New York, 1960, pp. 780.

# 8

# Numerical simulation of effect of non-spherical particle shape and bed size on hydrodynamics of packed beds

**Saeid Mohammadmahdi, Ali Reza Miroliaei***

*Department of Chemical Engineering, University of Mohaghegh Ardabili, Ardabil, Iran*

HIGHLIGHTS

GRAPHICAL ABSTRACT

- Flow channelling and vortex flow depend on particle shape, fluid velocity and bed porosity.

- The pressure drop of fluid with truncated cone particles is lower than the cone and cylindrical particles.

- Stationary points with cylindrical particles are more than the other particles.

- Vortex flow increases the pressure drop of fluid.

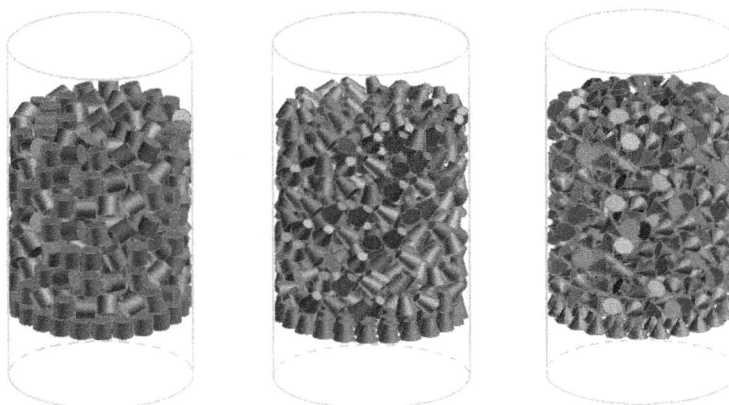

ARTICLE INFO

ABSTRACT

*Keywords:*
Flow channeling
Non-spherical particles
Pressure drop
Porosity
Vortex flow

Fluid flow has a fundamental role in the performance of packed bed reactors. Some related issues, such as pressure drop, are strongly affected by porosity, so non-spherical particles are used in industry for enhancement or creation of the desired porosity. In this study, the effects of particle shape, size, and porosity of the bed on the hydrodynamics of packed beds are investigated with three non-spherical particles namely cylindrical, cone, and truncated cone in laminar and turbulent flow regimes ($15 \leq \text{Re} \leq 2500$) using computational fluid dynamics. According to results obtained from the simulations, it was observed that flow channeling occurs in the parts of the bed that are not well covered by particles, which is more near the wall. CFD simulations showed that the vortex flow around the cylindrical particles is more than the cone and truncated cone particles and are caused by increasing the pressure drop of fluid in the bed. It was also found that the particles creating less porosity in the bed, due to their shape, are caused by increasing the pressure drop of fluid. The numerical results showed good agreement with available empirical correlations in the literature.

* Corresponding author:  ;  E-mail address: armiroliaei@uma.ac.ir

## 1. Introduction

There are several processes in industries, such as chemical, petrochemical and refinery industries, which require packed bed contactors and reactors. A packed bed is usually a hollow tube filled with layers of catalyst particles. The particles in the bed can be different shapes such as spherical, cylindrical, cubic, cone, truncated cone, etc. In many cases, non-spherical particles are selected due to the operating conditions and suitable distribution of fluid in the bed. Figure 1 shows a packed bed with spherical particles. Since packed beds are designed for interaction and increasing collision between materials, hydrodynamics and its related issues such as pressure drop of fluid, flow regimes and some of the incoming forces on the particles have an essential role in the performance of packed bed reactors. In fact, pressure drop of fluid in packed beds is strongly influenced by porosity, and for this reason non-spherical particles are used in industries to create the desired porosity. Due to safety and economic conditions, bed to particle diameter ratio of packed bed is also selected in the ranges of $3 < N = D/d_p < 8$ [1,2].

There are many numerical and experimental studies on the hydrodynamics of packed beds with the spherical particles. For example, the pressure drop and drag coefficient in square channels were studied by Calis et al. [3]. Their results showed good agreement with LDA measurements. Atmakidis and Kenig [4] investigated the wall effect on pressure drop in packed beds. They compared the CFD results with the empirical correlations of Zhavoronkov et al. and Reichelt. Reddy and Joshi [5] investigated CFD modeling of pressure drop and

drag coefficient in fixed beds with spherical particles. They stated that drag coefficient obtained from the CFD simulations become closer to the empirical equation of Ergun as the bed to particle diameter ratio increased due to reducing the effects of wall friction. The shape effects on the packing density of frustums were studied by Zhao et al. [6]. Their studies showed that the optimal aspect ratio of truncated cones is 0.8 and increases as the radii ratio increases. Also, they proposed a correlation between the packing density and shape parameters. Allen et al. [7] studied the effects of particle shape, size distribution, packing arrangement and roughness of particles on the packed bed pressure drop. Their results showed that the particle shape, packing arrangement and surface roughness of particles affect the pressure drop. Rong et al. [8] investigated fluid flow in packed beds with different spheres using a parallel lattice-Boltzmann model. The effects of size ratio and volume fraction on the fluid flow and drag force were studied. Their results showed that the dispersion of particles affects flow distribution and fluid-particle interaction forces. Also, they suggested a correlation to calculate the drag force. Experimental study and numerical simulation of pressure drop in a packed bed with arbitrarily shaped particles were carried out by Vollmari et al. [9]. They indicated that simulations are in good agreement with experiments depending on the particle shape and size and is often better in comparison with empirical correlations. Bu et al. [10] considered the flow transitions in three different structured packed beds, such as simple cubic, body center cubic and face center cubic packing forms, using electrochemical techniques. They observed three different flow regimes in the packed beds, i.e. laminar, transition and turbulent flow regimes. Also, they explained that flow regimes in packed beds depend on the arrangement of the particles. Du et al. [11] studied the porosity and pressure drop in packed beds experimentally and statistically. Their analysis showed that the experimental data and the predicted equation for particles with different sizes have good agreement together. Pressure drop in slender packed beds was investigated by Guo et al. [12]. They found that pressure drop in packed beds depends on the bed structure, as a minor change in the bed structure creates a notable pressure drop even though the beds have the same porosity.

In this research, the effects of particle shape and bed size on pressure drop of fluid in packed beds with non-

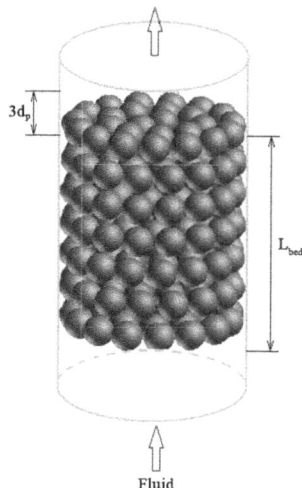

**Fig. 1.** Schematic of a packed bed with spherical particles.

spherical particles such as cylindrical, cone, and truncated cone particles are investigated to achieve a suitable distribution of fluid flow and lower pressure drop in the bed. The validation of the CFD simulation results is carried out with proposed empirical correlations in the literature.

## 2. Empirical correlations for pressure drop prediction in packed beds with non-spherical particles

The empirical equation of Ergun [13] is used to predict the pressure drop in the packed beds with spherical particles. This correlation is applicable in a wide range of flow regimes. The estimated pressure drop according to Ergun equation depends on the properties of the bed and fluid such as bed porosity and particle diameter, fluid flow rate, viscosity and density of fluid as follows:

$$\frac{\Delta P}{L} = \frac{150\mu(1-\varepsilon)^2}{\varepsilon^3 d_p^2 \varphi^2} u_s + \frac{1.75(1-\varepsilon)\rho}{\varepsilon^3 d_p \varphi} u_s^2 \tag{1}$$

In above equation $\mu$ and $\rho$ are the dynamic viscosity and density of fluid, $u_s$ is the superficial velocity of fluid, $\varepsilon$ is the porosity of bed, $d_p$ and $\varphi$ are the diameter and sphericity of particles, respectively.

The sphericity of particles is defined as the ratio between the surface area of the volume equivalent sphere and the surface area of the particle:

$$\varphi = \left[ 36\pi \cdot \frac{V_p^2}{A_p^3} \right]^{\frac{1}{3}} \tag{2}$$

The correction of Ergun equation for non-spherical particles has been carried out by some researchers. Some of these relations are shown in Table 1.

The coefficients of (3) and (7) equations have been mentioned in Tables 2 and 3, respectively.

Table 2. Coefficients in equation (3) [10].

| Particle shape | $K_1$ | $k_1$ | $k_2$ |
|---|---|---|---|
| Cylindrical | 190 | 2 | 0.77 |
| All particles | 155 | 1.43 | 0.83 |

Table 3. Coefficients in equation (7) [6].

| Particle shape | a | b | c |
|---|---|---|---|
| Cubic | 240 | 10.8 | 0.1 |
| Cylindrical | 216 | 8.8 | 0.12 |

## 3. CFD modeling

### 3.1. Characteristics of particles and beds

Packed beds were designed with three different particles types, of cylindrical, cone, and truncated cone, in different ratios of low bed to particle diameter. It was assumed that the size of particles is constant and the parameters of A and B were also defined for the dimensions of particles. A is the ratio of the height of the particles to the larger particle's diameter (A $=L_p/d_p$), and B is the ratio of the smaller particle's diameter to larger particle's diameter (B $= d/d_p$). Therefore, the values of A and B for the cylindrical, cone, and truncated cone particles are A=1 and B=1, A=1 and B=0, and A=1 and

Table 1. Empirical correlations for calculating pressure drop with non-spherical particles.

| | | |
|---|---|---|
| Eisfeld and Schnitzlein [10] | $\dfrac{\Delta P}{L} = K_1 \dfrac{\mu_f(1-\varepsilon)^2}{\varepsilon^3 d_p^2} u_s M^2 + \dfrac{(1-\varepsilon)\rho_f}{\varepsilon^3 d_p} u_s^2 \dfrac{M}{B_w}$ | (3) |
| | $M = 1 + \dfrac{4d_{pe}}{6D(1-\varepsilon)} \quad . \quad B_w = \left[ k_1 (\dfrac{d_p}{D})^2 + k_2 \right]^2$ | (4) |
| Nemec and Levec [11] | $\dfrac{\Delta P}{L} = \dfrac{150\mu(1-\varepsilon)^2}{\varepsilon^3 d_{sv}^2 \varphi^{3/2}} u_s + \dfrac{1.75(1-\varepsilon)\rho}{\varepsilon^3 d_{sv} \varphi^{4/3}} u_s^2$ | (5) |
| Singh et al. [12] | $\dfrac{\Delta P}{L} = 4.4666 \cdot (\dfrac{\rho_f \cdot u_s \cdot d_e}{\mu_f})^{-0.2} \cdot \varepsilon^{-2.945} \cdot \varphi^{0.696} \cdot e^{11.85(\log \varphi)^2} \cdot \dfrac{\rho_f \cdot u_s^2}{d_e}$ | (6) |
| Allen et al. [6] | $\dfrac{\Delta P}{L} = \left[ \dfrac{a}{Re_{Duct}} + \dfrac{b}{Re_{Duct}^c} \right] \cdot \dfrac{(1-\varepsilon)}{\varepsilon^3} \cdot \dfrac{\rho_f \cdot u_s^2}{8} \cdot \dfrac{\sum A_p}{\sum V_p}$ | (7) |
| | $Re_{Duct} = \dfrac{4 \cdot \rho_f \cdot u_s^2}{\mu_f \cdot (1-\varepsilon)} \cdot \dfrac{\sum V_p}{\sum A_p}$ | (8) |

B=0.5, respectively. Table 4 describes the characteristics of the particles and the beds.

The geometry of designed beds with the different shapes are shown in Figure 2.

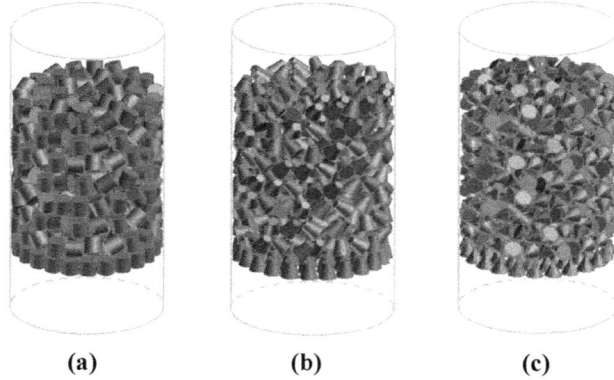

(a)                    (b)                    (c)

**Fig. 2.** Geometry of designed beds: (a) packed bed with cylindrical particles, (b) packed bed with truncated cone particles, and (c) packed bed with cone particles.

### 3.2. Governing equations

Momentum and continuity equations are used in order to investigate fluid flow through the packed beds. The continuity equation is defined as follows [2]:

$$\frac{\partial \rho}{\partial t} + \nabla \cdot (\rho u) = S_m \tag{9}$$

$S_m$ is the source term that is equal to zero in our simulations.

The equation for conservation of momentum is:

$$\frac{\partial (\rho u)}{\partial t} + \nabla \cdot (\rho u u) = -\nabla p + \nabla \cdot (\rho(\vartheta + \vartheta_t)\nabla u) + \rho g_i \tag{10}$$

In the above balance $\rho g_i$ is the gravitational force. $\vartheta$ and $\vartheta_t$ are the kinetic viscosity in laminar and turbulent flow regimes, respectively. The kinetic viscosity in the turbulent flow regime depends on two parameters, i.e. turbulent kinetic energy, k, and dissipation rate, ε.

The RNG k-ε model was used in the turbulent flow regime [17]. The parameters of this model are calculated from the following transport equations:

$$\frac{\partial (\rho k)}{\partial t} + \nabla \cdot (\rho k u) = \rho \vartheta_t (\nabla u)^2 + \nabla \cdot (\rho \alpha_k \vartheta_t \nabla k) - \rho \varepsilon \tag{11}$$

$$\frac{\partial (\rho \varepsilon)}{\partial t} + \nabla \cdot (\rho \varepsilon u) = \nabla \cdot (\rho \alpha_\varepsilon \vartheta_t \nabla \varepsilon) + \tag{12}$$

$$C_{1\varepsilon} \frac{\varepsilon}{k} \rho \vartheta_t (\nabla u) - C_{2\varepsilon} \rho \frac{\varepsilon^2}{k} - R_\varepsilon$$

In these equations, $C_{1\varepsilon}$ and $C_{2\varepsilon}$ are equal to 1.48 and 1.68, respectively. Also, $\alpha_k = \alpha_\varepsilon = 1.393$, $\vartheta_t$ and $R_\varepsilon$ are defined as follows:

$$\vartheta_t = C_\mu \frac{k^2}{\varepsilon} \tag{13}$$

$$R_\varepsilon = \frac{C_\mu \rho \eta^3 \left(1 - \eta/\eta_0\right) \varepsilon^2}{1 + \beta \eta^3} \tag{14}$$

Here, $C_\mu = 0.0845$, $\eta = Sk/\varepsilon$, $\eta_0 = 4.38$ and $\beta = 0$.

S is the modulus of the mean rate-of-strain tensor:

$$S = \sqrt{S_{ij} S_{ij}} \tag{15}$$

The finite volume method, ANSYS FLUENT software, was chosen for solving momentum and

**Table 4.** Characteristics of particles and beds.

| Particle shape | Dimension (mm) | | | Particle volume (mm³)×10⁻² | Sphericity | Bed volume (mm³)×10⁻⁵ | Porosity | N= D/d_sv |
|---|---|---|---|---|---|---|---|---|
| | $L_p$ | $d_p$ | d | | | | | |
| | 23 | 23 | 23 | 95.56 | 0.87 | 8.34 | 0.576 | 4.17 |
| | 23 | 23 | 23 | 95.56 | 0.87 | 28.1 | 0.625 | 6.26 |
| | 23 | 23 | 23 | 95.56 | 0.87 | 94.4 | 0.602 | 9.39 |
| | 23 | 23 | 11.5 | 55.74 | 0.84 | 8.34 | 0.722 | 4.36 |
| | 23 | 23 | 11.5 | 55.74 | 0.84 | 28.1 | 0.699 | 6.54 |
| | 23 | 23 | 11.5 | 55.74 | 0.84 | 94.4 | 0.708 | 9.82 |
| | 23 | 23 | 0 | 31.85 | 0.77 | 8.34 | 0.710 | 5.24 |
| | 23 | 23 | 0 | 31.85 | 0.77 | 28.1 | 0.741 | 7.86 |
| | 23 | 23 | 0 | 31.85 | 0.77 | 94.4 | 0.748 | 11.80 |

continuity equations. The air was assumed as the fluid through the packed beds in the simulations. The boundary conditions are as follows:

- Steady state flow;
- Incompressible fluid;
- Constant velocity in the bed inlet;
- Constant pressure (1 atm) in the bed outlet; and
- Non-slip condition for the walls and the surface of particles.

The SIMPLE algorithm was used for coupling velocity and pressure. Second order upwind discretization method was also applied to increase the accuracy of the results. The convergence criterion was maintained to achieve a very low level of the residual, about $10^{-5}$ in all equations.

## 4. Results and discussion

### 4.1. Mesh generation

The most important step in the simulation is mesh generation. In fact, mesh geometry should be designed in such a way that changing a number of meshes doesn't affect results, and the geometry must be independent of the mesh. For example, a grid independence study with unstructured tetrahedral mesh was carried out in the packed bed with cylindrical particles of N = 6.26 with five different mesh sizes, i.e. 5, 4, 3, 2.5 and 2 mm. The pressure drop was evaluated for different mesh sizes and the optimum mesh was selected. It was observed that the pressure drop varies from 8.7%, 5.9%, 3.2% and 0.29% when the grid size is changed from 5 to 4, 4 to 3, 3 to 2.5 and 2.5 to 2 mm, respectively. As it can be seen from Table 5, in the grid size 2.5 mm the pressure drop is independent of mesh size. Therefore, the grid size of 2.5 mm was selected for our simulations of a packed bed with cylindrical particles. The results of the mesh independence study for cylindrical, truncated cone, and

**Table 5.** Grid independence results for a packed bed with cylindrical particles in N=6.26.

| Mesh size (mm) | 5 | 4 | 3 | 2.5 | 2 |
|---|---|---|---|---|---|
| $\Delta P \times 10^{-3}$ (pa) | 8.54 | 9.36 | 9.95 | 10.28 | 10.31 |
| Pressure drop variations | | 8.7% | | | |
| | | | 5.9% | | |
| | | | | 3.2% | |
| | | | | | 0.29% |

cone have been stated in Tables 5, 6 and 7, respectively.

**Table 6.** Grid independence results for packed bed with truncated cone particles in N=6.54.

| Mesh size (mm) | 4 | 3.5 | 3 | 2.5 | 2 |
|---|---|---|---|---|---|
| $\Delta P \times 10^{-3}$ (pa) | 7.16 | 7.79 | 8.06 | 8.23 | 8.37 |
| Pressure drop variations | | 7.98% | | | |
| | | | 3.38% | | |
| | | | | 2.1% | |
| | | | | | 1.65% |

**Table 7.** Grid independence results for a packed bed with cone particles in N=7.86.

| Mesh size (mm) | 4 | 3.5 | 3 | 2.5 | 2 |
|---|---|---|---|---|---|
| $\Delta P \times 10^{-3}$ (pa) | 6.61 | 7.81 | 8.92 | 9.10 | 9.78 |
| Pressure drop variations | | 15.36% | | | |
| | | | 12.4% | | |
| | | | | 1.9% | |
| | | | | | 6.9% |

### 4.2. Velocity profiles

Velocity profiles in the packed beds with cylindrical, cone, and truncated cone particles and a bed volume of $94.9\times10^5$ mm$^3$ in the turbulent flow regime have been shown as contour and vector in Figures 3 and 4. As shown in these figures, when the fluid enters the packed bed it passes through a porous media created by the particles. From velocity contours in Figure 3, it is known that in the special regions of the bed where the distance between the particle-particle and particle-wall is low, the fluid velocity increases because of a lower passing surface for fluid flow. It is also seen that the channeling phenomenon occurs in the parts of the bed that are not well covered by particles. On the other hand, the flow channeling was observed more near the wall.

After the collision of the fluid with particles in the bed the fluid velocity decreases (see Figure 3) and the region where the fluid velocity is close to zero is called the stationary point. The stationary points were seen more in the packed beds with cylindrical particles.

As it is shown in Figure 4, a vortex flow is created in the bed. It occurs in areas that particles are close each other. The vortex flow was also observed in parts of the bed outlet. This type of flow was seen less in the packed beds with the cone and truncated cone particles because

**(a)**

**(b)**

**(c)**

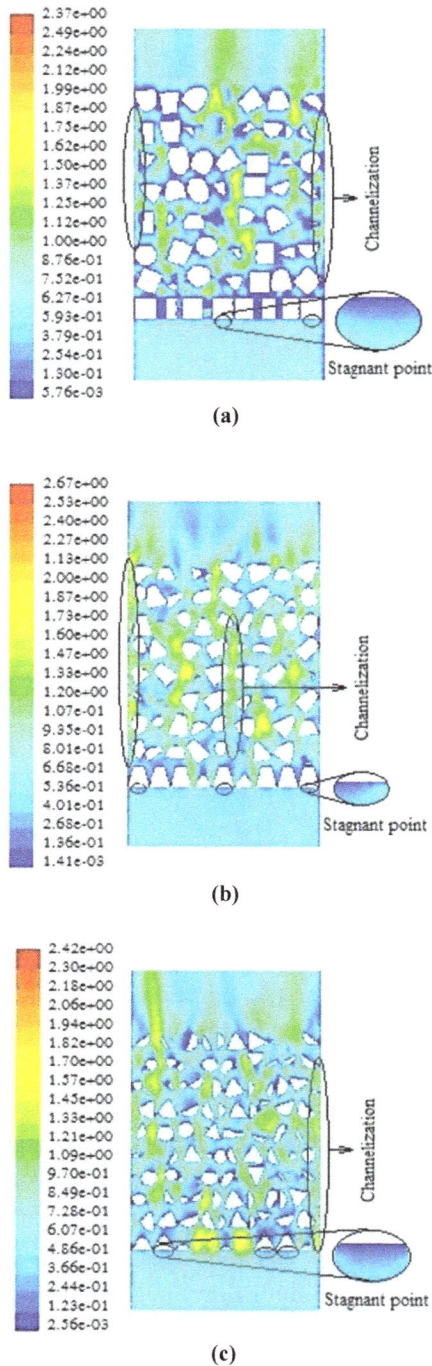

**Fig. 3.** Velocity contour in the packed beds with $V_b$ =94.9 ×$10^5$ mm$^3$ and $V_f$ = 0.5 m/s: a) cylindrical particles, b) truncated cone particles, and c) cone particles.

of the shape and surface of these particles.

Some of the most important parameters causing the turbulent flow in packed beds include particle shape, fluid velocity and bed porosity. So the characteristics of flow, such as flow channeling and vortex flow, can affect the pressure drop of fluid into the bed which will be discussed in the following sections.

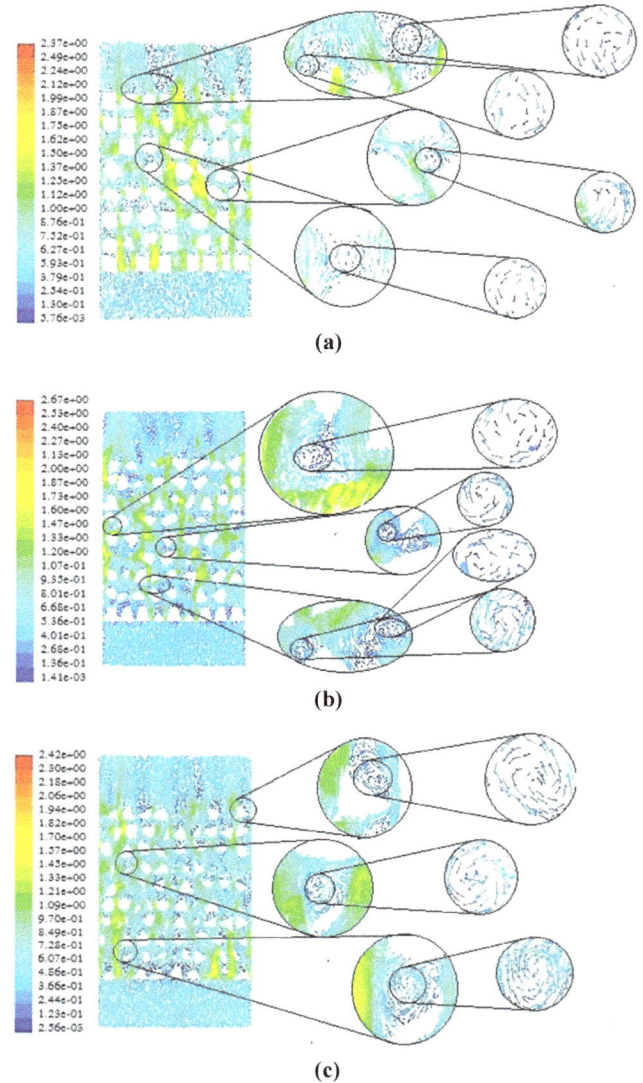

**(a)**

**(b)**

**(c)**

**Fig. 4.** Velocity vectors in the packed beds with $V_b$ =94.9 ×$10^5$ mm$^3$ and $V_f$ = 0.5 m/s: a) cylindrical particles, b) truncated cone particles, and c) cone particles.

## 4.3. Pressure drop

Pressure drop in packed beds is the most important parameter as the heat and mass transfer are strongly relevant to pressure drop. Therefore, study of effective parameters on pressure drop in packed beds is required. The CFD obtained pressure drop results for cylindrical, cone, and truncated cone particles in laminar and turbulent flow regimes are shown in Figures 5, 6 and 7, respectively. The validation of the CFD simulations was carried out with empirical correlations of Eisfeld and Schnitzlein [14], Nemec and Levec [15], and Allen et al. [7]. As it can be seen in these figures, increasing both length and diameter of the beds caused increasing pressure drop of fluid in the bed because the fluid needs

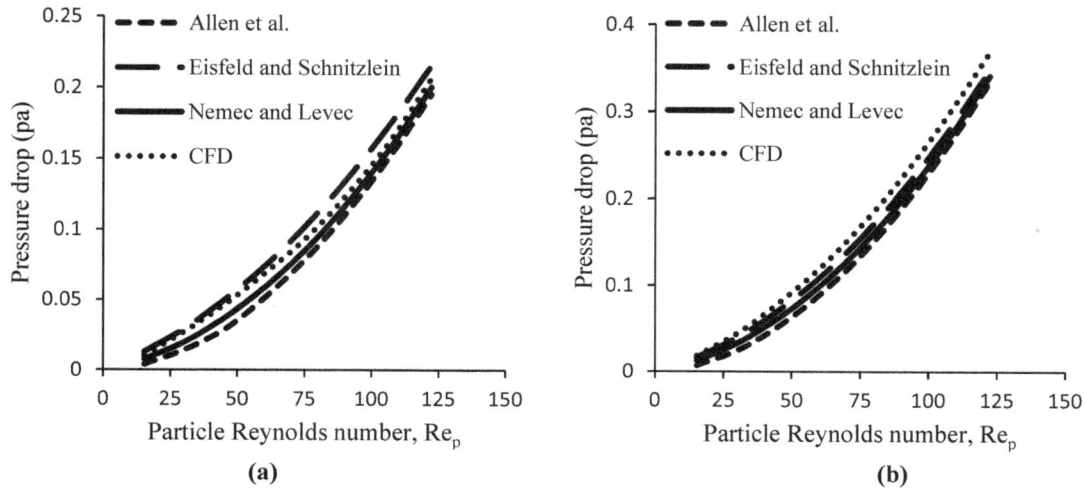

**Fig. 5.** Comparison of CFD pressure drop with empirical correlations for the packed bed with cylindrical particles in laminar flow: a) N=4.17 and b) N=9.39.

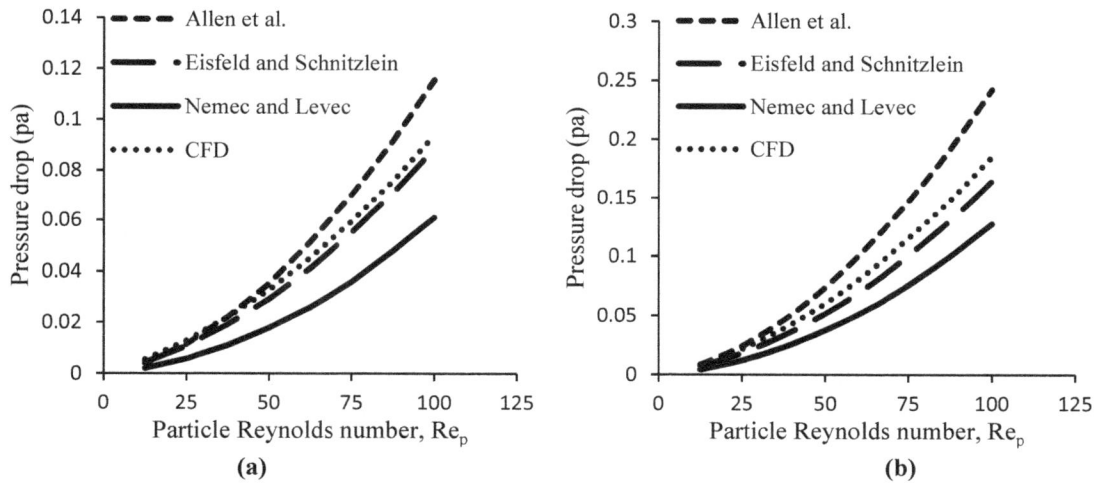

**Fig. 6.** Comparison of CFD pressure drop with empirical correlations for the packed bed with cone particles in laminar flow: a) N=5.24 and b) N=11.8.

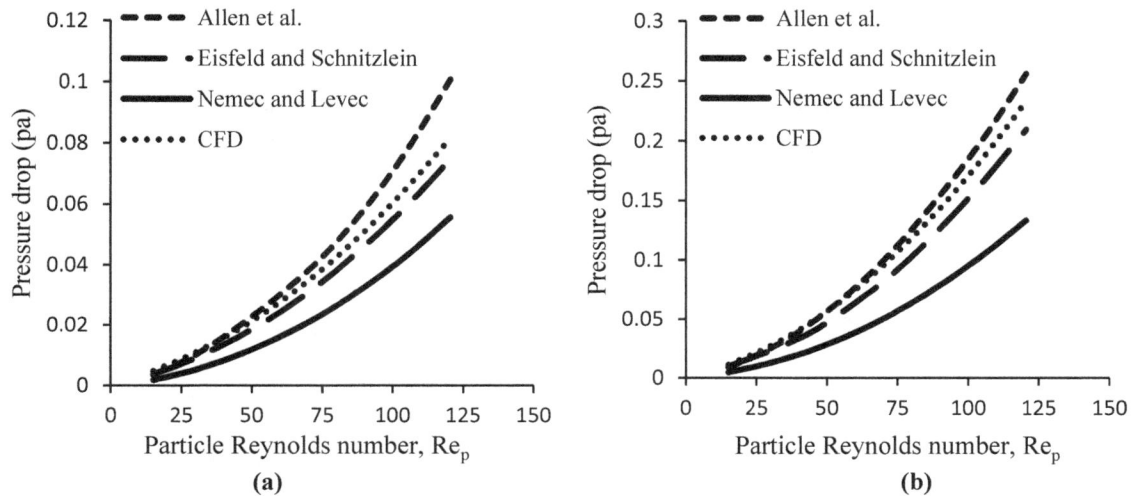

**Fig. 7.** Comparison of CFD pressure drop with empirical correlations for the packed bed with truncated cone particles in laminar flow: a) N=4.36 and b) N=9.82.

to pass a longer path in the bed. It is also observed that the pressure drop of fluid at the same Reynolds number for the truncated cone particles is lower than the cone and cylindrical particles.

Comparison of CFD pressure drop variations and empirical correlations in the turbulent flow regime for packed beds with different particles is carried out in Figures 8, 9 and 10, respectively. As mentioned before, the pressure drop values for cylindrical particles are more than the cone and truncated cone particles because of the eddy and vortex flows.

### 4.4. Effect of bed porosity on the pressure drop

The porosity of the bed is one of the parameters that most affects the pressure drop of fluid. A comparison between created pressure drop in the packed beds with cylindrical, cone, and truncated cone particles at different ratios of bed to particles diameter, according to Table 4, is shown in Figures 11 and 12. As seen in these figures, the pressure drop variations of fluid in the packed bed with cylindrical particles are more than the cone and truncated cone particles. It is also observed that the pressure drop of fluid in the packed beds with cone and truncated cone particles are close each other. The main reason for the pressure drop in the beds with different types of particles is that the bed porosity is created according to the shape of the particles. As can be seen in Figures 11 and 12, the pressure drop in the beds increases as the porosity of the bed decreases because as the porosity of the bed decreases the fluid passes through a more twisted path into the bed which causes the pressure drop of fluid to increase. It is also known that the particles having lower porosity, e.g. cylindrical particles, create more pressure drop in the bed.

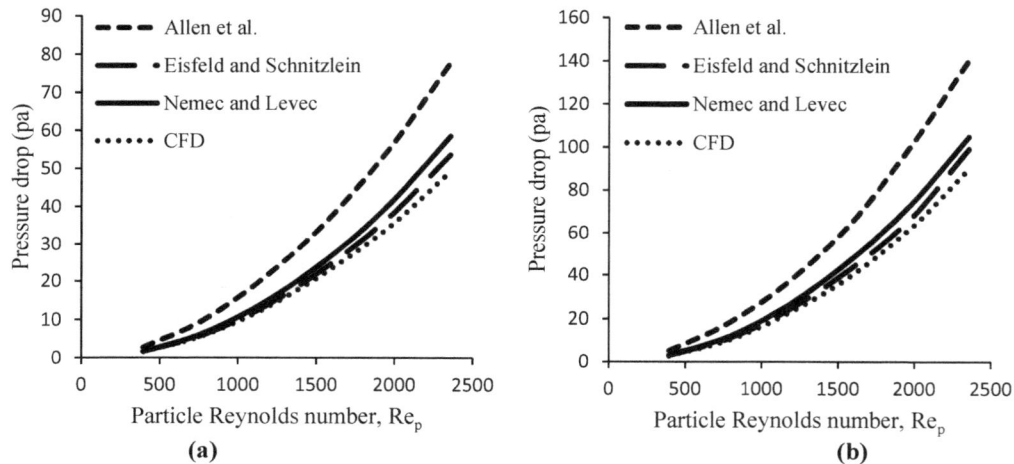

**Fig. 8.** Comparison of CFD pressure drop with empirical correlations for the packed bed with cylindrical particles in turbulent flow: a) N=6.26 and b) N=9.39.

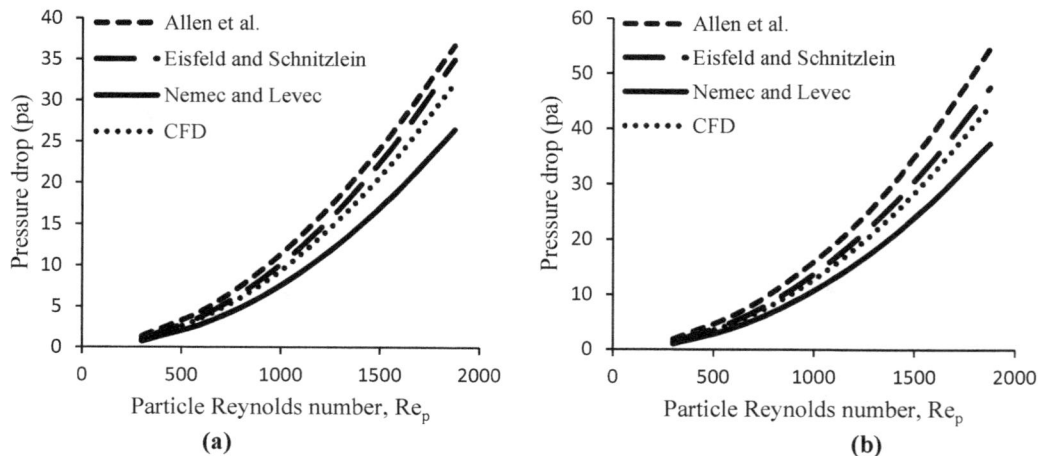

**Fig. 9.** Comparison of CFD pressure drop with empirical correlations for the packed bed with cone particles in turbulent flow: a) N=7.82 and b) N=11.8.

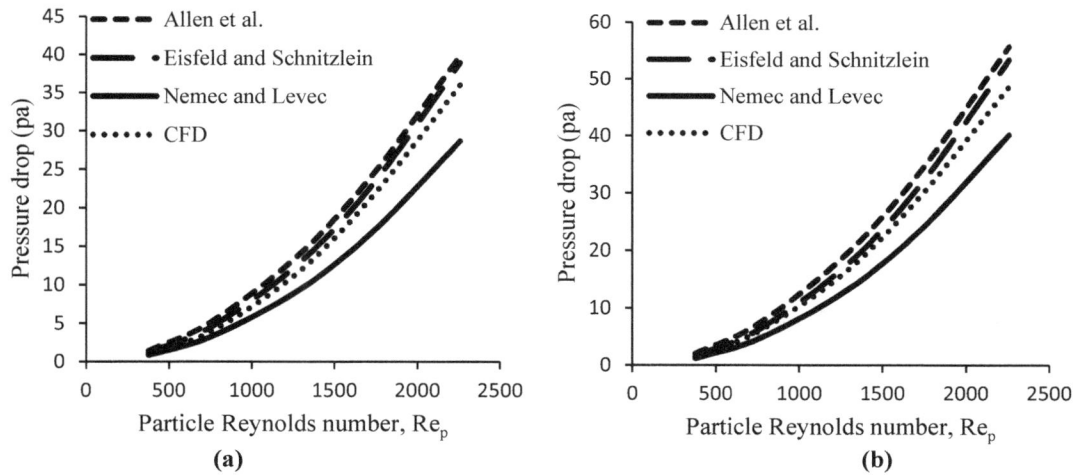

**Fig. 10.** Comparison of CFD pressure drop with empirical correlations for the packed bed with truncated cone particles in turbulent flow: a) N=6.52 and b) N=9.82.

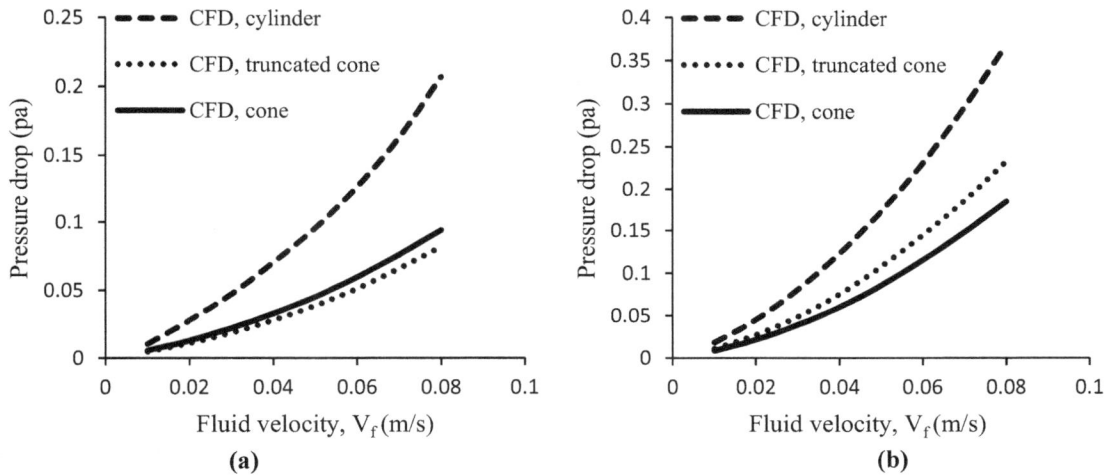

**Fig. 11.** Pressure drop variations vs. fluid velocity for cylindrical, truncated cone, and cone particles in a laminar flow regime: a) $V_b$ = $8.34 \times 10^5$ mm$^3$ and b) $V_b$ = $94.9 \times 10^5$ mm$^3$.

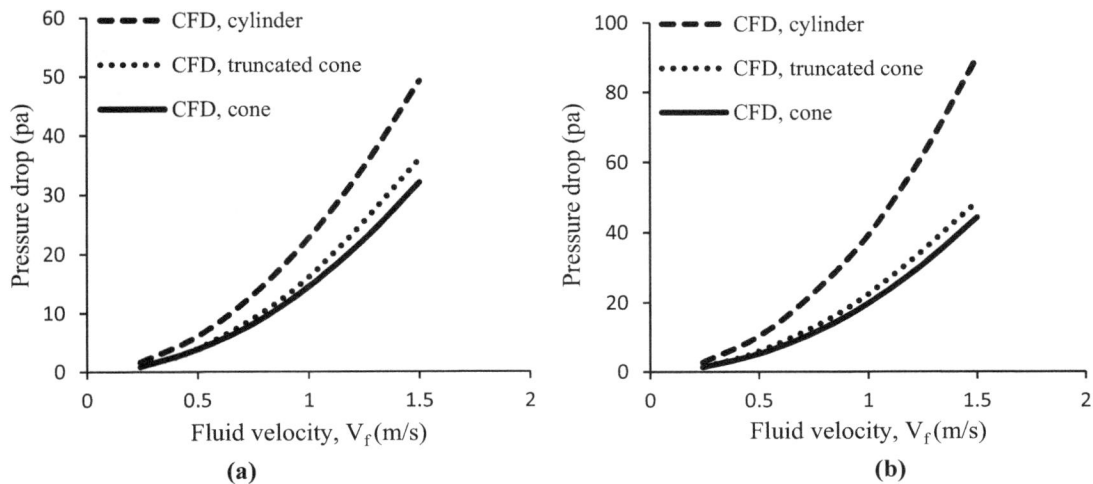

**Fig. 12.** Pressure drop variations vs. fluid velocity for cylindrical, truncated cone and cone particles in a turbulent flow regime: a) $V_b$ = $8.34 \times 10^5$ mm$^3$ and b) $V_b$ = $94.9 \times 10^5$ mm$^3$.

## 5. Conclusion

In this study, characteristics of fluid flow, such as flow channeling and vortex flow, and the effect of particle shape on the pressure drop of fluid in packed beds with a low bed to particle diameter ratio were investigated. Three types of particles cylindrical, cone, and truncated cone were selected. The CFD simulation results were validated using empirical correlations from the literature. According to the contours and vectors of fluid flow in all beds, it was seen that the channeling phenomenon occurs in some regions of the bed because of inadequate cover between particles and particle-wall. It was also observed that vortex flow with cylindrical particles is more than the cone and truncated cone particles. As result, we found the flow channeling and vortex flow properties of fluid flow depend on the shape of particles, fluid velocity and bed porosity.

Numerical results showed that the pressure drop of fluid in the packed bed with truncated cone particles is lower than the cylindrical and cone particles. The main reason for the pressure drop of fluid with different particles is the bed porosity, which is created according to the shape of the particles; but in equal porosity conditions pressure drop depends on eddies, vortex flow, and other forces (such as drag force) that are different due to the particle shape. As shown in the simulation results, the RNG k-ε model is appropriate for simulation and provides acceptable results in a turbulent flow regime.

## Nomenclature

| $A_p$ | Particle surface area (m²) |
|---|---|
| $A_w$ | Wall correction term |
| a, b, c | Constants in Eq. (7) |
| $B_w$ | Wall correction term |
| D | Bed diameter (m) |
| $d_p$ | Large diameter of particle (m) |
| D | Small diameter of particle (m) |
| $d_{sv}$ | Equivalent surface volume diameter, $d_{sv}=6.V_p/A_p$ (m) |
| G | Gravitational acceleration (m/s²) |
| $K_1$ | Constant in Eq. (3) |
| $k_1, k_2$ | Constants in Eq. (4) |
| $L_p$ | Length of particles (m) |
| L | Length of bed (m) |

| M | Wall correction term |
|---|---|
| N | Bed to particle diameter ratio, N =D/$d_{sv}$ |
| $n_p$ | Number of particles in the bed |
| $Re_{duct}$ | Duct Reynolds number, $Re_{duct} = \dfrac{2 . \rho . u_s . d_p}{3 . \mu (1-\varepsilon)}$ |
| $Re_p$ | Particle Reynolds number, $Re_p = \dfrac{\rho . u_s . d_p}{\mu}$ |
| $u_s$ | Superficial velocity (m/s) |
| $V_p$ | Particle volume (m³) |
| $V_b$ | Bed volume (m³) |

## Greek symbols

| ΔP | Pressure drop (Pa) |
|---|---|
| ε | Porosity, $\varepsilon = 1 - \dfrac{n_p V_p}{V_b}$ |
| $\mu_f$ | Fluid dynamic viscosity (kg/m.s) |
| $\rho_f$ | Fluid density (kg/m³) |
| φ | Particle sphericity |
| $\vartheta_t$ | Turbulence kinetic viscosity |
| K | Turbulence kinetic energy |
| ε | Rate of dissipation |

## References

[1] J. Ancheyta, J.A.D. Munoz, M.J. Maceas, Experimental and theoretical determination of the particle size of hydrotreating catalysts of different shapes, Catal. Today, 109 (2005) 120-127.

[2] A.G. Dixon, M. Nijemeisland, E.H. Stitt, Packed tubular reactor design using computational fluid dynamics, Adv. Chem. Eng. 3 (2006) 307-389.

[3] H.P.A. Calis, J. Nijenhuis, B.C. Paikert, F.M. Dautzenberg, C.M. van den Bleek, CFD modelling and experimental validation of pressure drop and flow profile in a novel structured catalytic reactor packing, Chem. Eng. Sci. 56 (2001) 1713-1720.

[4] T. Atmakidis, E.Y. Kenig, CFD-based analysis of the wall effect on the pressure drop in packed beds with moderate tube/particle diameter ratios in the laminar flow regime, Chem. Eng. J. 155 (2009) 404-410.

[5] R.K. Reddy, J. B. Joshi, CFD modeling of pressure drop and drag coefficient in fixed beds: Wall effects, Particuology, 8 (2010) 37-43.

[6] J. Zhao, S. Li, P. Lu, L. Meng, T. Li, H. Zhu, Shape

influences on the packing density of frustums, Powder Technol. 214 (2011) 500-505.

[7] K.G. Allen, T.W. von Backstrom, D.G. Kroger, Packed bed pressure drop dependence on particle shape, size distribution, packing arrangement and roughness, Powder Technol., 246 (2013) 590-600.

[8] L.W. Rong, K.J. Dong, A.B. Yu, Lattice-Boltzmann simulation of fluid flow through packed beds of spheres: Effect of particle size distribution, Chem. Eng. Sci. 116 (2014) 508-523.

[9] K. Vollmari, T. Oschmann, S. Wirtz, H. Kruggel-Emden, Pressure drop investigations in packings of arbitrary shaped particles, Powder Technol. 271 (2015) 109-124.

[10] Sh. Bu, J. Yang, Q. Dong, Q. Wang, Experimental study of flow transitions in structured packed beds of spheres with electrochemical technique, Exp. Therm. Fluid Sci. 60 (2015) 106-114.

[11] W. Du, N. Quan, P. Lu, J. Xu, W. Wei, L. Zhang, Experimental and statistical analysis of the void size distribution and pressure drop validations in packed beds, Chem. Eng. Res. Des. 106 (2016) 115-125.

[12] Z. Guo, Zh. Sun, N. Zhang, M. Ding, J. Wen, Experimental characterization of pressure drop in slender packed bed ($1 < D/d < 3$), Chem. Eng. Sci. 173 (2017) 578-587.

[13] S. Ergun, Fluid flow through packed columns, Chem. Eng. Prog. 48 (1952) 89-94.

[14] B. Eisfeld, K. Schnitzlein, The influence of confining walls on the pressure drop in packed beds, Chem. Eng. Sci. 56 (2001) 4321-4329.

[15] D. Nemec, J. Levec, Flow through packed bed reactors: 1. Single-phase flow, Chem. Eng. Sci. 60 (2005) 6947-6957.

[16] R. Singh, R.P. Saini, J.S. Saini, Nusselt number and friction factor correlations for packed bed solar energy storage system having large sized elements of different shapes, Sol. Energy, 80 (2006) 760–771.

[17] ANSYS FLUENT 12.0.16, Theory Guide. ANSYS Inc., 2009.

# 9

# Optimization of SiC particle distribution during compocasting of A356-SiC$_p$ composites using D-optimal experiment design

**Hamed Khosravi[1,*], Reza Eslami-Farsani[2], Mohsen Askari-Paykani**

[1] *Department of Materials Engineering, Faculty of Engineering, University of Sistan and Baluchestan, Zahedan, Iran*

[2] *Faculty of Materials Science and Engineering, K.N. Toosi University of Technology, Tehran, Iran*

[3] *Department of Materials Engineering, Tarbiat Modares University, Tehran, Iran*

## HIGHLIGHTS

- Compocasting processing of A356-SiC$_p$ composites was studied.

- Simultaneous effects of process parameters on SiC distribution were studied.

- D-optimal design of experiment was used for optimization.

## GRAPHICAL ABSTRACT

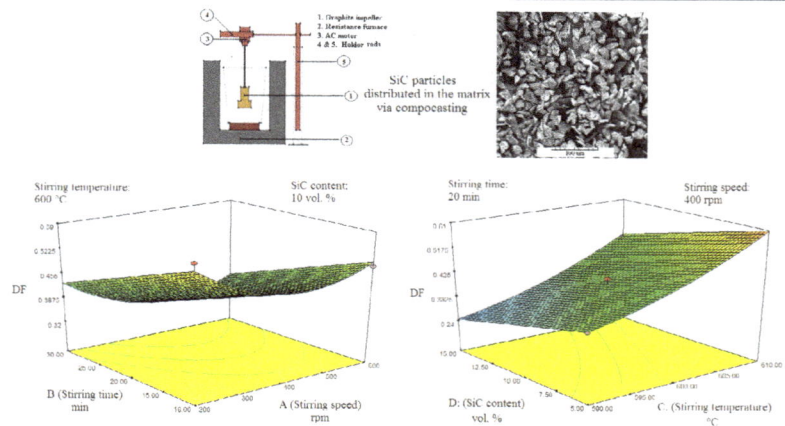

The smaller value of DF is indicative of more uniform distribution of SiC particles.

## ARTICLE INFO

*Keywords:*
A356-SiC$_p$ composite
Compocasting
Optimization
Modeling
D-optimal experiment design

## ABSTRACT

This paper presents an experimental design approach to the process parameter optimization for compocasting of A356-SiC$_p$ composites. Toward this end, parameters of stirring temperature, stirring time, stirring speed and SiC content were chosen and three levels of these parameters were considered. The D-optimal design of experiment (DODE) was employed for experimental design and analysis of results. In the experimental stage, different 20 μm-sized SiC particle contents (5, 10 and 15 vol %) were introduced into semisolid-state A356 aluminium alloy. Semisolid stirring was carried out at temperatures of 590, 600 and 610 °C with stirring speeds of 200, 400 and 600 rpm for 10, 20 and 30 min. The effect of these parameters on the distribution of the SiC particles within the matrix, represented by distribution factor (*DF*), was investigated. The smaller value of *DF* is indicative of the more uniform distribution of the SiC particles in the matrix. It was observed that the SiC particle content of 15 vol %, stirring temperature of 590 °C, stirring speed of 500 rpm, and stirring time of 30 min were the optimum parameter values producing the best distribution of the SiC particles in the matrix. The statistical test revealed that the main effect of the stirring temperature is the most significant factor.

* *Corresponding author:* ; *E-mail address: hkhosravi@eng.usb.ac.ir*

# 1. Introduction

It is well known that metal matrix composites (MMCs) are characterized by high specific strength, high specific stiffness, low thermal expansion coefficient and high wear resistance [1,2]. In particular, aluminium-based MMCs have gained extensive applications in automotive and aerospace industries due to their specific characteristics. Silicon carbide (SiC) has become the main type of reinforcement used for these materials. SiC exhibits good thermal conductivity and chemical compatibility with aluminium, creating a strong bond between particle and matrix [2-5]. MMCs can be fabricated via numerous processes mainly powder metallurgy and casting techniques. The casting process is a cost effective method while powder metallurgy is costly. Among the casting techniques, stir casting is the most frequently used route for production of particulate MMCs. However, it is associated with some inherent problems arising mainly from both the apparent non-wettability of ceramic reinforcing particles by liquid aluminium alloys and the density differences between the two phases [6,7]. In order to overcome some of these drawbacks that result in non-uniform distribution of the reinforcement within the matrix alloy, extensive interfacial reactions, and formation of brittle phases at the particle/matrix interface as well as a high level of porosity new semi-solid processing techniques have been considered for manufacturing of these MMCs [8-10].

Compcasting is a semi-solid processing route in which the ceramic reinforcing particulates are added to the semi-solid matrix alloy via mechanical stirring and then cast in a mold for solidification. This technique is superior because of its simplicity, flexibility and low cost, and is considered to be the best method for preparation of large quantities of composites at low cost [11,12].

From the available literatures on MMCs, it is obvious that the size, distribution and volume fraction of the reinforcement phase as well as the matrix properties are the main factors affecting the overall mechanical and physical properties [6,7,10,11].

One of the main challenges associated with the cast MMCs is to achieve a homogeneous distribution of reinforcement within the matrix alloy. In order to achieve the optimum properties of the MMCs, the distribution of the reinforcing particles in the matrix alloy should be uniform and the porosity levels need to be minimized. A non-homogeneous particle distribution often arises as a result of agglomeration, settling, and segregation of ceramic particles during the processing of these composite materials. The particle distribution has a significant effect on the mechanical properties of MMCs. For example, clustered particles act as crack initiation sites and have a negative influence on the mechanical properties of composite materials. Clustered particle arrangements significantly reduce the failure strain of composites. To obtain a homogeneous distribution of reinforcing particles in the cast particulate MMCs, several factors such as the good wettability of the particles with the molten alloy, proper mixing, reinforcement size, reinforcement content, mold temperature and solidification rate should be considered [10,11,13-16].

The key processing parameters affecting the final microstructure of the solidified slurry during compocasting processing are stirring time, stirring temperature and stirring speed. From an industrial point of view, it is essential to find out the best combination of compocasting parameters to attain the best mechanical and physical properties.

In general, an experiment is an observation which leads to characteristic information about a studied object [17]. One of the most common and classical approaches employed by many experimenters is one-factor-at-a-time (OFAT), in which one factor is varied while all other variables or factors in the experiment are fixed. The success of this approach depends on guesswork, luck, experience and intuition. Moreover, this type of experimentation requires large resources to obtain a limited amount of information about the process [17-21]. In many situations, in view of the high cost of experimentation, the number of observations is kept to a minimum [22-24]. With design of experiment (DOE) this number is kept as low as possible and the most informative combination of the factors is chosen [22,23]. Hence, DOE is an effective and economical solution. The aim of this so-called design is to optimize a process or system by performing each experiment and to draw conclusions about the significant behavior of the studied object from the results of the experiments. In recent years, the use of D-optimal design of experiment (DODE) in industrial experimentation has grown rapidly, due in part, to the fact that the methodology is now being introduced in standard DOE text books [19-

21], and also because facilities for constructing DODE have become generally available.

On the other hand, unlike standard classical designs, such as factorials and fractional factorials, DODE is usually not orthogonal [17,18]. This type of design is always an option regardless of the type of model the experimenter wishes to fit (for example, first order, first order plus some interactions, full quadratic, cubic, etc.) or the objective specified for the experiment (for example, screening, response surface, etc.) [18]. DODE is straight optimization based on a chosen optimality criterion and the model that will be fit. The optimality criterion used in generating DODE results in minimizing the generalized variance of the parameter estimates for a pre-specified model. As a result, the 'optimality' of a given DODE is model dependent [17,18]. That is, the experimenter must specify a model for the design before the computer can generate the specific treatment combinations. Given the total number of treatment runs for an experiment and a specified model, the computer algorithm chooses the optimal set of design runs from a candidate set of possible design treatment runs. This candidate set of treatment runs usually consists of all possible combinations of various factor levels that one wishes to use in the experiment.

Although some researchers have already utilized DOE to optimize the different types of casting routes [24], no effort has yet been made to perform this optimization on the com-casting process. In the present work, an attempt has been made to develop a model for predicting the uniformity in SiC particle within the matrix as a function of key input parameters in the compocasting processing of A356-SiC$_p$ composites.

## 2. Experimental

### 2.1. Materials and experimental procedure

Al-A356 with a nominal chemical composition, as given in Table 1, formed the matrix and SiC particles (average size 20 μm) with 5, 10 and 15 % volume fractions were used as the reinforcement phase. A356 aluminium alloy is a hypoeutectic Al-Si alloy and its relatively broad semisolid interval (32 °C) makes it suitable for semisolid processing. The SEM micrograph of the SiC powder is shown in Figure 1. SiC particles were artificially oxidized in air at 1000 °C for 120 min to allow a layer of SiO$_2$ to form on them and improve

their wettability with molten aluminium. This treatment helps the incorporation of the particles while reducing undesired interfacial reactions [11].

**Table 1.** Chemical composition (wt %) of A356 Alloy.

| Si | Mg | Mn | Zn | Cu | Fe | Al |
|------|------|------|------|------|------|---------|
| 6.93 | 0.38 | 0.23 | 0.26 | 0.25 | 0.11 | Balance |

The aluminium alloy matrix composites were synthesized by the compocasting method. Figure 2a shows a schematic representation of the compocasting apparatus used in this study. In the first stage, 1 kg of A356 aluminium alloy was put in a graphite crucible and melted at 750 °C by an electric resistance furnace. Two calibrated thermocouples were inserted into the melt and the furnace to measure their temperatures. SiC particles were preheated at 600 °C in a stainless steel crucible. Given density values for Al and SiC (2.7 and 3.2 g/cm³), the crucible charge was determined to obtain the A356-SiC$_p$ composite samples with different SiC contents. The semi-solid stirring process was carried out by a graphite impeller (Figure 2b) [25] at temperatures of 590, 600 and 610 °C for 10, 20 and 30 min. Three different stirring speeds of 200, 400 and 600 rpm were also utilized. In the last stage, the slurry was heated up to 660 °C and stirred at this temperature for another 8 min. Casting was done in a cylindrical shaped steel mold (40 mm internal diameter and 30 mm in height), preheated at 400 °C.

The prepared samples were subjected to standard metallographic procedures and examined via an "Olympus-BX60M" light microscope.

Fig. 1. SEM image of SiC particles.

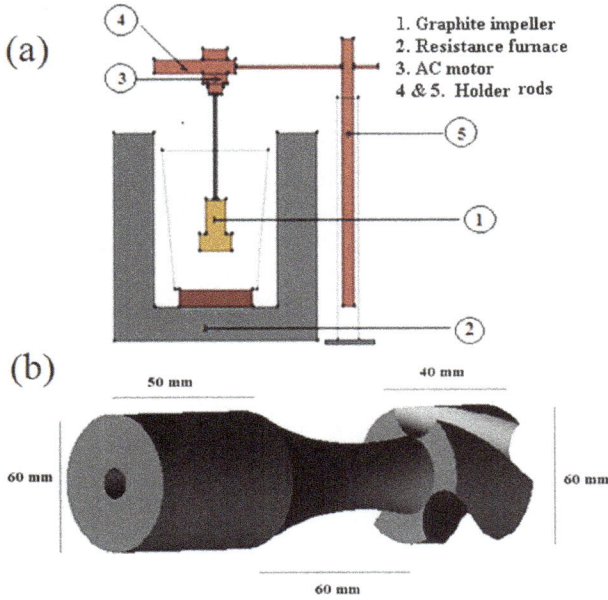

1. Graphite impeller
2. Resistance furnace
3. AC motor
4 & 5. Holder rods

**Fig. 2.** (a) Schematic representation of the compocasting apparatus and (b) graphite impeller used in this study.

The distribution of the SiC particles within the matrix alloy was characterized by calculating the distribution factor (DF) defined by Eq. (1) [26].

$$DF = \frac{S.D.}{A_f} \qquad (1)$$

in which $A_f$ is the mean value of the area fraction of the SiC particles measured on 100 fields of a sample and S.D. is its standard deviation. A non-uniform microscopic distribution of the reinforcing phase within a sample is reflected as a relatively high value of DF.

### 2.2. Experimental design and statistical analysis

To explore the effect of the operation factors on the response (DF) in the region of investigation, a DODE at three levels was performed. Stirring speed (rpm, A), stirring time (min, B), stirring temperature (°C, C) and SiC content (vol %, D) were selected as independent factors. The range of values and coded levels of the factors are given in Table 2.

A polynomial equation (Eq. 2) was used to predict the response (DF, Y) as a function of independent factors and their interactions. An interaction is the failure of the one factor to produce the same effect on the response at different levels of another factor [20]. In this work, there were four independent factors; therefore, the response for the quadratic polynomials becomes:

**Table 2.** Independent Factors and their Levels for DODE of compocasting process.

| Independent factors | Unit | level | | |
|---|---|---|---|---|
| | | -1 | 0 | 1 |
| Stirring speed (A) | rpm | 200 | 400 | 600 |
| Stirring time (B) | min | 10 | 20 | 30 |
| Stirring temperature (C) | °C | 590 | 600 | 610 |
| SiC content (D) | vol % | 5 | 10 | 15 |

$$Y = \beta_0 + \sum \beta_i x_i + \sum \beta_{ii} x_i^2 + \sum \sum \beta_{ij} x_i x_j \qquad (2)$$

where $\beta_0$, $\beta_i$, $\beta_{ii}$, $\beta_{ij}$ are the constant, linear, square and interaction regression coefficient terms, respectively, and xi and xj are the independent factors (A, B, C or D).

Design-Expert 7 (State-Ease, Inc., Trial version) software was used for multiple regression analysis, analysis of variance (ANOVA), and analysis of ridge maximum of data in the response surface regression (RSREG) procedure. The goodness of fit of the model was evaluated by the coefficient of determination $R^2$ and its statistical significance was checked by the F-test.

### 3. Results and discussions

This study demonstrates the effect of stirring speed, stirring time, stirring temperature and SiC content for the optimization of the compocasting route. Hence, the knowledge about the process is relatively limited, and the design is used to obtain 38 design points within the whole range of four factors for experiments. The designs and the response (DF (Y)) are given in Table 3. Following the experiments, the response surface is approximated by DODE.

The results of the DODE are presented in an ANOVA table (Table 4) with a confidence interval (CI) of 95% (P < 0.05) for the model. In statistics, CI is a kind of interval estimate of a population parameter and is used to indicate the reliability of an estimate. The level of confidence of CI would indicate the probability that the confidence range captures this true population parameter given a distribution of samples [17-21]. By considering a half normal plot and a normal plot (not shown here), four main effects and their squares all with CI = 95% were selected as significant factors for modeling. The effect of a factor is defined as the change in response produced by a change in the level of a factor. This

**Table 3.** DODE tests and the response for compocasting process.

| Standard Order | Run Order | Factor 1 A | Factor 2 B | Factor 3 C | Factor 4 D | Response DF (by Experiment) |
|---|---|---|---|---|---|---|
| 1 | 26 | 200 | 20 | 600 | 5 | 0.47 |
| 2 | 28 | 600 | 30 | 610 | 10 | 0.52 |
| 3 | 8 | 200 | 10 | 590 | 15 | 0.51 |
| 4 | 15 | 600 | 20 | 610 | 5 | 0.41 |
| 5 | 7 | 400 | 30 | 610 | 15 | 0.28 |
| 6 | 25 | 200 | 30 | 610 | 10 | 0.34 |
| 7 | 22 | 400 | 10 | 600 | 15 | 0.52 |
| 8 | 29 | 600 | 30 | 600 | 5 | 0.44 |
| 9 | 13 | 200 | 20 | 600 | 15 | 0.34 |
| 10 | 33 | 600 | 20 | 590 | 5 | 0.66 |
| 11 | 32 | 200 | 20 | 590 | 15 | 0.40 |
| 12 | 34 | 600 | 20 | 610 | 15 | 0.36 |
| 13 | 18 | 200 | 30 | 610 | 15 | 0.35 |
| 14 | 31 | 200 | 20 | 590 | 10 | 0.65 |
| 15 | 19 | 200 | 10 | 600 | 15 | 0.41 |
| 16 | 23 | 600 | 10 | 610 | 15 | 0.60 |
| 17 | 3 | 200 | 10 | 610 | 15 | 0.67 |
| 18 | 17 | 200 | 30 | 610 | 5 | 0.64 |
| 19 | 27 | 200 | 10 | 600 | 5 | 0.65 |
| 20 | 24 | 600 | 10 | 600 | 10 | 0.50 |
| 21 | 12 | 200 | 10 | 600 | 10 | 0.59 |
| 22 | 35 | 600 | 30 | 590 | 15 | 0.22 |
| 23 | 9 | 400 | 30 | 590 | 15 | 0.17 |
| 24 | 6 | 400 | 20 | 610 | 15 | 0.46 |
| 25 | 2 | 400 | 30 | 610 | 15 | 0.40 |
| 26 | 37 | 600 | 20 | 590 | 10 | 0.35 |
| 27 | 30 | 400 | 30 | 600 | 5 | 0.39 |
| 28 | 36 | 400 | 20 | 600 | 10 | 0.42 |
| 29 | 38 | 400 | 30 | 600 | 10 | 0.33 |
| 30 | 20 | 200 | 30 | 590 | 5 | 0.41 |
| 31 | 5 | 400 | 10 | 610 | 5 | 0.65 |
| 32 | 16 | 400 | 10 | 590 | 10 | 0.41 |
| 33 | 11 | 400 | 10 | 610 | 10 | 0.60 |
| 34 | 1 | 400 | 20 | 590 | 5 | 0.38 |
| 35 | 14 | 400 | 10 | 590 | 5 | 0.48 |
| 36 | 4 | 200 | 30 | 590 | 15 | 0.29 |
| 37 | 10 | 200 | 10 | 590 | 5 | 0.55 |
| 38 | 21 | 200 | 10 | 590 | 10 | 0.48 |

**Table 4.** ANOVA with CI = 95% for model and factors.

| Source | Sum of squares | Degree of freedom | Mean square | F value | P-value Prob> F | |
|---|---|---|---|---|---|---|
| Model | 0.60 | 5 | 0.12 | 281.69 | <0.0001 | Significant |
| $A$ | 0.019 | 1 | 0.019 | 45.71 | <0.0001 | |
| $B$ | 0.14 | 1 | 0.14 | 327.98 | <0.0001 | |
| $C$ | 0.29 | 1 | 0.29 | 687.10 | <0.0001 | |
| $D$ | 0.12 | 1 | 0.12 | 292.61 | <0.0001 | |
| $A^2$ | 0.040 | 1 | 0.040 | 93.53 | <0.0001 | |
| Residual | 0.014 | 32 | 0.0004261 | | | |
| Corrected Total | 0.61 | 37 | | | | |

is frequently called a main effect because it refers to primary factors of interest in the experiment [21].

ANOVA results for *DF* show a significant model with adequate precision of 61.818. Adequate precision compares the range of the predicted values at the design points to the average prediction error; on the other hand, adequate precision measures the signal to noise ratio and a ratio greater than 4 is desirable [18]. Here, the value of the ratio is greater than 4, so it represents an adequate model (Eq. (3)) for predicting the results within the design space without doing any further experiments.

$$Y=0.42-0.030A-0.037B+0.11C-0.069D+0.070A^2 \quad (3)$$

The quality of fittings of the equations was expressed by the coefficient of regression "Adjusted R-squared" or in a better way by "Predicted R-squared". The "Adjusted R-squared" values indicate variability in the observed response values which can be explained by the experimental factors and their interactions. The closer "Predicted R-Squared" and "Adjusted R-Squared" values are to 1, the better the fit [27]. The "Predicted R-squared" of 0.9684 is in reasonable agreement with the "Adjusted R-squared" of 0.9743. The model F-value of 281.69 implies that the model is significant ($F_{model}$ = 281.69 >> $F_{table}$ (5,32) = 2.530) and there is only a 0.01% chance that a "Model F-value" could occur due to noise. F-value is the test for comparing the variance associated with that term with the residual variance. It is the mean square for a term divided by the mean square for the residual. This term should be as large as possible [18]. Tables of F-value (a,b) for different confidence intervals exist in statistical references [17]. Where, the first number in parenthesis is the parameter or model degree of freedom and the second one is the error's

(residuals) degree of freedom. To categorize the parameter or the model as a significant value, calculated F-value must be more than its value in the statistical tables. If the calculated value of F is greater than that in the F table at a specified probability level, a statistically significant factor or interaction is obtained [20]. The lack of fit of the F-value for the response showed that the lack of fit is not significant ($p > 0.05$) relative to the pure error. This model (Eq. (3)) can be used to navigate the design space.

The quadratic regression coefficients obtained by employing a least squares method technique to predict quadratic polynomial models for the *DF* (*Y*) are given as Eq. (3). For *Y*, the linear term and the quadratic terms (without interaction terms) of *A*, *B*, *C* and *D* were significant ($P < 0.05$).

Sum of squares (*SS*) of each factor quantifies its importance in the process and as the value of the *SS* increases the significance of the corresponding factor in the undergoing process also increases. As shown in ANOVA table (Table. 4), the effect of *C* (stirring temperature) is the strongest and then *B*, *D*, $A^2$ and *A*, respectively. If we consider the model equation in actual terms, one can find that the effect of $A^2$ and *C* is positive (synergistic effect). However, *A*, *B* and *D* have a negative (antagonism) effect on *DF*. To decrease *DF*, the positive effect should be descending and negative effect should be ascending.

Significant factors in the fitted model (Eq. (3)) were chosen as the axes for the 3D surface plots (Figures 3a and 4a) and contour plots (Figures 3b and 4b). In a contour plot (base plots in the 3D plots), curves of equal response values are drawn on a plane whose coordinates represent the levels of the independent factors. Each contour represents a specific value for the height of the

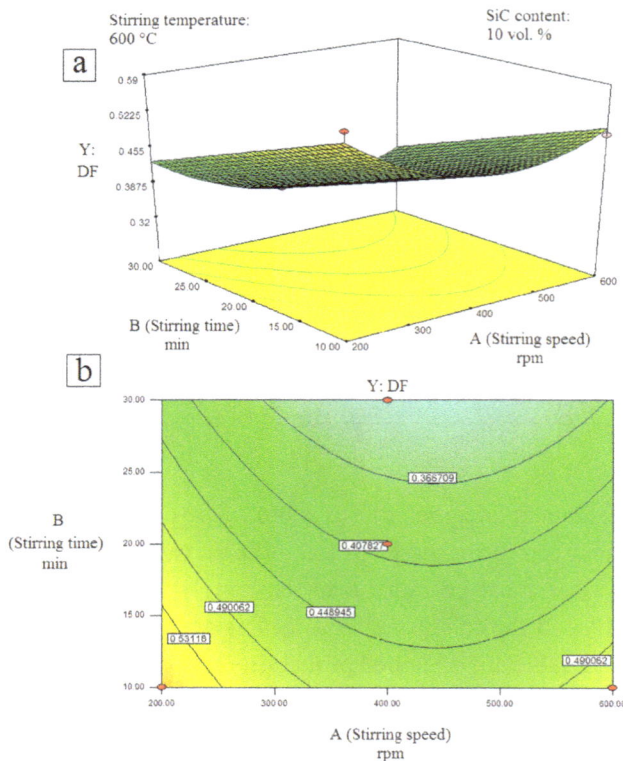

**Fig. 3.** a) Response surface and b) contour plots for the effect of the stirring time and stirring speed on the *DF*.

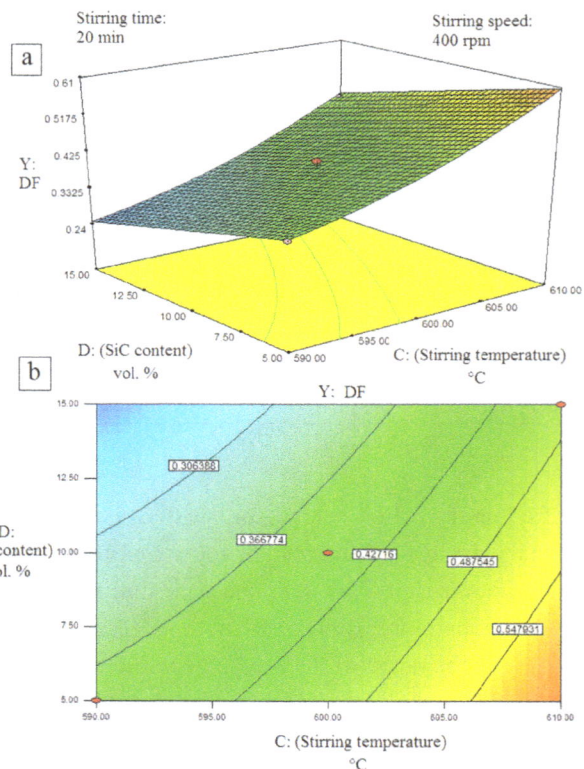

**Fig. 4.** a) Response surface and b) contour plots for the effect of the stirring temperature and SiC content on the *DF*.

surface above the plane defined for a combination of the levels of the factors.

From Figures 3 and 4, *DF* decreases by decreasing the amount of stirring temperature and increasing the amount of SiC content and stirring time. On the other hand, the relationship between the stirring speed and DF was almost parabolic. This trend is in good agreement with the trend of factor effects. The observed values were reasonably close to the predicted ones as shown in Figure 5.

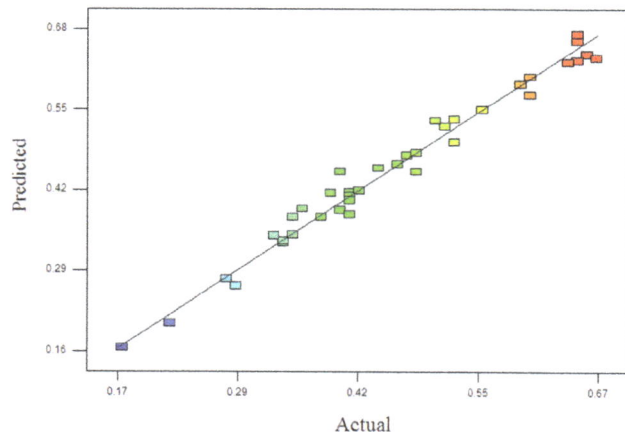

**Fig. 5.** Predicted vs. actual plot of *DF*.

The normality of the data can be checked by plotting a normal probability plot of the residuals. If the data points on the plot fall fairly close to a straight line, then the data are normally distributed [21]. The normal probability plot of the residuals for SF (not shown here) depicted that the data points were fairly close to the straight line and this indicates that the experiments come from a normally distributed population.

The confirmation experiments were conducted in three different conditions. The results are listed in Table 5. If the average of the results of the confirmation is within the limits of the CI, then the significant factors as well as the appropriate levels for obtaining the desired results are properly chosen [17-21]. From Table 5, the experimental responses are in 95% CI range and this model can be used to navigate within the design space.

In order to test the validity of the optimized conditions given by the model, an experiment was also carried out with parameters as suggested by the model. The conditions used in the confirmatory experiment are given in Table 6. The *DF* value at the optimal condition was found to be 0.16 (Table 6), which is consistent with the model. Therefore, the formulated model is acceptably valid. It should be noted that the smaller value of *DF*

**Table 5.** Results of confirmation tests.

| | | Stirring speed (rpm) | Stirring time (min) | Stirring temperature (°C) | SiC content (vol. %) | DF |
|---|---|---|---|---|---|---|
| 1 | Model | 400 | 20 | 590 | 10 | 0.30 |
| | Confirmation test | | | | | 0.32 |
| 2 | Model | 200 | 30 | 600 | 5 | 0.51 |
| | Confirmation test | | | | | 0.53 |
| 3 | Model | 400 | 20 | 610 | 10 | 0.53 |
| | Confirmation test | | | | | 0.52 |

**Table 6.** DF at optimal conditions.

| Parameter | Stirring speed | Stirring time | Stirring temperature | SiC content | DF |
|---|---|---|---|---|---|
| Model | 500 rpm | 30 min | 590 °C | 15 vol % | 0.16 |
| Confirmation test | | | | | 0.19 |

is indicative of the more uniform distribution of the SiC particles in the matrix [11]. Figure 6 demonstrates the optical micrographs of the composite samples fabricated by different compocasting process parameters and SiC contends for confirmation and optimal condition tests.

Figure 3 shows that DF of the SiC particles decreases with increasing the semisolid stirring time, representing a more homogenous SiC distribution within the matrix. At lower stirring time (10 min), in some zones the matrix is free from SiC particles and in other regions clustering of the SiC particles is visible. This shows that this stirring time is insufficient for obtaining an

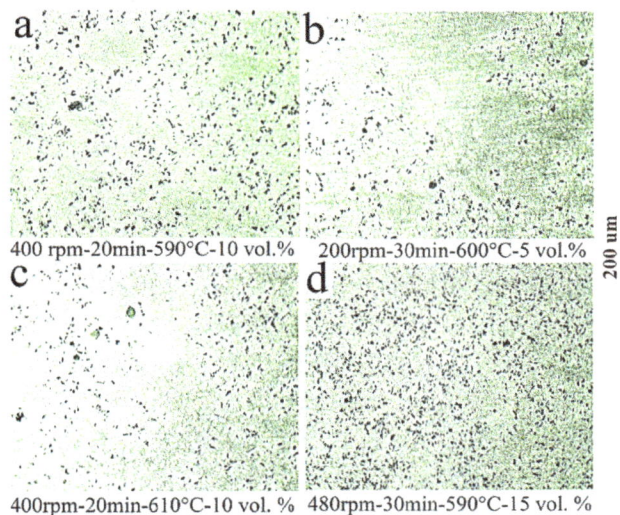

400 rpm-20min-590°C-10 vol.%    200rpm-30min-600°C-5 vol.%

400rpm-20min-610°C-10 vol. %    480rpm-30min-590°C-15 vol. %

200 um

**Fig. 6.** Optical micrographs of the A356-SiC$_p$ composites fabricated by the different compocasting process parameters and SiC contents (a-c) confirmation tests and (d) at optimal condition.

acceptable SiC distribution in the matrix. Higher stirring time results in a better distribution of the particles. From Figure 4, it can be seen that decreased stirring temperature resulted in a more homogeneous distribution of these particles within the matrix, as indicated by smaller DF values. This means that by increasing the semisolid stirring temperature, a less homogeneous distribution of the SiC particles is obtained in the matrix alloy. Decreasing the stirring temperature from 610 to 590 °C (at the fixed stirring speed and stirring time of 400 rpm and 20 min, respectively) leads to a 45% decrease in the DF value, which is attributed to the increased viscosity of the semisolid slurry. According to the equilibrium binary Al-Si diagram, A356 aluminium alloy solidifies at a broad temperature interval (32 °C) between 583 to 615 °C. This alloy consists of 45%, 35% and 18% solid fractions in semisolid slurry at 590, 600 and 610 °C, respectively [28]. This shows that the viscosity of the alloy increases as the semisolid temperature decreases. The restricted movement of the particles within the slurry during semisolid stirring prevents the SiC particles from settling as a consequence of the increased effective viscosity; consequently, a more uniform particle distribution is obtained. The presence of a solid phase in the semisolid slurry can also help the breakdown of the SiC clusters during stirring.

The results of this study show a remarkable improvement in the uniformity of the SiC particle distribution (as reflected by the decreased DF) when the stirring speed of 500 rpm was used (Figure 3 and Table 6). Particle clustering is observable at a relatively low stirring speed (i.e. 200 rpm), and in some regions the matrix is free of SiC particles (Figure 6b). By increasing the stirring speed to 500 rpm, a better distribution of the SiC particles within the matrix alloy is obtainable. These results are in agreement with some related studies [10,12], and can be attributed to the increase of shear forces applied by increasing the stirring speed, which can improve the uniformity of the SiC particle distribution as a result of a larger vortex within the slurry. On the other hand, the higher stirring speed (from 500 to 600 rpm) imposed a considerable non-uniformity in the SiC particle distribution, which can be attributed to the increased agitation severity of the slurry, resulting in clustering of the SiC particles.

The effect of the SiC content on the uniformity of the particles distribution within the matrix is given in Figure 4. From this figure, improvement in the uniformity of the SiC particle distribution is obtainable when the particle content increases. This can be attributed to the (a) restricted movement of particles within the melt during solidification as a consequence of the increased effective viscosity of the slurry and (b) finer matrix microstructure as a result of increased barriers for growth of $\alpha$-Al phase.

## 4. Conclusion

Compocasting processing of Al-A356-SiC$_p$ composites was studied and modeled using the D-optimal design of experiment (DODE). The effects of compocasting process parameters (stirring temperature, stirring time and stirring speed) as well as SiC content on the uniformity in the particle distribution were investigated. The conclusions drawn from the results can be summarized as follows:

1. The optimum values of stirring temperature, stirring time and stirring speed were found to be 590 °C, 30 min and 500 rpm, respectively.

2. The correlation coefficient ($R^2$) of the regression model was 0.97, which confirms the excellent accuracy of the model.

3. The most important factor affecting the SiC distribution within the matrix alloy was found to be the stirring temperature.

4. The uniformity in the SiC distribution improved by increasing the SiC content and stirring time and decreasing the stirring temperature. A remarkable improvement in the uniformity of the SiC particle

distribution was achieved when the stirring speed of 500 rpm was used.

## References

[1] N. Chawla, K. K. Chawla, Metal matrix composites, Springer, New York, 2006.

[2] H. Beygi, M. Shaterian, E. Tohidlou, M.R. Rahimipour, Development in wear resistance of Fe-0.7Cr-0.8Mn milling balls through in situ reinforcing with low weight percent TiC, Adv. Mat. Res. 413 (2012) 262-269.

[3] B.K. Vinoth, J.J.T. Winowlin, T.P.D. Rajan, M. Uthayakumar, Dry sliding wear studies on SiC reinforced functionally graded aluminium matrix composites, Proceedings of the Institution of Mechanical Engineers, Part L: J. Mater. Design Appl. 30 (2016) 182-189.

[4] O. El-Kady, A. Fathy, Effect of SiC particle size on the physical and mechanical properties of extruded Al matrix nanocomposites, Mater. Design, 54 (2014) 348-353.

[5] H. Khosravi, F. Akhlaghi, Comparison of microstructure and wear resistance of A356-SiC$_p$ composites processed via compocasting and vibrating cooling slope, T. Nonferr. Metal. Soc. 25 (2015) 2490-2498.

[6] S.T. Kumaran, M. Uthayakumar, S. Aravindan, S. Rajesh, Dry sliding wear behavior of SiC and B4C-reinforced AA6351 metal matrix composite produced by stir casting process. Proceedings of the Institution of Mechanical Engineers, Part L: J. Mater. Design Appl. 230 (2016) 484-491.

[7] S.A. Sajjadi, H. R. Ezatpour, M. Torabi Parizi, Comparison of microstructure and mechanical properties of A356 aluminium alloy/Al$_2$O$_3$ composites fabricated by stir and compo-casting processes, Mater. Design, 34 (2012) 106-111.

[8] B. Abbasipour, B. Niroumand, M. Monir-Vaghefis, Compocasting of A356-CNT composite, T. Nonferr. Metal. Soc. 20 (2010) 1561-1566.

[9] K.H.W. Seah, S.C. Sharma, M. Krishna, Mechanical properties and fracture mechanism of ZA-27/TiO$_2$ particulate metal matrix composites, Proceedings of the Institution of Mechanical Engineers, Part L: J. Mater. Design Appl. 217 (2003) 201-206.

[10] H. Zhang, L. Geng, L. Guan, L. Huang, Effects of SiC particle pretreatment and stirring parameters on

the microstructure and mechanical properties of SiC$_p$/Al-6.8Mg composites fabricated by semi-solid stirring technique, Mat. Sci. Eng. A - Struct. 528 (2010) 513-518.

[11] F Akhlaghi, A. Lajevardi, H. M. Maghanaki, Effects of casting temperature on the microstructure and wear resistance of compocast A356/SiC$_p$ composites: a comparison between *SS* and *SL* routes, J. Mater. Process. Tech. 155-156 (2004) 1874-1880.

[12] S.A. Sajjadi, M. Torabi-Parizi, H.R. Ezatpour, A. Sedghi, Fabrication of A356 composite reinforced with micro and nano Al$_2$O$_3$ particles by a developed compocasting method and study of its properties, J. Alloy. Compd. 511 (2012) 226-231.

[13] A. Ourdjini, K. Chew, C. Khoo, Settling of silicon carbide particles in cast metal matrix composites, J. Mater. Process. Tech. 116 (2001) 72-76.

[14] L. V. Vugt, L. Froyen, Gravity and temperature effects on particle distribution in Al-Si/SiC$_p$ composites, J. Mater. Process. Tech. 104 (2000) 133-144.

[15] M. Gupta, L. Lu, S. E. Ang, Effect of microstructural features on the aging behavior of Al-Cu/SiC metal matrix composites processed using casting and rheocasting routes, J. Mater. Sci. 32 (1997) 1261-1267.

[16] A. Cetin, A. Kalkanli, Effect of solidification rate on spatial distribution of SiC particles in A356 alloy composites, J. Mater. Process. Tech. 205 (2008) 1-8.

[17] M. J. Anderson, P. J. Whitcomb, DOE simplified: practical tools for effective experimentation, New York, Productivity Inc., 2000.

[18] Software helps Design-Expert Software, Version 7.1, User's guide, Technical Manual, Stat-Ease Inc., Minneapolis, 2007.

[19] J. Antony, Design of experiments for engineers and scientists, Oxford, Heinemann, 2003.

[20] D.C. Montgomery, Design and analysis of experiment, Wiley, New York, 1997.

[21] A. Dean, D. Voss, Design and analysis of experiments, Springer text in statistics, Springer-Verlag, New York, 1999.

[22] A.K. Sahoo, S. Pradhan, Modeling and optimization of Al/SiC$_p$ MMC machining using Taguchi approach, Measurement, 46 (2013) 3064-3072.

[23] N. Mandal, B. Doloi, B. Mondal, R. Das, Optimization of flank wear using Zirconia Toughened Alumina (ZTA) cutting tool: Taguchi method and regression analysis, Measurement, 44 (2011) 2149-2155.

[24] H. Khosravi, R. Eslami-Farsani, M. Askari-Paykani, Modeling and optimization of cooling slope process parameters for semi-solid casting of A356 Al alloy, T. Nonferr. Metal. Soc. 24 (2014) 961-968.

[25] B. Rahimi, H. Khosravi, M. Haddad-Sabzevar, Microstructural characteristics and mechanical properties of Al-2024 alloy processed via a rheocasting route, Int. J. Min. Met. Mater. 22 (2015) 1-9.

[26] R. Rahmani, F. Akhlaghi, Effect of extrusion temperature on the microstructure and porosity of A356-SiC$_p$ composites, J. Mater. Process. Tech. 187-188 (2007) 433-436.

[27] S. Neseli, S. Yaldiz, E. Turkes, Optimization of tool geometry parameters for turning operations based on the response surface methodology, Measurement, 44 (2011) 580-587.

[28] H. Khosravi, H. Bakhshi, E. Salahinejad, Effects of compocasting process parameters on microstructural characteristics and tensile properties of A356-SiC$_p$ composites, T. Nonferr. Metal. Soc. 24 (2014) 2482-2488.

# 10

# Crushing analysis of the industrial cage mill and the laboratory jaw crusher

Reza Zolfaghari, Mohammad Karamoozian*

*School of Mining, Petroleum and Geophysics, Shahrood University of Technology, Shahrood, Iran*

## HIGHLIGHTS

- Industrial cage mill creates better mineral liberation of middling than the jaw crusher.

- Grinding the middle product with a cage mill results in a better yield than jaw crusher.

- The rate of fines produced through the jaw crusher is less than the cage mill.

## GRAPHICAL ABSTRACT

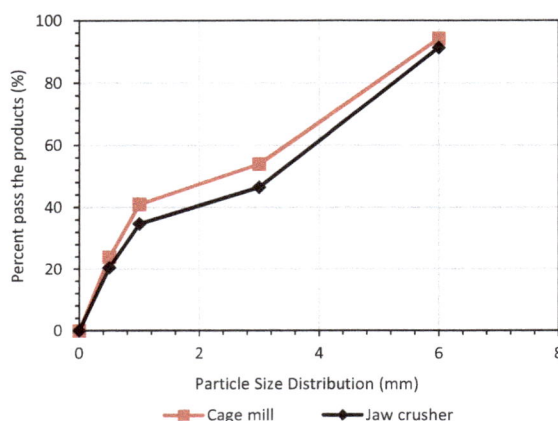

## ARTICLE INFO

*Keywords:*

Comminution
Middle coal
Washability analysis
Cage mill
Jaw crusher

## ABSTRACT

Many research studies have been conducted on the liberation of locked minerals using a crusher and comparing this device with the other devices. This paper reviews the liberation of middle coal by different methods of crushing force. In the Tabas coal washing plant, particles of 0.5-50 mm size are processed through the heavy media method (using 3 Tri-flo separators) and particles of 0-0.5 mm size are processed using the flotation method (using 6 column flotation cells). A Tri-flo separator with a diameter of 700 mm and the capacity of 120 tons per hour is used for the cleaning of 6-50 mm raw coal particles. The study was conducted using a laboratory jaw crusher and a cage mill with a specific comminution ratio, both crushing forces were analyzed with the same distribution and mechanism of production of fines. In this study, grading and washability characteristics of a representative sample of middle product were reviewed and the dimensions of the ash were measured for each section. Intermediate product crushing using a laboratory jaw crusher and an industrial cage mill were conducted at up to 5 mm size and 50 percent of final speed. The amount of coal released after each section grading was determined by a sinking and floating test for size +0.5 mm and release analysis and ash testing for smaller dimensions of -0.5, these tests were conducted for each section product dimension. The results indicated that utilizing a cage mill is more effective than a laboratory jaw crusher, resulting in 11-percent more yield with 12 ash. The rate of fines produced through the laboratory jaw crusher is less than the industrial cage mill.

* Corresponding author:   ;  E-mail address: m.karamoozian@shahroodut.ac.ir

# 1. Introduction

Coal preparation in the late 1890's and early 1900's began in the United States. At that time, the coal separation process was conducted by hand and mechanical operations [1]. Over the past century, the role of coal and its importance in the world economy was remarkable. In 1860, coal was so important to the world that it allocated 60 percent of the total value of all minerals. However, with the arrival of oil and gas the use of coal as an alternative fuel in the world became rarer. Iran is ranked twenty-sixth in the world in terms of coal reserves, the largest of which are the Tabas reserves.

In the past, only coarse fragments of coal were recovered after extraction, and due to the lack of appropriate technologies small coal particles would be transferred into tailing damps. Later, the demand for a product with a uniform distribution particle size and the need for acceptable degrees of liberation in coal crushing caused coal crushing devices to be developed at the same rate as washing processes [2]. The initial load of coal washing plants is usually derived from underground and ground mines. Occasionally, a 10 inch coal might be extracted. In mineral processing plants the size reduction process is performed using crushers and mills [1]. The extraction method in the Tabas Parvadeh coal mines is underground mining. Because of the nature of mass and extraction method of coal, the size of extracted materials is different.

In comparison with the released metals, coal fines generated from coal have a relatively larger size. Controlling the size of coal fines in the crushing procedure would be very effective. The coal petrology of middle and raw coal has significant differences indicating there are noticeably different characteristics on their surfaces [7,8]. For coal mining and extraction coal should first be crushed, and a controlled crushed particle size would be useful to feed the processing plant. In addition, there is the need to use the right equipment to reduce the fines and to reduce the contact surface with rock minerals. To further develop the system in this paper we analyzed changes to the operating and systematic parameters and investigated the effect of the two parameters of known speed and added water.

Coal comminution is the final process of the grinding operation. The process in the comminution phase involves reduction of the particle size, impact, and abrasion. The main goal of crushing is to obtain the appropriate degree of liberation. The highest rate of energy consumption in the processing plant is attributed to crushing. In practice comminution is performed along with event impact, in which both free and locked particles are present. Among these types of particles, the locked particles are suitable for comminution. Impact creating on the border between mineral particles results in the most ideal situations. Many experts have studied and investigated the fracture mechanics for coal, which is a brittle and fragile material. Pressure, impact and cutting are the main steps in coal crushing devices. Middle product crushing in the Tabas coal washing plant is conducted using a cage mill. A cage mill is a rotating crushing system in which a multiple grinding plates moves in retrograde motion, and where the cages move together with the same speed but in the opposite direction. As the material is passed through the device and is crushed from one step to the next, the impact velocity increases. Preferably, coarser particles will be crushed. Figure 1 shows the structure in this device.

The material enters the internal cage through a slot. These devices are mainly used for brittle materials such as coal and salt. The main mechanism of crushing occurs in the cage mill [3]. The movement of particles can be controlled through design of the crushing rods in the device. By using more rows, more coal with a size of less than 75 micros is achievable. The mill size and design is based on impact in terms of feed parameters (volume of the sample and feed particle moisture), product parameters (shape and distribution), and some system parameters (rate of wear parts and specific machinery) [9].

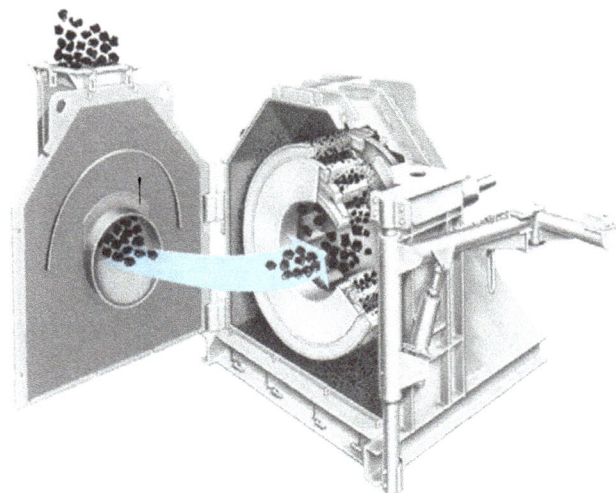

Fig. 1. Schematic of a cage mill and two cages [3].

Coal preparation plants generally do not reduce the size too much. Fractures in the mass of coal during processing result in the production of fines, depending on the nature of the coal and processing plant [4].

Gravity separation techniques are used for various materials such as sulfide minerals, e.g. galena, and coal in sizes smaller than 50 microns. The use of this method has increased in recent years because of the increased cost of chemicals necessary in the flotation operation, simplicity of installation, and low environmental pollution. Although these methods are known as gravity methods because of their special mass, classification of shape and dimension play important roles in these methods [11,12]. Analyses of sinking and floating are based on the floating particle density. Particles in the analyses of sinking and floating in each section consists of two even parts, namely ash and burnt material. The ratio of these two parts is important for calculation of the degree of liberation in coal. The degree of liberation related to the ash could be measured.

Nowadays, the demand for environment-friendly products, with respect to environmental regulations and requirements, has increased; in this case that refers to the recovery and quality of the product in relation to the coal distribution in the minerals [5]. Due to this coal concentrate production costs will increase because of the high costs of crushing. Nowadays, electrical disintegration (ED) equipment is a new technology for crushing coal. In this technology the failure mechanism of action is selective; however, they are not generally used due to the operational costs [10].

In the Tabas coal washing plant the cage mill is located in a key part of the plant, and its halt would affect the whole circuit break down. In the event of the necessity for extreme repairs of the cage mill, the plant's goal was to replace the cage mill with another device which has more availability and also produces less fine particles in the product. In this research a jaw crusher was selected to investigate the subject in batch scale. Finally, the purpose of crushing is to increase the liberation in a

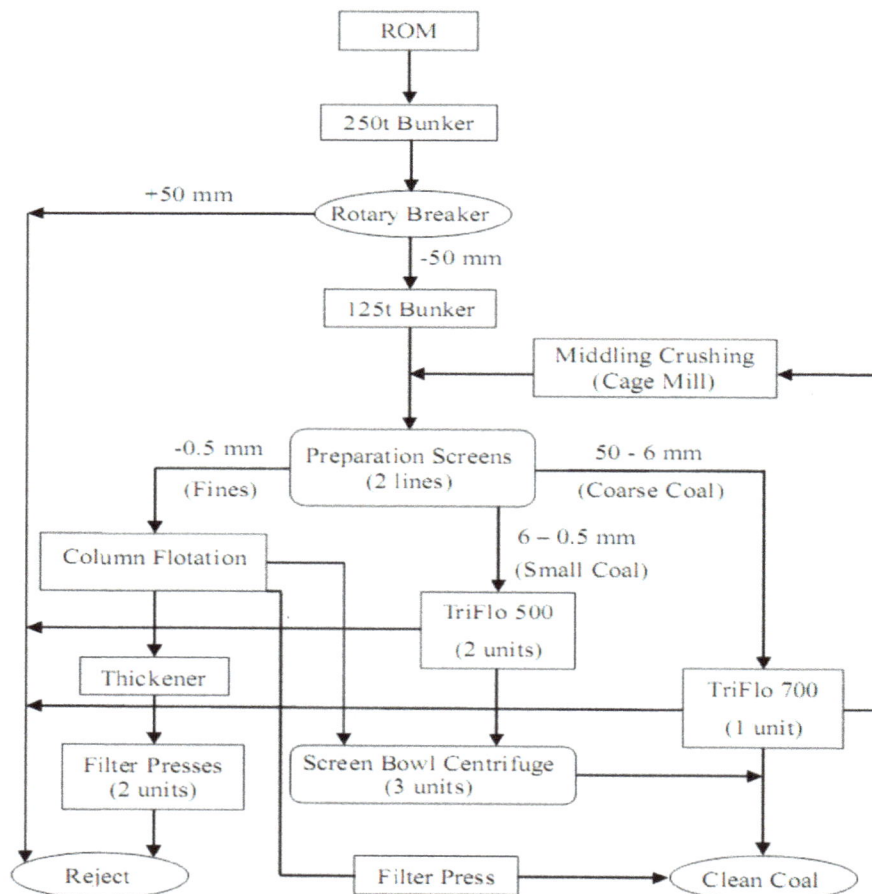

**Fig. 2.** The process flowsheet of the Tabas Coal Preparation Plant (TCPP) [13].

mineral with the minimum rate of size reduction. This goal needs selective fractures in coal. In this research, middle load crushing was conducted using a cage mill and jaw crusher with different force.

## 2. Material and methods

### 2.1. Sampling

The effect of size reduction in coal recovery can be expressed by the curved washability test capability. In this paper, liberation of middle coal using the industrial cage mill and laboratory jaw crusher were examined with the same degree of liberation according to the Tabas coal washing plant flow sheet (Figure 2).

According to the original plant design a Tri-flo 700 model of DWS700, which is used in the heavy media section for condensing the coal from the size of 6 to 50 mm, is used for two different densities in two separate parts of the plant. Each section consists of a cylindrical enclosure. The material ejected from the output 2 section comes out as middle material with 30-40 ash (Figure 3). With respect to the relatively high tonnages of this material in the Tabas coal washing circuit (30 ton per hour) appropriate grinding of coal and an increase in the degree of liberation will result in a reduction in the amount of waste of the valuable product. One of the sections includes the densities of +1.5-1.7 g/cm$^3$, and constitutes a 10 percent share of the total feed. The size of this section is usually less than 50 mm. With respect to theoretical criteria, 400 kg of sample were provided. The samples were collected in a flow middle load in a shift of 5 hours. In this study, crushing was examined on the middle load. Therefore, the analysis of a representative sample was conducted before and after grinding. The products with +0.5 mm were analyzed through sinking and floating as well as ash percentage. For -0.5 mm products, release analysis and ash percentage were conducted.

## 3. Results and disscusion

### 3.1. The middle coal crushing

Fluctuations in the size and difficulty of the feed are the most important factors in grinding circuit disruption. If there is an increase in the size or hardness feed, coarser grading will be achieved unless the feed is

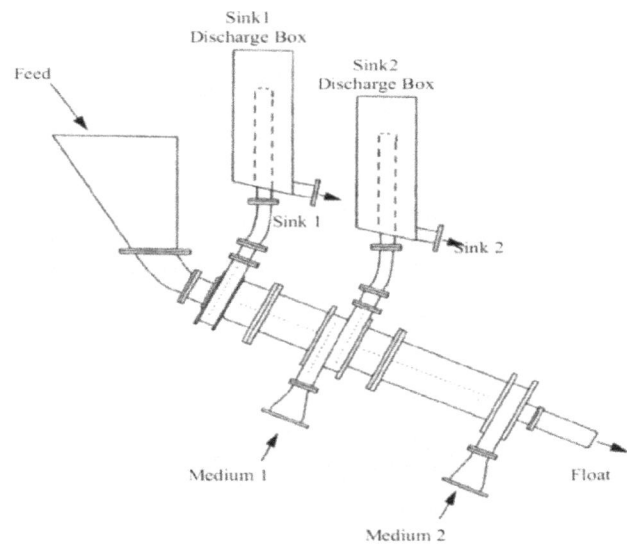

Fig. 3. Schematic diagram of a tri-flo separator [13].

reduced. Grading and ash feed is shown in Table 1.

As Table 1 shows, an increase in the size of minerals results in the increase in the ash percentage.

The cage mill model 40B2C4R with 4 cages and the jaw crusher model BM2 were used for crushing the middle load in the coal washing plant circuit. The cage mill and the jaw crusher were activated in the form of an open circuit. The sieve analysis is shown in Figure 4. According to this diagram, the two devices have almost the same liberation degrees.

### 3.2. Washability test of the middle product

Washability analysis of the middle product was investigated using yield-ash curves. The analysis of washability of crusher feed results are shown in Table 2.

According to the Table 2, there is a noticeable conflict between coal and waste in the coarser size. In this study, for ease of comparison between before and after

Table 1. Gradation and middle ash of the tri-flo separation in TCPP.

| Size (mm) | W (%) | Ash (%) |
|---|---|---|
| (+50) | 7.66 | 35.6 |
| (+25-50) | 10.7 | 36.7 |
| (+12-25) | 28.14 | 34.1 |
| (+6-12) | 26.2 | 35.7 |
| (+3-6) | 23.19 | 33.5 |
| (-3) | 4.11 | 38.3 |
| Total | 100 | 34.9 |

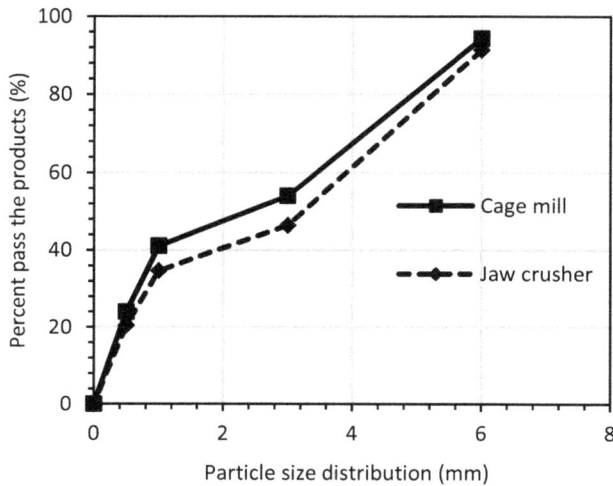

**Fig. 4.** Comparison of industrial cage mill and the laboratory jaw crusher for the degree of liberation of the middle coal after crushing.

crushing conditions in each sample of the concentrates, the middle and waste are separated and compared with each other. As can be seen in the table, 94 percent of the total middle sample consists of middle product with 33.83 ash, which is considered a high share. The purpose of grinding is mainly to reduce this section.

### 3.3. Washability analysis of products

Crushed products were granulated into 5 classes: +6, (-6+3), (-3+1), (-1-0.5) and -0.5 mm. The sinking and floating test was conducted in 6 different fractions. The Washability test Chart showed that the industrial cage mill resulted in the best yield (Figure 5).

For products with different forces, final yield was achieved with 12 ash. Liberation of middle products after crushing with the different devises is shown in

**Table 2.** Washability test results of the middle load of the tri-flo separation in TCPP.

| Feed | W (%) | Ash (%) |
|------|-------|---------|
| Yield (%) | 1 | 12 |
| Coal | 0.6 | 10.59 |
| Middle | 94.72 | 33.83 |
| Reject | 4.68 | 60.13 |
| Total | 100 | 34.92 |
| Coarse (%) | | 78.32 |
| Small (%) | | 21.68 |
| Fine (%) | | 0 |
| Total (%) | | 100 |

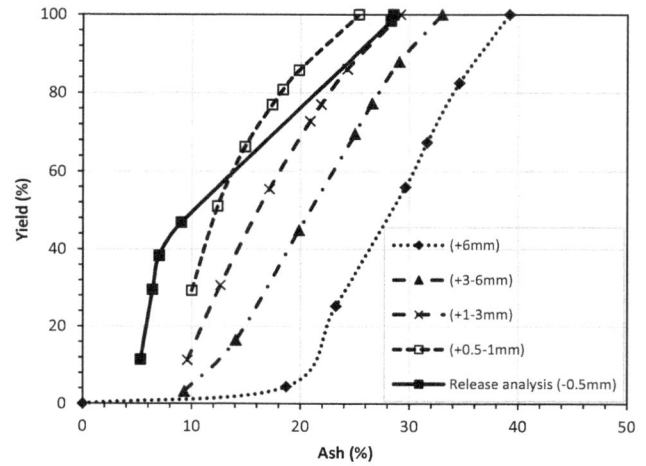

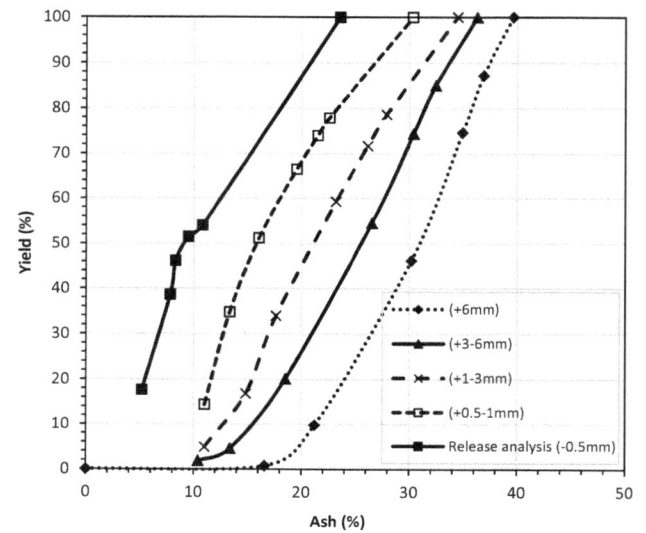

**Fig. 5.** Diagram of the sink-float analysis of the industrial cage mill and the laboratory jaw crusher.

Table 3. The base was +0.5 mm, as it was for feed. Floating 1.3 and 1.4 were considered as coal and sinking 1.8 was considered as waste. And the density of 1.5 and 1.7 was considered as middle. The table shows the contribution of each size with respect to building screening of the coal washing plant and what share after crushing will be allocated to each part of the plant (Figure 6).

As can be seen, if ash 12 is chosen as the criterion for the comparison of different force of devises yield after crushing, the yield would increase using the cage mill while the jaw crusher results in a decrease in the yield because of the excessive grinding of materials. This in turn leads to an increase in ash; but the laboratory jaw crusher yield is reduced because of the lack of material and fines production. Thus, we conclude that grinding

Feed

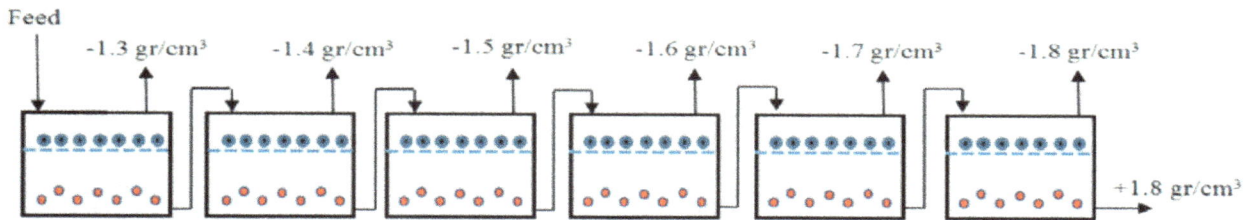

-1.3 gr/cm³    -1.4 gr/cm³    -1.5 gr/cm³    -1.6 gr/cm³    -1.7 gr/cm³    -1.8 gr/cm³

+1.8 gr/cm³

**Fig. 6.** Schematic diagram of the sink-float experiment [13].

**Table 3.** Washability test middle samples results and the amount of concentrates, middle and waste in the middle materials (after crushing).

| | Jaw crusher | | Cage mill | |
|---|---|---|---|---|
| | W (%) | Ash (%) | W (%) | Ash (%) |
| Yield (%) | 17.5 | 12 | 28.5 | 12 |
| Coal | 19.5 | 11.3 | 30.5 | 11.5 |
| Middle | 66.0 | 33.9 | 58.5 | 34.9 |
| Reject | 14.3 | 57.2 | 10.9 | 59.3 |
| Total | 100 | 32.8 | 100 | 30.5 |
| Coarse (%) | 9.09 | | 5.7 | |
| Small (%) | 69.3 | | 70.3 | |
| Fine (%) | 21.5 | | 23.9 | |
| Total (%) | 100 | | 100 | |

the middle product with an industrial cage mill results in the best yield rate. This would cause the release rate to increase to about 30.5 percent of coal ash, which in turn results in a 3 percent increase in the total yield of the plant. The fluctuation of ash products is due to fluctuation of the tri-Flo device.

### 3.4. Mechanism of fine

Milled products were categorized in dimensions of -0.5 mm in 4 classes: + 500, (-500+300), (-300+150), (-150+75) and -75 microns. The -75 micron products were considered as fine, results of which are shown in Table 4.

**Table 4.** Determine the amount of fine in the middle materials (after crushing with different forces).

| | Cage mill | | Jaw crusher | |
|---|---|---|---|---|
| | W (%) | Ash (%) | W (%) | Ash (%) |
| Fine | 4.89 | 25.60 | 4.01 | 25.1 |
| Ash (total) | 30.5 | | 32.8 | |

It is observed that increasing the ash decreases the amount of produced fine in line. At different forces, the rate of fine produced in the jaw crusher is less than the industrial cage mill. The crushing sequences in the jaw crusher were less than the cage mill. Thus, the particles were exposed to less impact; therefore, they produce less fine particles.

### 3.4. Fracture mechanism of these two methods

The freedom degree of the products comminuted by the jaw crusher and the cage mill are contrasted under similar size distribution. Combined effects of crushing, splitting and bending are applied to realize size reduction by the jaw crusher. In these fragmentation forces, crushing is the dominant force which urges particles to be separated through boundaries. When an irregular particle is crushed by crushing the product falls into two distinct size ranges: coarse particles resulting from the induced tensile failure and fines produced from either compressive failure near the points of loading or by shear at projections as shown in Figure 7a [7].

Nevertheless, impact is the main force utilized by cage mills. The particles are then struck by the subsequent cage rows before exiting through the bottom of the mill, contact with materials and stress concentration causes mineral liberating through the interface as shown in Figure 7b.

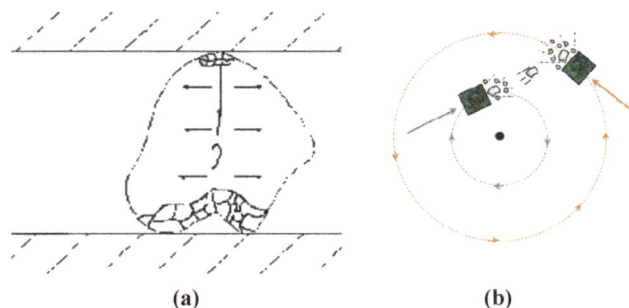

(a)      (b)

**Fig. 7.** Size reduction mechanism of (a) jaw crushing and (b) cage milling [3,7].

## 4. Conclusion

By contrasting the effects of different kinds of fragmentation forces on mineral liberation, the industrial cage mill which utilizes crushing as the main force creates better mineral liberation of middling than the laboratory jaw crusher with pressure as its main force.

We conclude that grinding the middle product with a cage mill results in a better yield rate than the laboratory jaw crusher. This would cause the release rate to increase to about 30.5% of coal ash, which in turn results in a 3% increase in the total yield of the plant.

In addition, the rate of fines produced through the laboratory jaw crusher is less than the industrial cage mill.

## References

[1] A. Noble, G.H. Luttrell, A review of state-of-the-art processing operations in coal preparation, Int. J. Mining Sci. Tech. 25 (2015) 511-521.

[2] D.L. Khooury, Coal cleaning technology N.D.C., Park. Ridge, N.J., 1981, pp. 34-46.

[3] F. Rodriguez, M. Ramirez, R. Ruiz, F. Concha, Scale-up procedure for industrial cage mills, Int. J. Miner. Process. 97 (2010) 39-43.

[4] J. Hao, H. Zhang, K. Yang, C. Lu, J. Chen, Y. Li, Effect of different milling processes on the mineral distribution in a coal powder, Int. J. Mining Sci. Tech. 22 (2012) 237-242.

[5] E.T. Oliver, J. Abbott, N.J. Miles, Liberation characteristics of a coal middlings, Coal Prep. 16 (1995) 167-178.

[6] T. Oki, H. Yotsumoto, S. Owada, Calculation of degree of mineral matter liberation in coal from sink-float separation data, Miner. Eng. 17 (2004) 39-51.

[7] W. Xie, Y. He, X. Zhu, L. Ge, Y. Huang, H. Wang, Liberation characteristics of coal middlings comminuted by jaw crusher and ball mill, Int. J. Mining Sci. Tech. 23 (2013) 669-674.

[8] W. Zou, Y. Cao, Z. Zhang, J. Liu, Coal petrology characteristics of middlings from Qianjiaying fat coal mine, Int. J. Mining Sci. Tech. 23(5) (2013) 777-782.

[9] G. Unland, Y. Al-Khasawneh, The influence of particle shape on parameters of impact crushing, Miner. Eng. 22 (2009) 220-228.

[10] M. Ito, S. Owada, T. Nishimura, T. Ota, Experimental study of coal liberation: electrical disintegration versus roll-crusher comminution, Int. J. Miner. Process. 92 (2009) 7-14.

[11] J.W. Leonard, Coal Preparation, 5th ed., Society for Mining, Metallurgy & Exploration Inc., 1990.

[12] B.A. Wills, Mineral processing technology, 7th ed., Butterworth Heinemann Publisher, 2006.

[13] R. Dehghan, M. Aghaei, Evaluation of the performance of tri-Flo separators in Tabas (Parvadeh) coal washing plant, Res. J. Appl. Sci. Eng. Tech. 7 (2014) 510-514.

# Analyzing the effect of gradation on specific gravity and viscosity of barite powder used in excavation mud

**Mohamad Alizade Pudeh[1], Esmaeil Rahimi[1,*], Mehran Gholinejad[1], Amirhossein Soeezi[2]**

[1] Department of Mining Engineering, Islamic Azad University, South Tehran Branch, Tehran, Iran

[2] School of Mining, College of Engineering, University of Tehran, Tehran, Iran

## HIGHLIGHTS

- The excavating companies consume more barite powder when the specific gravity of excavation mud is diminished.

- It is very important to select the appropriate material to adjust the viscosity and density.

- The dimensions of the particles get finer; the specific gravity is changed and increased.

## GRAPHICAL ABSTRACT

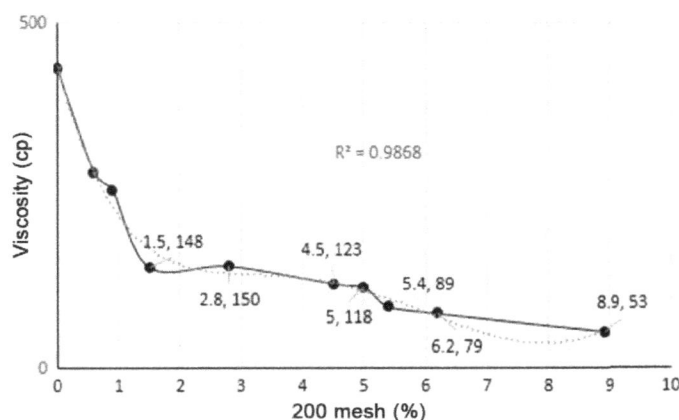

## ARTICLE INFO

*Keywords:*

Barite
Excavating mud
Mesh
Specific weight
Viscosity

## ABSTRACT

Barite is an important additive material to increase the weight of excavation mud. It has been used extensively in the excavation mud industry. The barite particle sizes used in excavation mud is very important. There is a direct relationship between the amount on the sift of 200 and 325 mesh. By reducing particle sizes of the 200 and 325 mesh, the specific gravity of the barite power increased slightly. It could be concluded that increasing the size of the barite powder particle can reduce density, but this should be considered as a laboratorial error. Thus, granulation has a great influence on the viscosity of barite powder. According to previous studies dimensions below 325 mesh are very influential on viscosity. It can be said excavating companies consume more barite powder when the specific gravity of excavation mud is diminished because barite powder can enhance specific gravity but the viscosity goes up. There will be more problems in the excavation mud if viscosity increases.

* Corresponding author:   ; E-mail address: SE_rahimi@azad.ac.ir

## 1. Introduction

The main task of drilling mud is to carry the drill bit to the head well [1]. In fact, the mud will pump from the reservoir into the well [2-3]. Barite is one of the most important additive materials to increase the weight of excavation mud. It has been used extensively in this industry. According to different statistics, approximately 6 million tons of barite is used in the excavating mud industry. When Barite powder was ground it demonstrated an inverse relationship between viscosity and gradation in the lab. In fact, increasing the size of barite powder particles can reduce viscosity of its mud. Barite is identified a strategic mineral existing in nature with several applications specially in excavating mud [2-4]. The specific gravity of excavating mud is 4.2 or more, its stiffness is 2.5 to 3.5 in Moss stiffness measurement, and the size of the particles is less than 74 micron. The mud can support and fill the well around the drilling area. Therefore, the concentration and viscosity of the drilling mud should be sufficient to control the pressure of the wall. It should be noted, water, air and oil fluids are used to make drilling mud. Air is the cheapest and most accessible fluid used up to now [6]. In some cases, in order to prevent the loss of drilling performance, ilmenite, manganese oxide, barite and silica particles are used on a Nano scale. So, it is important to select the appropriate material to adjust the viscosity and density. The specific gravity and viscosity of barite powder are very effective parameters in the type of drilling mud. Depending on the amount of powder, barite powder can have variable values [7-10].

In the current essay we investigate and optimize the parameters of specific gravity and viscosity for drilling mud. Therefore, by changing the amount of barite powder aggregation and analyzing the obtained data the ideal aggregate for proper operation of drilling mud can be determined.

## 2. Methodology

The feed is crushed to less than 6 mesh with a jaw crusher. After homogenization the powder was divided into 10 samples of 2 kg weight. Then it was thoroughly control ground by milling at different times (Figure 1). It should be noted, time was differed for grain control. In order to produce the desired grain, the crushed matter was controlled with 200 and 325 mesh screens.

The test parameters are: pH at normal temperature, pH at 82 °C, specific gravity, the remaining matter on the sift 200 mesh, the remaining matter on the sift 325 mesh, the solid particles dissolved in the water, the apparent viscosity before adding the plaster, after adding the real plaster and the alkaline amount of the soil metals dissolved as $Ca^{2+}$.

### 2.1. The instructions and evaluating methods of barite powder

Barite is a weight increasing substance with the specific gravity of 4.2 $gr/cm^3$. 75PCF is used as the specific gravity increasing substance in non-productive layers like Gachsaran. The barite powder used in the excavating mud should have the following global standards (API 13A).

### 2.1.1. Determining the pH of barite powder

35 gr of barite powder is mixed with 350 ml of water and its pH is measured by the pH paper or the pH meter. The mixture is heated to 82 °C and its pH is measured again.

### 2.1.2. The specific gravity of barite powder

The specific gravity is calculated by the Le Chatelier's method. About 200 gr of barite is dried in the oven and poured in the Le Chatelier container until the zero line. Then, it is placed in a bath at a stable 31.6 °C in order to register the maximum petroleum expansion, after half an hour (primary volume) the dried barite is poured into a metallic cylinder until it is completely filled. The

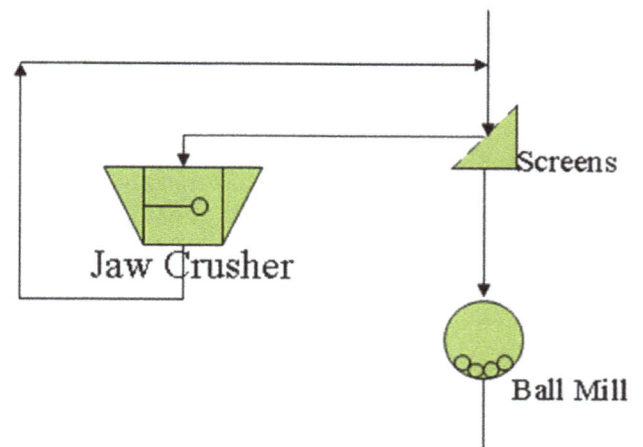

**Fig. 1.** Crushing and screening of feed.

cylinder and the barite are weighed and registered as the primary substance. After an hour, the maximum temperature of the petroleum is 31.6 °C. The Le Chatelier container is moved out of the bath and the dried barite is added to it. Thus, the height of the petroleum is 18 cm. The Le Chatelier container is again placed in the bath. The metallic cylinder with the remained barite is weighed again and registered as the secondary weight. After an hour the volume of Le Chatelier volume is read and registered as the secondary volume.

### 2.1.3. Determining the gradation by washing using sift

Sifts 200 and 325 meshes were used and then 100 gr of barite was added and washed by water at 15 psi pressure until clear water passed through the sift. The sifts should be dried in the oven. The barite content of each of them is weighed and registered based on the weight percentage.

25 gr of barite is added to 50 cm$^3$ of distilled water in an Erlenmeyer flask and then stirred for 5 min. Next, it remains still for 5 min and then is filtered by filter paper 42. Addition to 50 cm$^3$ of warm distilled water and the above mentioned steps are repeated. The third time, 50 cm$^3$ of cold distilled water is poured on the solid particle left in the Erlenmeyer flask. After three stages, about 150 cm$^3$ of water will be in the glass container. This container is put on the evaporating machine to evaporate the water content. Next, the glass container is next put into the oven to until completely dried. The solid sediments remaining in the glass container is weighed. Thus, the weight of the solid particles dissolved in the water is obtained by subtracting the weight of the empty glass container from the weight of the sediment inside. The calculated number is multiplied by 4 to achieve the percentage of the solid particles dissolved in water.

### 2.1.4. Determining the apparent viscosity

The amount of the required barite could be calculated by Eq. (1).

$$\text{Amount of the required barite} = \frac{D \times V(d_2 - d_1)}{D - d_2} \quad (1)$$

where $D$ is the specific weight of the barite (4.2), $V$ is the volume of distilled water for the test (250 cm$^3$), $d_2$ is the maximum weight of the specific weight by increasing

the barite (2.5 gr/cm$^3$) and $d_1$ is the specific weight off the distilled water (1 gr/cm$^3$).

The amount of the intended barite is 927 gr which is poured into 250 cm$^3$ of distilled water. The weight is measured by the mud scale which should be 2.5 gr/cm$^3$. This mud is put in a closed glass container for 24 hrs. Then, the mud is mixed for 10 min and the $\theta_{600}$ is read and the apparent viscosity is calculated by a viscometer (Eq. (2)).

$$\text{Apparent viscosity} = \theta_{600}/2 \quad (2)$$

Then, 2.3 gr of plaster is added to the mud, it is stirred for 10 min and again the and the apparent viscosity are calculated.

### 2.1.5. Determining the amount of dissolved soil alkaline metals

100 gr of barite is added to 100 cm$^3$ of distilled water. Then, it is stirred on the shaker for 10 min three times and then let stand still for another 10 min. The liquid phase is filtered by filter paper 42 and the water is trapped in the glass container. 50 cm$^3$ distilled water, 5-6 drops of indicator EBT, and 10 cm$^3$ of clear water added into a Erlenmeyer flask. Then, it is put on a magnetic stirrer and titrated by the 400 ppm (EDTA) of Versonite solution to change the red color of the solution into blue one. The percentage of dissolved soil alkaline metals could be determined by using Eq. (3).

$$\text{Alkaline earth metal (\%)} = \frac{\text{Versonite consumption} \times 40}{10} \quad (3)$$

## 3. Result and discussion

Barite is one of the most strategic minerals in nature with many applications in drilling mud. Barite is a weight increasing substance which is used to a large extent in the base water.

This article investigated and optimized the effective parameter such as barite aggregation, bpecific gravity and proper viscosity. It was planned to crush the barite in a gradation higher and lower than the standard used by excavation companies to scrutinize the complete trend of viscosity and specific gravity changes. Our results indicated that the coarser gradation corroded faster than other samples. The gradation changes of the samples have been made by sifts of 200 and 325 mesh in order to be analyzed.

Therefore, Figure 2(a) shows the amount of barite gradient on 200 mesh grains per specific gravity. Also, the effect of aggregates on 325 mesh on specific gravity is measured as shown in Figure 2(b).

As can be seen in Figure 2(a), reducing the size of the grains increases the amount of the specific weight. The reason for these changes can be a loss of porosity between the particles. In other words, by increasing the crushing and the smaller the size of the particles, the hollow spaces between the particles will be eliminated. According to the specific gravity formula, the volume of particles decreases and ultimately increases the specific weight.

Figure 2(b) shows that reducing the size of the grains increases the amount of specific weight. The amount of 2% of the remaining material on the silicon can increase the density up to 4.22 $gr/cm^3$. By increasing the amount of material up to 8%, the density can be increased to 8 $gr/cm^3$.

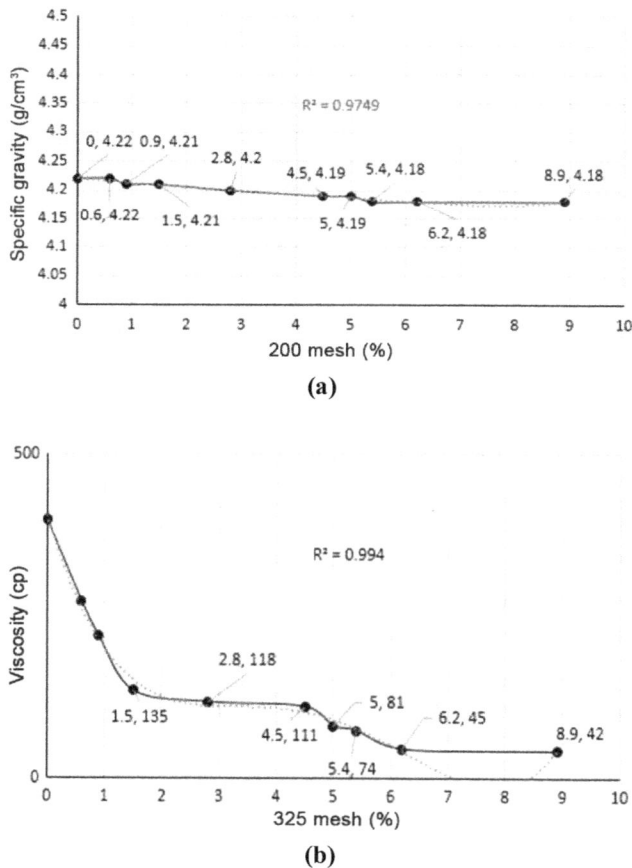

**(a)**

**(b)**

**Fig. 2.** The effect of graining on specific gravity (82 °C) of (a) 200 mesh and (b) 325 mesh.

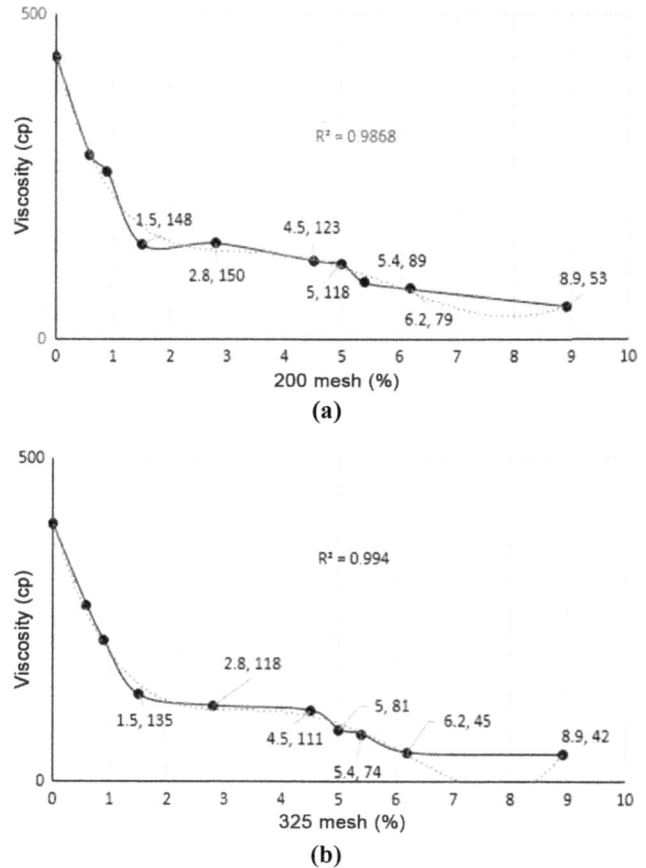

**(a)**

**(b)**

**Fig. 3.** The effect of grains on specific gravity of (a) 200 mesh and (b) 325 mesh.

The effect of graining on the viscosity on 200 mesh is shown in Figure 3(a). The mesh size on the mesh 325 mesh is also measured in Figure 3(b).

Figure 3(a) shows that reducing the particle size or decreasing grain size can increase the viscosity. The results show that if 2.8% of the material remains on the 200 mesh, it can produce a viscosity of 118 cp. Increasing the material on the screens to 9% can decrease viscosity to 42 cp. Figure 3(b) shows increasing the size of the particle increases the viscosity. The results show that 5% of the material remaining on the 325 mesh screen can produce a viscosity of 118 cp. So, increasing the available materials on the surface of the screen to 9% can decrease the viscosity to 53 cp.

## 4. Conclusion

Barite is identified as one of the strategic mineral existing in nature and it has several applications specially in excavating mud. In some cases, in order to prevent the loss of drilling performance, ilmenite, manganese oxide, barite and silica particles are used on a nano scale. Therefore, it is very important to select the appropriate material to adjust the viscosity and density. A drastic increase in the size of particles or the gradation reduction can lead to high viscosity. It is hard for excavating companies to transmit the excavating mud by pumps. In contrast, it is difficult to treat excavating mud after the cycle of excavating mud movement. Specific gravity variations with a percentage on 200 mesh screen can be estimated by increasing the particle size. On the other hand, changes in the particle size of 325 mesh have a greater effect on the specific weight than changes in the particle size of 200 mesh. More normal numbers and slopes are observed in the changes on 200 mesh than the percentage of the changes on 325. The dimensions of the particles get finer; the specific gravity is changed and increased.

## References

[1] M. Ghazi, G. Duplay, R. Hadjamor, M. Khodja, H. Amar, Z. Kessaissia, Life-cycle impact assessment of oil drilling mud system in Algerian arid area, Resour. Conserv. Recy. 55 (2011) 1222-1231.

[2] S. Elkatatny, H. Nasr-El-Din, Removal of water-based filter cake and stimulation of the formation in one step using an environmentally friendly chelating agent, Int. J. Oil Gas Coal T. 7 (2014) 169-181.

[3] B. Benayada, N. Habchi, M. Khodja, Stabilization of clay walls during drilling in southern Algeria, Appl. Energ. 2 (2003) 51-59.

[4] A. Tehrani, A. Popplestone, T. Ayansina, Barite sag in invert-emulsion drilling fluids, in: Offshore Mediterranean Conference and Exhibition, 25-27 March, Ravenna, Italy (2009).

[5] T. Ofei, C. Bavoh, A. Rashidi, Insight into ionic liquid as potential drilling mud additive for high temperature wells, J. Mol. Liq. 242 (2017) 931-939.

[6] A.S. Apaleke, A.A. Al-Majed, M.E. Hossain, Drilling fluid: state of the art and future trend, in: SPE Latin America and Caribbean Petroleum Engineering Conference, 16-18 April, Mexico City, Mexico (2012).

[7] S. Elkatatny, A. AlMoajil, A. Texas, Evaluation of a new environmentally friendly treatment to remove $Mn_3O_4$ filter cake, Paper SPE 156451, in IADC/SPE Asia Pacific Drilling Technology Conference (APDT), July 9-11, Tianjin, China (2012) pp. 1-13.

[8] B. Bageria., M. Mahmouda., A. Abdulraheema., S. Al-Mutairib, S. Elkatatnya, R. Shawabkeha, Single stage filter cake removal of barite weighted water-based drilling fluid, J. Petrol. Sci. Eng. 149 (2017) 476-484.

[9] D. Xue, R. Sethi, Viscoelastic gels of guar and xanthan gum mixtures provide long-term stabilization of iron micro- and nanoparticles, J. Nanopart. Res. 14 (2012) 1239-1252.

[10] M.E. Hossain, A.A. Al-Majed, Fundamentals of Sustainable Drilling Engineering, John Wiley & Sons, Canada, 2015.

# 12

# A new method for the preparation of pure topiramate with a micron particle size

**Bahman Hassanzadeh**[*], **Farajollah Mohanazadeh**

*Department of Chemical Technology, Iranian Research Organization for Science and Technology (IROST), Tehran, Iran*

## HIGHLIGHTS

- A micron-sized topiramate sample was synthesized by using of Triton X-10.

- The purification and reduction of the size of topiramate were performed in one step.

- The kinetic solubility of topiramate was improved by size reduction.

## GRAPHICAL ABSTRACT

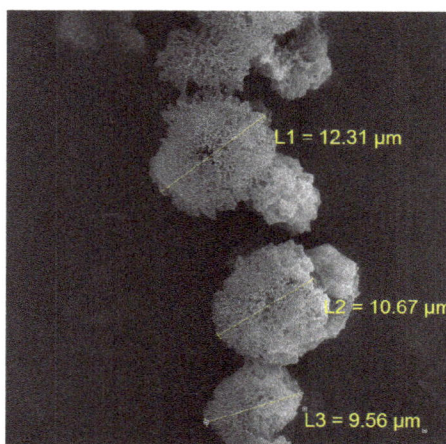

## ARTICLE INFO

*Keywords:*

Topiramate
Particle size
Micronized drugs
Solubility

## ABSTRACT

Crude topiramate was prepared from the reaction between diacetonefructopyranose, sulfuric diamide, and 2-picoline. During a controlled processing the presence of Triton-X-100, crude topiramate in a methanol-water solvent, was converted to pure micron size topiramate particles. The factors affecting the purification and micronization of topiramate, such as solvent and anti-solvent type and concentration ratio, temperature, and mixing speed, were investigated. A mechanism for the preparation of topiramate was proposed based on the results of changes in the concentration of the raw materials and an investigation of the intermediates. In addition, the solubility rate of topiramate with different particle sizes has been determined.

* Corresponding author:  ; E-mail address: Hzbahman@yahoo.com

# 1. Introduction

Dissolution of drug is an important rule of drug absorption and bioavailability [1-3]. The relationship between particle size, bioavailability, and dissolution is well documented. Bioavailability is often directly dependent on the particle size for solid delivery systems because it controls dissolution solubility characteristics. It is well know that a fine particle size has more solubility. At present, most new drugs and 40% of drugs currently used in industry have poor solubility [1].

Several techniques have been used to overcome the solubility problems of poorly soluble drugs [4-10]. These techniques include amorphous formulations, salt/pro-drug formation, co-solvents, and complexation. In recent years, the use of micronized particles has been reported in pharmaceutical applications to increase the solubility rate of poorly soluble drugs [11-14]. Bottom up and top down are the two technologies resulting in micronized drugs or nanoparticles. In bottom up technology, the micronized or nano drug is precipitated by an anti-solvent method. The drug is dissolved in a solvent in the presence of a surfactant, and then precipitated by the addition of an anti-solvent as a micronized form [15-17]. In top down technology the particle is crushed by mailing or a high pressure homogenizer [18]. In these technologies the particle size is reduced by pressure, scrubbing, and abrasion which maybe damage the drug [19]. The particle size range for direct-compression tablets is a very important factor. In direct-compression tablets, the drug should have suitable properties of compaction behavior and powder flow [20]. Although particle size reduction increases solubility, it also increases the density of matter and the accumulation of particles. It should be kept in mind that the accumulation and particle density does not neutralize the positive effect of increased solubility [21]. Therefore, the use of micronized materials is more useful than nano-scale materials in the formulation of tablets [21].

Topiramate is a sulfamate-substitute of the natural occurring monosaccharide D-fructose with anticonvulsant or antiepileptic properties. It is designated chemically as 2,3:4,5-bis-O-(1-methylethylidene)-β-D-fructopyranosesulfamate (Figure 1). It is a white or almost white crystalline powder with a bitter taste. Topitamate was invented by Maryanoff [22]. The solubility of topiramate is low in water (9.8 mg/mL) [23]. Topiramate is known as an anticonvulsant or antiepileptic drug [24]. It is also used to prevent migraine headaches [25]. Topiramate can also help people lose weight [26]. Topiramate is efficacious in the treatment of alcohol use disorders and is an alternative treatment to FDA-approved medications [27].

The preparation of active pharmaceutical intermediates (APIs) in various shapes and sizes is always considered. Topiramate is a new drug whose use is expanding, so far a grinding method has been used to prepare its various sizes. Usually grinding of materials is associated with weight loss of the product, destruction of the chemical structure due to heat and pressure, and consumption of energy [19]. In this work, we investigated the preparation of micronized topiramate using the sedimentation method.

# 2. Materials and Methods

## 2.1. Materials

Diacetonefructopyranose (99.21%) was purchased from Grindlays Pharmaceuticals Pvt. Ltd., India. Sulfuric diamide, 2-picoline, toluene, ammonia, and hydrochloric acid 37% were obtained from Aldrich. All compounds were used as received without further purification. Topiramate was synthesized in the lab.

## 2.2. Instruments

A high-performance liquid chromatography system (2 pumps waters 510, detector IR waters 410), FT-IR

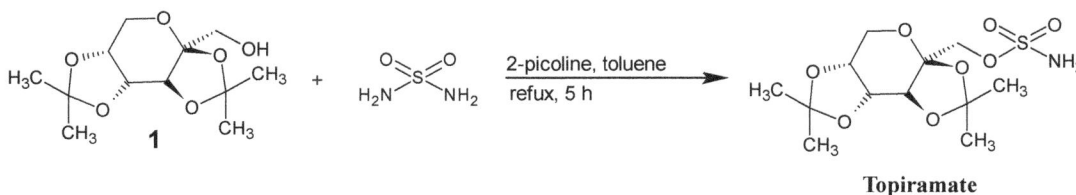

**Fig. 1.** Preparation of topiramate.

spectrometer (Philips PU 9624), scanning electron microscope (Tescan), and laser scattering particle size analyzer (Mastersize 3000, Malvern) were used to characterize purity and mesh size of the micronized topiramate.

## 2.3. Preparation of topiramate

In a three necked round bottom flask equipped with a magnetic stirrer, condenser, and thermometer, sulfuric diamide (9.6 gr, 0.1 mole), and 2-picoline (9.3 gr, 0.1 mole) were mixed in toluene (60 ml) and stirred at 50 °C. After 15 minutes, a two-layered mixture was obtained. Diacetonefructopyranose (10 gr, 0.1 mole) was added to the mixture. The reaction mixture was refluxed for 5 h. The reaction was cooled to room temperature. An aqueous solution of sodium hydroxide (4 gr in 60 ml of water) was added to the reaction mixture. The aqueous layer was separated and neutralized with HCl 37% (8.22 ml). The resulting crude product was filtered, washed with water (2×20 ml), and dried in oven at 50 °C for 3 h (24.45 gr, 72%).

## 2.4. Preparation of micronized topiramate

In a three necked round bottom flask equipped with a magnetic stirrer, condenser, thermometer, and addition valve connected to the peristaltic pump, crude topiramate (10 gr, 0.03 mole) and Triton-X-100 (0.02 gr) were mixed with methanol (10 ml). The mixture was refluxed for 5 min. until a clear mixture was obtained. The clear mixture was cooled to 30 °C and stirred at 600 rpm. Water (20 ml) was added to the mixture at a rate of 1ml/min by the peristaltic pump. Under these conditions, the mixture was stirred for 30 min. The reaction temperature was adjusted to 10 °C for 30 min. and the mixture was stirred for an additional hour. The resulting powder was separated by filtration, washed with water (2×5 ml) and dried at 50 °C under vacuum (50 mmHg) for 3h. The yield and purity of the micronized product (12 µm) were, respectively, 83% and 99.54% (by HPLC). 1H-NMR (CDCl$_3$) δ: 1.34 (s, 3H, isopropylidene CH$_3$), 1.42 (s, 3H, isopropylidene CH$_3$), 1.48 (s, 3H, isopropylidene CH$_3$), 1.55 (s, 3H, isopropylidene CH$_3$), 3.76-3.80 (d, 1H, j=12.9 Hz, H-6a), 3.87-3.92 (dd, 1H, j=13.0, 1.8 Hz, H 6-b), 4.22-4.24 (d, 1H, j=10.9 Hz, H-1a), 4.25-4.26 (dd, 1H, j=7.9, 1.0 Hz, H-5), 4.29- 4.30 (d, j=2.7 Hz, H-3), 4.31-4.33 (d, 1H, j=10.9 Hz, H-1b), 4.59-4.63 (dd, 1H, j=2.6, 7.9 Hz, H-4), 5.15 (s, 2H, NH$_2$).

## 2.5. Kinetic solubility studies

### 2.5.1. Sample preparation

A synthetic sample of topiramate was milled. Three samples of topiramate with mesh sizes of 841-1000 µm, 297-344 µm, and 149-177 µm were separated by sieves No. 18, 20, 45, 50, 80 and 100. A fourth sample of micronized topiramate with a mesh size about 12 µm was produced at the lab.

### 2.5.2. Solubility measurements

An excess of each of the topiramate samples (micronized, 841-1000 µm, 297-344 µm, and 149-177 µm) was poured into 100 ml glass vials containing 50 ml of water. The vials were capped and sealed with parafilm®. They were then placed on the shaker platform and continuously agitated at 100 rpm in a water bath maintained at 25 °C. They were sampled at intervals of 15 min. The samples were centrifuged and the concentration of clear solutions was determined by HPLC.

## 3. Results and discussion

### 3.1. Preparation of topiramate

In this work, we tried to provide a micronized crystalline topiramate in order to increase topiramate solubility. Topiramate was prepared from the reaction of **1** (Figure 1), 2,3:4,5-bis-O-(1-methylethylidene)-β-D-fructopyranose (diacetone fructopyranose), with sulfuric diamide and 2-picoline in toluene.

Purification of topiramate is normally performed in different solvents or mixed solvents such as hexane/2-propanol [22], ethanol/water [22], 2-propanol [28], acetone/water [29], and ethylacetate/hexane [30]. Due to the cheapness of alcohols, we used a mixture of alcohols and water to purify the topiramate. There was no significant difference between methanol and ethanol in the experiments. Therefore, a mixture of one volume of methanol and two volumes of water relative to the weight of crude topiramate was used as solvent for purification of the topiramate. It should be noted that increasing the amount of methanol (as solvent) will reduce efficiency; and although increasing the amount of water (as anti-solvent) can increase efficiency, the purity of the product is reduced (Table 1).

Temperature is an important factor in the crystallization of topiramate. When the crude topiramate is dissolved in hot methanol and water and the mixture is slowly cooled down, a crystalline product with a coarse mesh is prepared [21]. Change in other parameters, such as stirring speed and anti-solvent addition rate, did not play a role in reducing particle size. According to reports [29], the use of surfactants affects the size of crystals formed. In this study, different amounts of a non-ionic surfactant, Triton-X-100, were used to produced micronized scale particles. The lowest amount of Triton-X-100 was 0.2% wt to crude topiramate. No uniformity was observed in particle size in amounts less than 0.2%. Mixing speed was the last factor to be investigated. The optimum mixing speed was found to be 600 rpm. The resulting micronized powder passed all the tests of USP 38 and had an average mesh size of 12 μm according to the results of SEM (Figure 2) and the laser scattering particle size analyzer.

## 3.2. Mechanism of reaction

Molar equivalent amounts of diacetonefructopyranose, sulfuric diamide, and 2-picoline were dissolved in toluene and refluxed. After an hour, the reaction stopped and the product was separated. The yield of reaction was 30%. The reaction was then repeated with half the amount of diacetonefructopyranose. This reaction efficiency was 30% and there was no change in the yield of product. In two other similar reactions, reducing sulfuric diamide and 2-picoline by half their initial values, the reaction efficiency was also halved. These

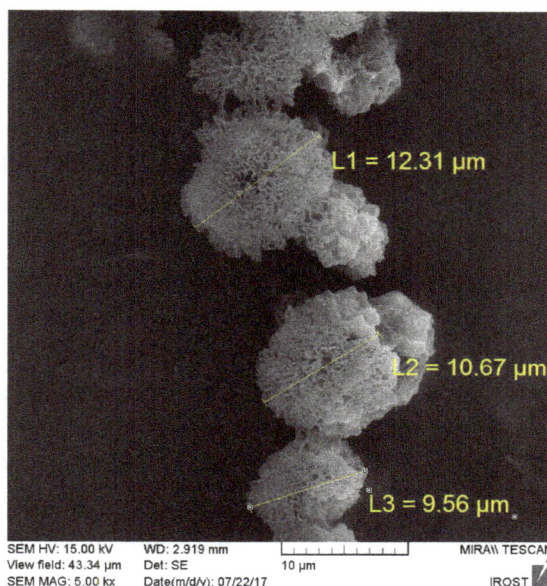

**Fig. 2.** The SEM image of micronized topiramate.

experiments showed that the concentration of sulfuric diamide and 2-picoline effect the efficiency of the reaction. The proposed mechanism for topiramate synthesis is shown in Figure 3. This suggests that initially an active intermediate **2** is generated by a reaction between sulfuric diamide and 2-picoline. This active intermediate **2** is attacked by diacetonefructopyranose and intermediate **3** is produced and 2-picoline is released. In the final step, intermediate **3** is fragmented into topiramate and ammonia. Thin layer chromatography (TLC) studies did not show any stable intermediate during the reaction.

**Table 1.** Amount of methanol and water used for purification of topiramate.

| Entry | Crude topiramate/methanol/water (w / v / v) | Yield (%) | Purity[1] (%) |
|-------|---------------------------------------------|-----------|---------------|
| 1 | 1 / 1 / 1 | 82 | 99.94 |
| 2 | 1 / 1.5 / 2 | 71 | 99.77 |
| 3 | 1 / 2 / 2 | 65 | 99.95 |
| 4 | 1 / 1 / 1.5 | 87 | 99.61 |
| 5 | 1 / 1 / 2 | 83 | 99.54 |

[1] Based on HPLC.

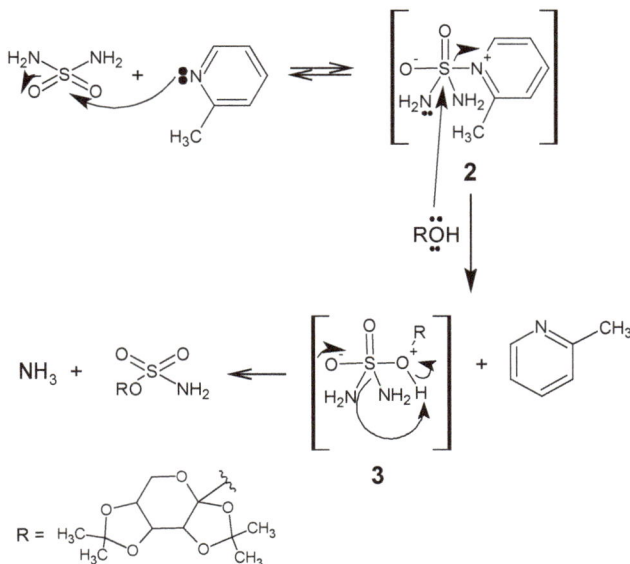

**Fig. 3.** The Proposed mechanism for topiramate synthesis.

*3.3. Solubility of micronized topiramate*

A sample of topiramate with an average particle size of 12 microns was generated using crystallization of crude topiramate in a methanol solvent with the aid of a surfactant. Three other samples of topiramate with different particle sizes (841-1000 μm, 297-344 μm, and 149-177 μm) were produced using topiramate screening. The samples were suspended in vials containing water and shaken at 25 °C. The contents of vials were filtered at 15-minute intervals and their concentration was measured by HPLC. As shown in Table 2 and Figure 4, The solubility time of topiramate (mg/ml) with a particle size of 12 μm is 45 minutes, which is half the time required for the solubility of the topiramate with a larger particle size (841-1000, 297-344, and 149-177).

**4. Conclusion**

Topiramate was prepared from the reaction of

**Table 2.** Topiramate solubility (mg/ml) variation with time for different particle sizes in water at 25 °C.

| Time (min) | Mesh size | | | |
|---|---|---|---|---|
| | 841-1000 | 297-344 | 149-177 | 12 |
| 15 | 1.4 | 2.1 | 2.7 | 4.5 |
| 30 | 2.1 | 3.0 | 3.9 | 9.6 |
| 45 | 3.9 | 4.3 | 5.1 | 9.8 |
| 60 | 5.2 | 6.1 | 6.5 | 9.8 |
| 75 | 6.3 | 7.6 | 7.9 | 9.8 |
| 90 | 8.7 | 9.4 | 9.8 | 9.8 |
| 105 | 9.5 | 9.8 | 9.8 | 9.8 |
| 120 | 9.8 | 9.8 | 9.8 | 9.8 |

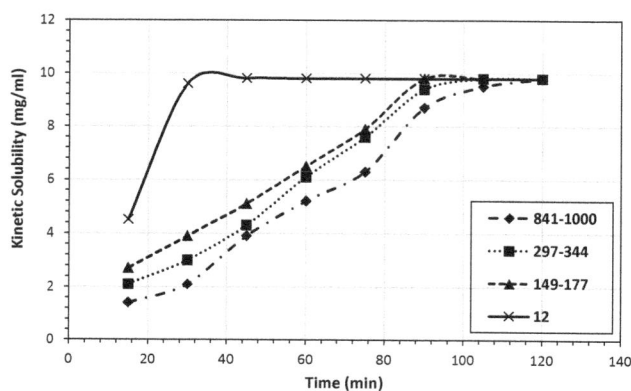

**Fig. 4.** Kinetic solubility of different particle size topiramate.

diacetonefructopyranose with sulfuric diamide and 2-picoline in toluene. Micronized topiramate was produced by a new non-grinding method. With the process control in this method, the purification and production of micronized particles were performed concurrently. The solubility time of the produced sample is half the time required to dissolve the samples containing larger particles of topiramate. The factors affecting the production of micronized topiramate, such as temperature, stirring speed, and surfactant effect, were investigated. Factors affecting the production of topiramate were also investigated, and based on this evidence a mechanism for this reaction was proposed. Based on the results of the reaction, diacetonefructopyranose does not play a role in the rate determining step. The reaction progress rate is dependent on the concentration of 2-picoline and sulfuric diamide.

**References**

[1] K. Gao, L. Ma, X. Wang, L. Zuou, X.F. Wang, Application of drug nanocrystal technologies on Oral drug delivery of poorly soluble drugs, Pharm. Res. 30 (2013) 307-324.

[2] J.B. Dressman, C. Reppas, In vitro-in vivo correlations for lipophilic, poorly water-soluble drugs, Eur. J. Pharm. Sci. 11 Suppl. 2 (2000) S73-S80.

[3] M. Wang, M. Thanou, Targeting nanoparticles to cancer, Pharmacol. Res. 62 (2010) 90-99.

[4] A.T.M. Serajuddin, Solid dispersion of poorly-soluble drugs: early promises, subsequent problems, and recent breakthroughs, J. Pharm. Sci. 88 (1999) 1058-1066.

[5] M.E. Davis, M.E. Brewster, Cyclodextrin-based pharmaceutics: past, present and future, Nat. Rev. Drug Discov. 12 (2004) 1023-1035.

[6] J. Breitenbach, Melt extrusion: from process to drug delivery technology, Eur. J. Pharm. Biopharm. 54 (2002) 107-117.

[7] S.S. Davis, C. Washington, P. West, L. Illum, G. Liversidge, L. Sternson, R. Kirsh, Lipid emulsions as dug delivery systems, Ann. N. Y. Acad. Sci. 507 (1987) 75-88.

[8] K. Kawakami, T. Yoshikawa, T. Hayashi, Y. Nishihara, K. Masuda, Microemulsion formulation for enhanced absorption of poorly soluble drugs II in-vivo study, J. Control Release, 81 (2002) 75-82.

[9] D.B. Fenske, A. Chonn, P.R. Cullis, Liposomal

nanomedicines: an emerging field, Toxicol. Pathol. 36 (2008) 21-29.

[10] V.P. Torchilin, Multifunctional nanocarriers, Adv. Drug Deliver. Rev. 64 (2012) 302-315.

[11] R.J. Aitken, M.Q. Chaudhry, A.B.A. Boxall, M. Hull, Manufacture and use of nanomaterials: current status in the UK and global trends, Occup. Med. 56 (2006) 300-306.

[12] K. Praveen, C. Singh, A study on solubility enhancement methods for poorly water soluble drugs, Am. J. Pharmacol. Sci. 14 (2013) 67-73.

[13] S.K. Poornachary, G. Han, J.W. Kwek, P.S. Chow, R.B.H. Tan, Crystallizing micronized particles of a poorly water-soluble active pharmaceutical ingredient: nucleation enhancement by polymeric additives, Cryst. Growth Des. 16 (2016) 749-758.

[14] A.R. Mokarram, A. Kebriaeezadeh, M. Keshavarz, A. Ahmadi, B. Mohabat, Preparation and in-vitro evaluation of indomethacin nanoparticles, DARU J. of Pharm. Sci. 18 (2010) 185-192.

[15] H.P. Thakkar, B.V. Patel, S.P. Thakkar, Development and characterization of nanosuspensions of olmesartan medoxomil for bioavailability enhancement, J. Pharm. Bioallied Sci. 3 (2011) 426-434.

[16] J.P.J. Dhaval, M.P. Vikram, R.J. Rishad, R. Patel, Optimization of formulation parameters on famotidine nanosuspension using factorial design and the desirability function, Int. J. PharmTech Res. 2 (2010) 155-161.

[17] P.R.J. Khadka, H. Kim, I. Kim, J.Y. Kim, H. Kim, J.M. Cho, G. Yun, J. Lee, Pharmaceutical particle technologies: An approach to improve drug solubility, dissolution and bioavailability, Asian J. Pharma. Sci. 9 (2014) 304-316.

[18] J. Leleux, R.O. Williams, Recent advancements in mechanical reduction methods: particulate system, Drug Dev. Ind. Pharm. 3109 (2013) 1-12.

[19] N. Rasenack, B. W. Muller, Micron-size drug particles: common and novel micronization techniques, Pharm. Dev. Technol. 9 (2004) 1-13.

[20] T. Yajima, S. Itai, H. Hayashi. K. Takayama, T. Nagai, Optimization of size distribution of granules for tablet compression, Chem. Pharm. Bull. 44 (1996) 1056-1060.

[21] B.Y. Shekunov, P. Chattopadhyay, H.H.Y. Tang, A.H.L. Chow, Particle size analysis in pharmaceutics: principles, methods and applications, Pharm. Res. 24 (2007) 203-227.

[22] C.A. Maryanoff, L. Scott, K.L. Sorgi, U.S. Patent No. 5,387,700 (issued Feb 7, 1995).

[23] a) Food and Drug Administration, Center for Drug Evaluation and Research, Application Number 020505s038s039, 020844s032s034lbl.
b) Örn ALMARSSON, M.L. Peterson, J. Remenar, EP 1,485,388A2 (Issued Dec. 15, 2004)

[24] B.E. Maryanoff, S.O. Nortey, J.F. Gardocki, R.P. Shank, S.P. Dodgson, Anticonvulsant O-alkyl sulfamates. 2,3:4,5-Bis-O-(1-methylethylidene)-beta-D-fructopyranose sulfamate and related compounds, J. Med. Chem. 30 (1987) 880-887.

[25] A. Ferrari, I. Tiraferri, L. Neri, E. Sternieri, Clinical pharmacology of topiramate in migraine prevention, Expert Opin. Drug Met. 7 (2011) 1169-1181.

[26] Food and Drug Administration, Center for Drug Evaluation and Research, Application Number 22580Orig1s000.

[27] B.A. Johnson, N. Ati-Daoud, Topiramate in the new generation of drugs: efficacy in the treatment of alcoholic patients, Curr. Pharm. Design, 16 (2010) 2103-2112.

[28] L.H. Wang, C.T. Huang, U.S. Patent No. 8,748,594 (issued Jun. 10, 2014).

[29] D.P. Balwant, L.P. Kumar, P.A. Kumar, H.B. Prafulbhai, WO/2007/108009 (issued Sep. 9, 2007).

[30] H.P. Chawla, A.S. Chowdhary, S.M. Patel, WO/2007/099388 (issued Sep. 07, 2007).

# 13

# Diagnosis of the disease using an ant colony gene selection method based on information gain ratio using fuzzy rough sets

**Mohammad Masoud Javidi***, **Sedighe Mansoury**

*Faculty of Mathematics and Computer, Department of Computer Science, Shahid Bahonar University of Kerman, Kerman, Iran*

## HIGHLIGHTS

- Gene selection as a preprocessing phase is very important in the diagnosis of diseases.

- By applying a two-stage gene selection method, the accuracy of detecting diseases process was increased.

- By detecting the genes which were statistically differentially abundant in different phenotypes, the genes that related to healthy or diseases were detected.

## GRAPHICAL ABSTRACT

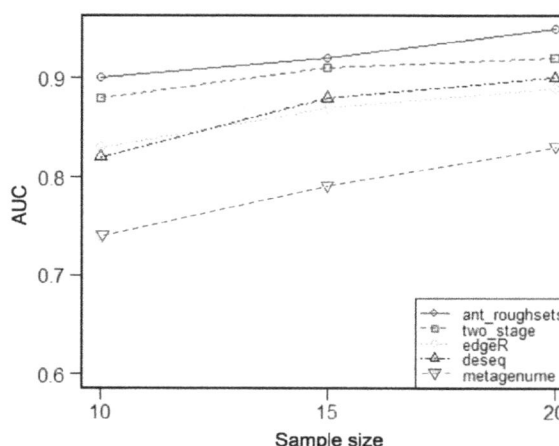

## ARTICLE INFO

*Keywords:*

Microarray
Gene selection
Phenotype
Fuzzy rough set
Ant colony optimization
Gain ratio

## ABSTRACT

With the advancement of metagenome data mining science has become focused on microarrays. Microarrays are datasets with a large number of genes that are usually irrelevant to the output class; hence, the process of gene selection or feature selection is essential. So, it follows that you can remove redundant genes and increase the speed and accuracy of classification. After applying the gene selection, the dataset is reduced and detection of differentially abundant genes facilitated with more accuracy. This will, in turn, increases the power of genes which are correctly detected statistically differentially abundant in two or more phenotypes. The method presented in this study is a two-stage method for functional analysis of metagenomes. The first stage uses a combination of the filter and wrapper gene selection method, which includes the ant colony algorithm and utilizes fuzzy rough sets to calculate the information gain ratio as an evaluation measure in the ant colony algorithm. The set of features from the first stage is used as input in the second stage, and then the negative binomial distribution is used to detect genes which are statistically differentially abundant in two or more phenotypes. Applying the proposed method on a microarray dataset it becomes clear that the proposed method increases the accuracy of the classifier and selects a subset of genes that have a minimum length and maximum accuracy.

*\* Corresponding author: ; E-mail address: javidi@uk.ac.ir*

# 1. Introduction

In the last two decades, the advent of the DNA microarray data set has stimulated a new movement of research into bioinformatics and machine learning. All cells have a nucleus and inside the nucleus, there is DNA. DNA has coding and encoding sections, the coding sections are known as genes. The genes in each individual have different abundances known as gene expression. Each gene performs essential work in any organism [1]. Advances in molecular genetic technologies, such as the micro-arrays of DNA, allow us to obtain a general view of the cell and we can observe expression of a large number of genes [2]. The general process of obtaining gene expression data from a DNA microarray is presented in Figure 1, where the dataset is formed for two classes of normal and diseased. In order to detect differentially abundant genes for different classes and to study their effects on diseases we need to analysis the gene expression dataset. The large dimensions of the dataset lead to statistical and analytical problems, and also, there are very small samples compared to the number of genes in the dataset. In addition, the presence of noise in the genes makes it difficult to detect the specific genes that cause the disease. The application of gene selection is a good approach to overcome to these problems. Using gene selection redundant and irrelevant genes are deleted; thereby, reducing processing time and also diminishing the interference of noisy or unwanted information

leading to incorrect classification, in other words the accuracy of the classifier is increased [3]. Gene selection methods can be divided into four categories: filter, wrapper, embedded and hybrid, which are shown in Table 1 [2]. After applying a gene selection method, diseases or tumors is done are determined by detecting differentially abundant genes in two or more phenotypes. Applying an appropriate method to correctly detect these genes is essential. Statistical procedures play a critical role in detecting differentially abundant genes. In this paper, a two stage method is proposed to determine whether a person is ill or healthy. In the first phase, the dimension of the dataset is reduced by applying an ant colony gene selection evaluation method based on information gain ratio that is calculated by fuzzy rough sets. Then in the second phase, a negative binomial distribution is used to determine the health or sickness. The proposed method can be applied to the comparison of more than two microbial conditions; two microbial conditions; so, our method can be applicable to more general situations.

# 2. Related works

In recent years, gene selection has received much attention. Many optimization algorithms of feature or gene selection have been presented to increase classification accuracy. The concept of gene selection is viewed as one of the most important techniques in Rough Set Theory. There are many feature selection methods that use rough sets. Inbarani et al. [4] presented a supervised hybrid feature selection algorithm based on particle swarm optimization (PSO) and rough sets. This method applies a positive region-based dependency measure to calculate the dependency of the decision feature on the conditional features, which is suitable only for smaller datasets. Chen et al. [5] present a rough set-based feature selection method using the Fish Swarm Algorithm. This algorithm uses a rough set-based dependency measure and thus is not suitable for large datasets. Park and Choi [6] proposed information-theoretic dependency roughness. This algorithm considers the information-theoretic attribute dependency degrees of categorical-valued information systems. The execution time of this method is not provided.

The majority of studies on rough sets have been focused on constructive approaches. In the Pawlak's

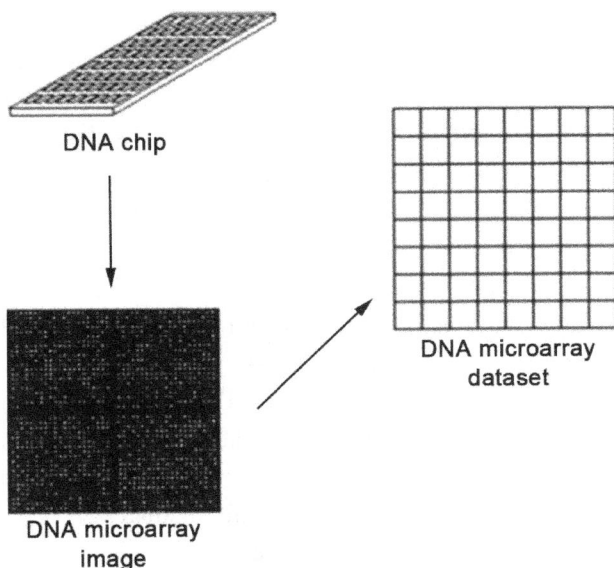

**Fig. 1.** General process of acquiring gene expression data from the DNA microarray.

**Table 1.** Comparison of General Schemes for Gene Selection Methods [2].

| Method | Advantages | Disadvantages |
|---|---|---|
| Filter method<br><br>filter → classifier<br><br>Gene selection space | ➢ Easily scaled to very high-dimensional datasets.<br>➢ Very fast and are computationally simple.<br>➢ Not dependent on any particular algorithm.<br>➢ Feature selection is to be carried out only once, and then different classifiers can be evaluated.<br>➢ Time complexity is O(n), which is low as Compared to other methods Simple. | ➢ Do not take into account the interaction with the classifier.<br>➢ Each feature is measured separately and thus does not take into account the feature dependencies.<br>➢ Lack of feature dependencies results in the degraded performance as compared to other techniques.<br>➢ Creates redundancy.<br>➢ Evaluates genes based on their individual scores ignores their relevance in combination with other genes. |
| Wrapper method<br><br>Gene selection space<br>Hypothesis space<br>classifier | ➢ Involve the interaction between model selection and feature subset search.<br>➢ Take feature dependencies into account.<br>➢ Implementing a wrapper method is quite easy and straightforward in supervised learning.<br>➢ Tests the predictive power of genes.<br>➢ Carries out exhaustive search, generating.<br>➢ optimal solutions. | ➢ These methods have to overfit with a higher risk than filter techniques.<br>➢ Wrapper methods are computationally intensive.<br>➢ Exponential time complexity.<br>➢ Doesn't take enough measures to eliminate redundancy. |
| Embedded method<br><br>Gene selection U hypothesis<br>classifier | ➢ Interacts with the classifier.<br>➢ Achieves computational complexity better than wrapper methods.<br>➢ Models feature dependencies.<br>➢ Tests the predictive power of genes fitting. | ➢ Classifier dependent selection.<br>➢ Prone to over-fitting. |
| Hybrid method | ➢ Can combine the advantages of various approaches. | ➢ Time complexity may increase. |

rough set model [7], the correlation relationship is a key concept. However, this correlation relationship is a very stringent condition that can restrict the application domain of the rough set model. To solve this problem a fuzzy similarity relation can be used to replace an equivalence relation, which was called the fuzzy rough set. Applying fuzzy rough sets in gene selection has received much attention. Gene or feature selection by fuzzy rough sets was first proposed by Wang et al. [8]. They evaluated the hypoxic resistance of a patient on the basis of the values of his blood pressure during a barocamera examination. The measurements were evaluated by the FRS criteria. Jenson and Shen [9] proposed a feature selection method which uses the dependency function to compute the importance of attributes by fuzzy rough sets. Pradipta and Partha [10]

proposed a feature selection where the fuzzy rough set was used to measure the relevance and significance of features. In [11] a method is presented that uses consistence degree as a critical value to reduce redundant attributes in a database. In this approach, a rule based classifier applying a generalized fuzzy-rough set model is proposed. This classifier is effective on noisy data. In [12] a feature selection method with fuzzy-rough and ant colony optimization, similar to our method, is provided. However, the entropy value is used in this method. The disadvantage of this method is that the optimal subset may not be properly selected, because in some cases the entropy criterion (a gene with many distinct values) is high causes the algorithm to selects this gene although it may not be the proper gene. In this paper we presented a method that applies the information gain ratio criteria as the evaluation measure, so it can select the proper genes.

Several statistical methods have been developed to compare various microbial communities in terms of detecting differentially abundant genes, e.g., SONs [13], XIPE-TOTEC [14], Metastates [15] and MEGAN [16]. However, these methods are designed to compare exactly two phenotypes. The Shot Gun Functionalize R [17] method is based on regression and is useful in data with more than two phenotypes; however, the disadvantage of this method is that it only works with discrete data that has a Poisson distribution. Poisson distribution is not flexible for discrete data that has high dispersion. In this paper, we proposed a hybrid gene selection method that uses ant colony and fuzzy rough sets in order to calculate information gain ratio as an evaluation criteria. After selecting an optimal subset of genes, this subset is used in negative binomial (NB) distribution. The NB distribution is widely used to model count data.

## 3. Some basic notations

In this section, we briefly describe the theory of rough set and also information measures in rough and fuzzy-rough sets theory. Rough set theory was proposed by Pawlak [7]. The concept of a rough set has been proposed as a new mathematical tool to deal with uncertain and imprecise data. This theory has been accepted from the beginning, and has been used in many fields of data analysis such as banking [18], economics and finance [19], medical imaging [20], medical diagnosis [21], and data mining [22].

### 3.1. Basic rough set notation

Let, $IS = <U, A, V, f>$, be an information system, where $U$ is a nonempty set of finite object, $A$ is a finite set of attributes or genes, and $V$ is the union of attribute domains, where $V_a$ is the set of values for the attribute $a$; $f: A \times U \rightarrow V$ is an information function that appropriate special values from the domains of attribute to object. If $P \subseteq A$, then an associated indiscernibility equivalence relation, $IND(P)$, is defined as [23]:

$$IND(P) = \{(x,y) \in U^2 \, \forall \mid a \in Pf(a,x) = f(a,y)\} \quad (1)$$

Since $IND(P)$ is a reflexive, symmetric, and transitive relation, it is an equivalence relation; therefore, $IND(P)$ can create a partition on $U$ that is denoted by $U|IND(P)$ or more simply $U|P$, and $[X]_P$ represents an equivalence class of $IND(P)$ containing $x$. The lower and upper estimates for $X \subset U$, respectively, are defined as follows [23]:

$$P \downarrow X = \{x \in U \mid [x]_P \subseteq X\} \quad (2)$$

$$P \uparrow X = \{x \in U \mid [x]_P \cap X \neq \varnothing\} \quad (3)$$

Based on the lower and upper estimates, the boundary regain is defined as follows [23]:

$$BND_P(X) = P \uparrow X - P \downarrow X \quad (4)$$

### 3.2. Information measures in rough set theory

Assume $X_i \in U|IND(P)$ and $X_j \in U|IND(Q)$ are partitions of $U$ which are induced by $P$ and $Q$, respectively. The probability distribution of $X_i$ is defined as follows and the probability distribution of $X_iX_j$ is defined as Eq. (6), where $|..|$ denotes the cardinality [23].

$$P(X_i) = \frac{|X_i|}{|U|} \quad (5)$$

$$P(X_iX_j) = \frac{|X_i \cap X_j|}{|U|} \quad (6)$$

**Definition 1**: If $IS = <U, A, V, f>$ is an information system, $B$ is a subset of $A$ and $X_i \in U|B$, then the Shannon's entropy $H(B)$ of $B$ is defined as [23]:

$$H(B) = -\sum_{i=1}^{n} P(X_i) \log P(X_i) = -\sum_{i=1}^{n} \frac{|X_i|}{|U|} \log \frac{|X_i|}{|U|} \quad (7)$$

**Definition 2**: In information system $IS = <U, A, V, f>$, the join entropy of $P$ and $Q$ is defined as [23]:

$$H(PQ) = H(P \cup Q) = -\sum_{i=1}^{n}\sum_{j=1}^{m} P(X_iX_j) \log P(X_iX_j)$$
$$= -\sum_{i=1}^{n}\sum_{j=1}^{m} \frac{|X_i \cap X_j|}{|U|} \log \frac{|X_i \cap X_j|}{|U|} \quad (8)$$

where $X_i \in U|B$, $X_j \in U|Q$, and $P,Q \subseteq A$.

**Definition 3**: The conditional entropy of $D$ with condition $B$ for decision system $DS = <U, C \cup D, V, f>$ is defined as [23]:

$$(D \mid B) = -\sum_{i=1}^{n} P(X_i) \sum_{j=1}^{m} P(X_j \mid X_i) \log P(X_j \mid X_i) \quad (9)$$

$$= -\sum_{i=1}^{n} \frac{|X_i|}{|U|} \sum_{j=1}^{m} \frac{|X_i \cap X_j|}{|U|} \log \frac{|X_i \cap X_j|}{|U|}$$

$$= -\sum_{i=1}^{n} \sum_{j=1}^{m} \frac{|X_i \cap X_j|}{|U|} \log \frac{|X_i \cap X_j|}{|U|}$$

where, $B$ is a subset of $C$, and $C$ is the condition attribute set; $X_i \in U|B$ and $X_j \in U|D$, where $D$ is the decision attribute.

**Definition 4**: The mutual information of $B$ and $D$ is defined as follows [23]:

$$I(B;D) = H(D) - H(D \mid B) \quad (10)$$

**Definition 5**: The gain of attribute $a \in C\text{-}B$ is defined as [23]:

$$Gain(a,B,D) = I(B \cup \{a\};D) - I(B;D) \quad (11)$$
$$= H(D \mid B) - H(D \mid B \cup \{a\})$$

**Definition 6**: The mutual information gain ratio of attribute $a$, is defined as [23]:

$$Gain\_Ratio(a,B,D) = \frac{Gain(a,B,D)}{H(\{a\})} \quad (12)$$
$$= \frac{I(B \cup \{a\};D) - I(B;D)}{H(\{a\})}$$

*3.3. Information measures in fuzzy-rough set theory*

In fuzzy rough sets, it is essential to define a fuzzy equivalence relation. $\tilde{R}$ is a fuzzy equivalence relation, if it satisfies:

Reflectivity: $\tilde{R}(x,y) = 1, \forall x \in X$

Symmetry: $\tilde{R}(x,y) = \tilde{R}(y,x), \forall x, y \in X$

Transitivity: $\tilde{R}(x,y) \geq \min_{y}\{\tilde{R}(x,y), \tilde{R}(y,z)\}$

$M(\tilde{R})$ represents a relation matrix for $x_i, x_j \in X$, that $\tilde{R}$ is a fuzzy equivalence relation defined on a nonempty finite set $X$.

$$M(\tilde{R}) = \begin{pmatrix} r_{11} & \cdots & r_{1n} \\ \vdots & \ddots & \vdots \\ r_{n1} & \cdots & r_{nn} \end{pmatrix} \quad (13)$$

Here, $r_{ij} \in [0,1]$ is the relation value of $x_i$ and $x_j$ that can be written as $\tilde{R}(x,y)$. For the crisp rough set model, if $x_i$ equals to $x_j$ with respect to the crisp equivalence relation $R$ then $r_{ij} = 1$; otherwise, $r_{ij} = 0$. A similarity function that has been used to calculate the equivalence relation is shown by Eq (14), where $x_i$ and $x_j$ are attribute values of two objects on attribute $a$; $a_{max}$ and $a_{min}$ are maximal and minimal values of attribute $a$, respectively [23].

$$r_{ij} = \begin{cases} 1 - 4 \times \frac{|x_i - x_j|}{|a_{max} - a_{min}|}, & \frac{|x_i - x_j|}{|a_{max} - a_{min}|} \leq 0.25 \\ 0 \end{cases} \quad (14)$$

Two important operators in the fuzzy equivalence relation that are useful for implementing fuzzy theory are defined by [23]:

$$\tilde{R} = \tilde{R}_1 \cup \tilde{R}_2 \Leftrightarrow \tilde{R}(x,y) = \max\left\{\tilde{R}_1(x,y), \tilde{R}_2(x,y)\right\}$$

$$\tilde{R} = \tilde{R}_1 \cap \tilde{R}_2 \Leftrightarrow \tilde{R}(x,y) = \min\left\{\tilde{R}_1(x,y), \tilde{R}_2(x,y)\right\}$$

**Definition 7**: The fuzzy partition of the universe $U$ generated by $\tilde{R}$, is defined as [23]:

$$U / \tilde{R} = \left\{[x_i]_{\tilde{R}}\right\}_{i=1}^{n} \quad (15)$$

Here, $\tilde{R}$ is a fuzzy equivalence relation and $[x]_{\tilde{R}}$ is the fuzzy equivalence class equal to $\frac{r_{i1}}{x_1} + \frac{r_{i2}}{x_2} + \ldots + \frac{r_{in}}{x_n}$.

**Definition 8**: The cardinality $[x]_{\tilde{R}}$ is defined as [23]:

$$|[x_i]_{\tilde{R}}| = \sum_{j=1}^{n} r_{ij} \quad (16)$$

**Definition 9**: Information quantity of the fuzzy attribute set or the fuzzy equivalence relation is defined as [23]:

$$H(\tilde{R}) = -\frac{1}{n} \sum_{i=1}^{n} \log \frac{|[x_i]_{\tilde{R}}|}{n} \quad (17)$$

**Definition 10**: The joint entropy of $B$ and $E$ is defined as [23]:

$$H(BE) = H(\tilde{R}_B \tilde{R}_E) = -\frac{1}{n} \sum_{i=1}^{n} \log \frac{|[x_i]_{\tilde{B}} \cap [x_i]_{\tilde{E}}|}{n} \quad (18)$$

where $FIS = <U, A, V, f>$ is a fuzzy information system, $A$ is the attribute set, and $B$ and $E$ are two subsets of $A$.

**Definition 11**: Let $FIS = <U, A, V, f>$ is a fuzzy decision system, $C$ is the condition attribute set, $D$ is the decision attribute and $B \subseteq C$. The condition entropy $D$ on condition $B$ can be calculated as follows [23]:

$$\tilde{H}(D \mid B) = -\frac{1}{n}\sum_{i=1}^{n}\log\frac{|[x_i]_{\tilde{B}} \cap [x_i]_{\tilde{D}}|}{|[x_i]_{\tilde{B}}|} \qquad (19)$$

In the above relation, $[x_i]_{\tilde{B}}$ and $[x_i]_{\tilde{D}}$ are fuzzy equivalence classes containing $x_i$ generated by $B$ and $D$, respectively.

**Definition 12**: The mutual information of $B$ and $D$ is defined as [23]:

$$\tilde{I}(B;D) = \tilde{H}(D) + \tilde{H}(B) - \tilde{H}(BD) \qquad (20)$$

**Definition 13**: In decision system FDS= $<U,C\cup D,V,f>$, $\forall a \in C\text{-}B$ the gain of attribute $a$, can be defined as [23]:

$$Gain(a,B,D) = \tilde{I}(B\cup\{a\};D) - \tilde{I}(B;D) \qquad (21)$$

**Definition 14**: According to the definition of 13, the mutual information gain ratio of attribute $a$, can be defined as [23]:

$$\begin{aligned} GainRatio(a,B,D) &= \frac{Gain(a,B,D)}{\tilde{H}(\{a\})} \qquad (22) \\ &= \frac{\tilde{I}(B\cup\{a\};D) - \tilde{I}(B;D)}{\tilde{H}(\{a\})} \end{aligned}$$

## 4. Proposed method

### 4.1. Gene selection phase

In this section, a new filter-wrapper approach for gene selection in fuzzy-rough sets is described. In this approach, filter phase employs a modified ACO search strategy which is able to do gene selection function as a multi-modal problem, and the wrapper phase includes a learning model that evaluates the chosen subsets of genes from the filter phase and select the best subset, then calculates pheromones changes in the selected subsets. Choosing the subsets of features with first and second maximum accuracies as candidate subsets for minimal data reductions is a contribution of this work; so each chosen minimal subset has a short length along with an acceptable accuracy value; consequently, the approach is able to satisfy both an increase the accuracy and a decrease in the length of reduced subsets, concurrently. In detail, in order to implement this approach we need

the feature selection problem space to be considered in the form of a complete non-directed graph. The nodes, indicating the genes and edges, represent the probability of choosing the next node. The algorithm starts with the production of k number of ants, which is half the number of genes. The following steps are followed to complete each ant's tour:

1- Initialize ants with random and different nodes.
2- For each ant k, consider set $S_K$ includes all the nodes without initial node, as accessible locations.
3- The ant k chooses the next node according to the transition rule that will be dealt with in the next section.
4- The selected node is removed from the $S_K$.
5- For each ant k, the third and fourth stage is repeated until $S_K$ is empty.
6- The best answer achieved is saved.

After each ant completes its tour, the pheromone is updated on the routes traversed from origin to destination, according to the algorithm explained in section 3-1-2. At the end of each iteration, the best observed solutions are kept; i.e. in each iteration, we consider the subsets of the genes that have maximum accuracy as the best candidate subsets. We preserve the subsets which have the first and the second maximum accuracies among all the best candidate subsets from the first iteration to the current. Then, we consider the minimal subsets from the preserved subsets as the best of all the iterations. Because the wrapper method utilizes a learning model, gene selection based on wrappers boosts the accuracy of the model; however, this method increases the order of mathematical complexity. In this method, instead of evaluating the genes separately, the subsets found by the filter are evaluated using the wrapper model to decrease the complexity. Output of wrapper model (accuracy of the classifier) is a criterion for the goodness evaluation of the subsets found. After the end of each run the best seen solution, from the first iteration until the current one is saved as an optimal solution. In addition to detecting high quality subsets of genes, finding more than one solution in one run is another advantage of this method compared to other methods.

### 4.1.1. Transition rule and gene deletion

The transition rule introduced in [24] is used for exploring the nodes' space. Node j, as a candidate for selection, is selected with probability 0.5 using the

$$p_{ij}^k = \begin{cases} 1, & j = \arg\max \left\{ \tau_{ij}^\alpha \eta_{ij}^\beta \right\} \\ 0, & otherwise \end{cases} \tag{23}$$

If an ant selects a new node, that node is removed from the set of available nodes, and if candidate node j is not selected that candidate node will also be removed from the set of available nodes. In this case, the following relation in the roulette wheel mechanism, as the probability of selecting the available nodes, is used to select the next node.

$$p_{ij}^k(t) = \begin{cases} \dfrac{[\tau_{ij}]^\alpha \cdot [\eta_j]^\beta}{\sum_{x \in S_K} [\tau_{ix}]^\alpha \cdot [\eta_x]^\beta}, & j \in s_k \\ 0, & otherwise \end{cases} \tag{24}$$

In both of the above equations, $\alpha = 0.5$ and $\beta = 1$, and the initial value of $\tau_j$ is equal to 0.1. By selecting each node, in the roulette wheel mechanism, that node and all nodes before it, $\eta_j = GainRatio\ (j, N_K, D)$ are calculated by Eq. (22) as heuristic information and $N_K$ is regarded as a set of selected nodes by ant k, and $\tau_{ij}$ is the pheromone value of edge ij.

### 4.1.2. Pheromones updating rules

After each individual ant created its own complete tour, the pheromone is updated on the path it travelled from the beginning to the end, as follows:

1- On each edge of the complete graph, the pheromone evaporates according to equation (25).
2- In each iteration, the pheromone on the path is updated according to equations (26) and (27).
3- In order to maintain the best answers, the pheromone on the best path in all of the repetitions is updated according to (28).

$$\tau^{new} = (1 - \rho).\tau^{old} \tag{25}$$

$$\Delta \tau_{ij} = \frac{\gamma'_{N_k}}{lenght(N_k)} \tag{26}$$

$$\tau_{ij}^{new} = \begin{cases} \tau_{ij}^{old} + \Delta \tau_{ij}, & if \quad ij \in BF \\ \tau_{ij}^{old} + \varphi * \Delta \tau_{ij}, & otherwise \end{cases} \tag{27}$$

$$\tau_{ij}^{new} = \tau_{ij}^{old} + \varphi * \gamma'_{N_k} \tag{28}$$

where $\varphi = 0.5$, $\rho = 0.2$ and BF is the best path traversed in the current iteration. $\gamma_{N_k}$ is the accuracy of the classifier as output of the learning model.

### 4.2. Differentially abundant feature detection stage

At this stage, the genes which are statistically differentially abundant in two or more conditions are detected. In real metagenomics count data, the variance is usually greater than the corresponding mean of the gene abundance. Negative binomial distribution (NB) is often used for high-dispersion data.

### 4.2.1. NB model

Suppose r of p genes are selected from the first stage. Let Y be the vector of the numbers of reads for gene i in all samples where i=1, 2, 3,..., r. Each element $(y_s)$ in the Y vector with a negative binomial distribution is modeled as follows:

$$f_Y(y_s; \mu_s, \theta) = \frac{\Gamma(y_s + \theta)}{\Gamma(\theta).y_s!} \cdot \frac{\mu_s^{y_s}.\theta^\theta}{(\mu_s + \theta)^{y_s + \theta}} \tag{29}$$

$E(y_s) = \mu_s$ is the mean and the variance is $var(y_s) = \mu_s(1 + \mu_s/\theta)$. The variance is quadratic in the mean. The negative binomial distribution model can also be modeled with the dispersion parameter, $\phi = 1/\theta$. In this case, the mean is equal to and the variance is $\mu_s(1 + \phi\mu_s)$. Initially, $\phi$ is greater than zero, and when $\phi \to 0$, the negative binomial distribution is reduced to the standard Poisson distribution with the parameter $\mu_s$. In the generalized linear model, the logarithmic link is the most appropriate method for linking the mean response $\mu$ in negative binomial distribution variable to a linear combiner of predictors $x$. For each gene i (i= 1, 2, 3, ..., r), $\log(\mu_s) = x_s^T \beta$ where $x_s^T$ is $1 \times K$, the line vector contains the indicative variables of the phenotypes, S=1, 2, ..., N, K represents the number of phenotypes and $\beta$ is the corresponding $K \times 1$ column vector of unknown regression parameters. Auxiliary variables can be introduced into a regression model based on the NB distribution via the relationship:

$$\log(\mu_s) = \sum_{j=1}^{K} x_{sj} \beta_{j-1} \tag{30}$$

In the negative binomial distribution model for mean $\mu_s = \exp(x_s^T \beta)$, $\beta$ and $\phi$ are estimated by maximizing the log likelihood function:

$$l(\beta, \phi; Y) = \sum_{s=1}^{N} \left\{ \log\left( \frac{\Gamma(y_s + \phi^{-1})}{\Gamma(\phi^{-1})} \right) - \log(y_s!) \right. \tag{31}$$

$$\left. -(y_s + \phi^{-1})\log(1 + \phi\mu_s) + y_s \log\phi + y_s x_s^T \beta \right\}$$

## 5. Results

To implement the proposed method, we use a data set with 20 samples and 1000 genes. This data set has two classes; one is healthy and the other expresses the sickness of the samples. Among these 10 samples are healthy and the rest are patient. In order to implement the proposed method, we utilize the R statistical software on a five-core computer that has 1 gigabyte of RAM. The proposed method has been implemented with a number of different samples of the data set and the results of these experiments were compared with four current reliable methods: Two-stage, edgeR, metagenomeSeq and DESeq. The results are expressed in terms of time, accuracy, ROC, AUC, PR Curve FDR, and the power in detection of the true differentially abundant genes. In addition, a criterion named, $\Psi_B$ as described below, is also examined in the results.

$$\psi_B = \frac{accuracy(B)}{length(B)} \qquad (32)$$

where $B$ is a subset of genes. By increasing the classification accuracy and reducing the length of the selected subset, $\Psi_B$ increases. This indicates an increase in the efficiency of the method, because the efficiency of the feature or gene selection is based on the accuracy and number of selected genes.

In Figure 2, the computational time of the proposed method for sample sizes of 10, 15, and 20, is compared with the other methods. According to Figure 2, the Runtime of the proposed method is less than DESeq and Metastates but longer than the rest of the methods. Therefore, the proposed method is not efficient in terms of run time. The high execution time of this algorithm is due to the implementation of the ant colony algorithm. Microarray data requires several nesting loops with a high repetition rate and the number of high repetitions is due to the fact that microarray data usually has a high number of genes. Another reason for the high execution time of this algorithm is the need for high-dimensional square matrices to calculate entropy and the information gain ratio, which are time consuming calculations.

In Table 2, the accuracy $\Psi_B$ and for the five methods are compared under various sample sizes. The subset obtained from the first phase in the proposed method is given as input to a SVM classifier and its accuracy is calculated. Regarding the $\Psi_B$ values, we find that the precision criterion, accuracy, alone is not a suitable

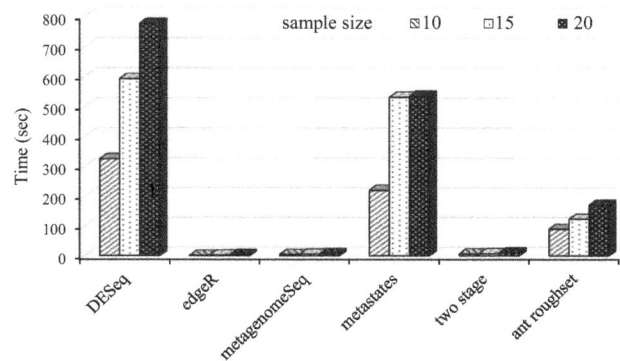

**Fig. 2.** Comparison of computational time (in second) for five methods under various sample sizes.

**Table 2.** Accuracy and $\Psi_B$ for the presented method for sample size of 10, 15, and 20.

| Sample size | Selected subset length | Accuracy (%) | $\Psi$ | p-value |
|---|---|---|---|---|
| 10 | 619 | 93 | 0.15 | 0.01074 |
| 15 | 527 | 97 | 0.17 | 0.001135 |
| 20 | 404 | 95 | 0.23 | 0.0002003 |

measure for evaluating the gene selection methods. So, in order to evaluate the methods, the $\Psi_B$ criterion must be used. Also, the p-value shows that the accuracy obtained from the proposed method is not random, because its value is much lower than the usual 0.05 threshold; therefore, the result is reliable.

The ROC curve is usually used to measure signal detection. It is created by plotting the true positive rate (TPR) or sensitivity versus the false-positive rate (FPR) [25]. Figures 3 and 4 display the ROC curves of the proposed method before applying the gene selection process for sample sizes of 10 and 20, Figures 5 and 6 indicates the ROC curves after the gene selection process for samples 10 and 20. In general in ROC curves, the closer the curve is to the diameter the weaker the classifier is in distinction, and as the ROC curve tends to be upwards and far from the diameter, the classifier is better in distinction showing that method is better. According to Figures 5 and 6, the ROC curves plotted for the SVM classifier in sample sizes of 10 and 20 are far from the diameter, which shows a good performance of the classifier, and Figures 3 and 4 are closer to the diameter, which shows the classifier does not have proper efficiency. A better way to express this (near the curve to the diameter) is to have a surface below the ROC curve (AUC). On the other hand, as the AUC is closer to 0.5, the weaker

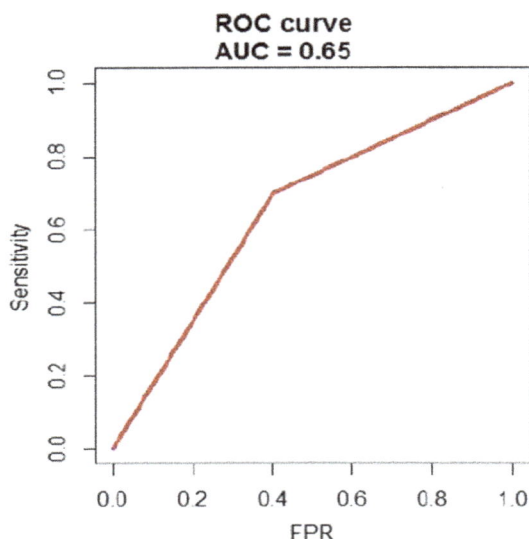

**Fig. 3.** ROC curves of the proposed method before applying the gene selection process for sample size of 20.

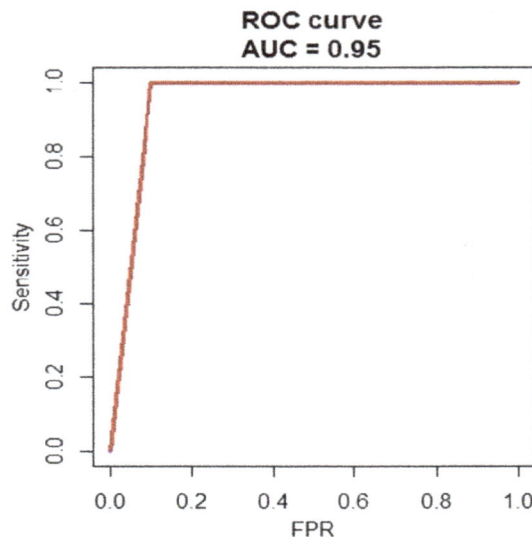

**Fig. 5.** ROC curves of the proposed method after applying the gene selection process for sample size of 20.

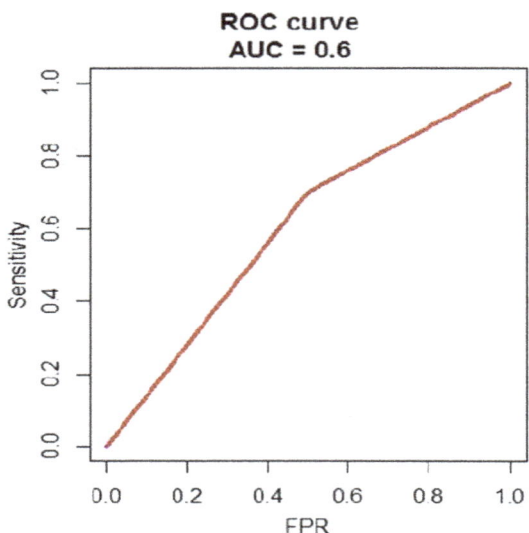

**Fig. 4.** ROC curves of the proposed method before applying the gene selection process for sample size of 10.

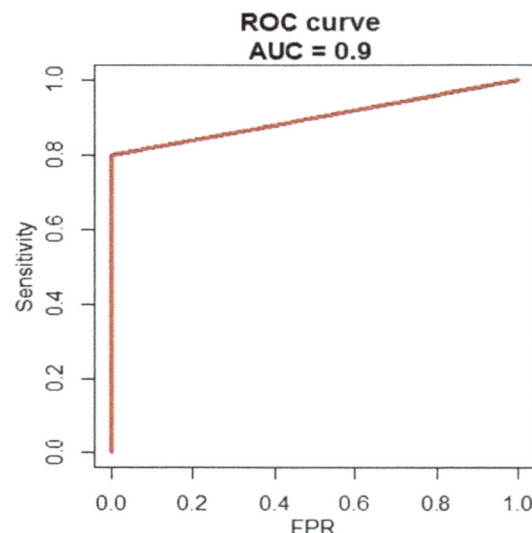

**Fig. 6.** ROC curves of the proposed method after applying the gene selection process for sample size of 10.

the classifier works in discrimination between the two groups, and whichever area is closer to one has a more favorable classification result. The AUC is independent of the various forms of population under investigation, and this is an advantage. In general, AUC represents the overall performance of the methods so that the greater the AUC value, the higher the performance of the method. Comparing the amount of AUC before and after the gene selection process, we found the overall performance of the both sample sizes 20 and 10, after the gene selection process, increased by 30%. It can also be realized that the number of samples has a significant effect on the classifier's performance, because a

comparison of the ROC curve of the 20-sample size and the 10-sample size, is higher and further from the diameter. Figure 7 shows the results of the AUC derived from the five methods in various sample size. Typically, the AUC increases as the sample size increases.

As seen in Figure 7, the proposed method has a higher AUC than the rest of the methods and this means that the proposed method has a higher overall performance. In a sample size of 10 the proposed method increased by 2% compared to the two stage method. In a sample size of 15, the proposed method is 1% better than the two-stage method, and in a sample size of 20 the proposed method, in overall performance or AUC, had a 3% growth.

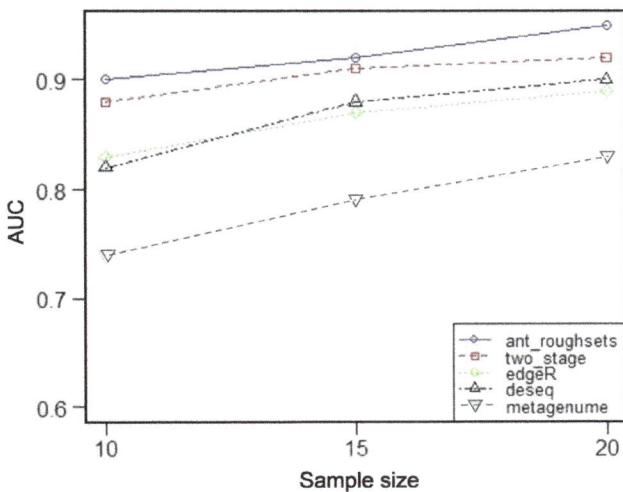

**Fig. 7.** Comparison of AUC curves.

Figures 8 and 9 indicate the precision and recall curves before applying the gene selection method, and Figures 10 and 11 show this curves after applying the gene selection method. In addition, this figure shows the plotted minimum and maximum PR curves and also a random PR curve to compare the performance of the method with the maximum, minimum and random mode. In this figure, the PR curve of the proposed method and the random PR curve are displayed by red and green, respectively. In general, precision and recall are opposite of each other but an ideal curve would have both equal to one. As seen in the curves, when the gene selection method is applied and the sample size increases the value of the recall increased, and when the precision rises the curve is more stable this means that the method has a higher performance. According to the curves, the amount of recall before applying the gene selection method in the sample size of 10 is approximately 0.59 and in the sample size of 20 it is nearly 0.63, which represents an increase of 4%. This amount of recall slowly diminished until the amount of precision reached a value of 1. The amount of recall, after applying a gene selection in a sample size of 10 is 1 and in the sample number of 20 it is approximately 90/0, this shows the recall is more stable in a sample size of 20. In addition, after selecting a gene, the PR curves do not have a significant difference between the maximum situations. Therefore, it can be concluded that classification is improved by applying the proposed method of gene selection.

Figure 12 shows the power in detection of the true differentially abundant genes for five methods at various levels of FDR for a sample size of 10. Typically,

Typically, increasing the amount of FDR increases the detection power rate. According to the diagram, we find that the proposed method has a higher detection power in determining the genes which are statistically differentially abundant in two or more phenotypes. In general, in the proposed method the detection power in determining genes with differentially abundant for a sample size of 10, on average, has been increased by 1.3% compared to the two-stage method. Figure 13 displays the power in detection of the true differentially abundant genes for the five methods at various levels of FDR for a sample size of 20. FDR for a sample size of 20. In this diagram, as in the previous diagram, we find that the proposed method has the highest detection power. On average, in proposed method for sample size of 20 the detection power has been increased by 1% compared to the two-stage method.

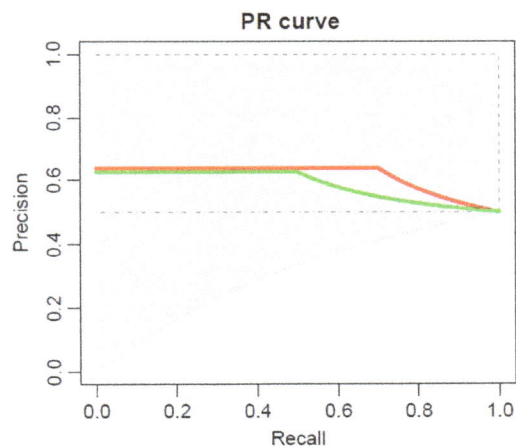

**Fig. 8.** The PR curves before applying the gene selection method for sample size of 10.

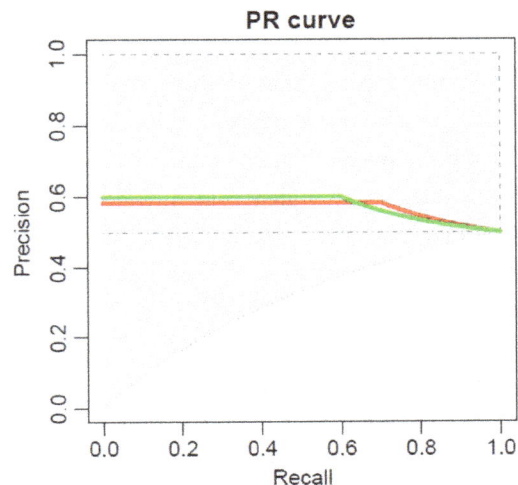

**Fig. 9.** The PR curves before applying the gene selection method for sample size of 20.

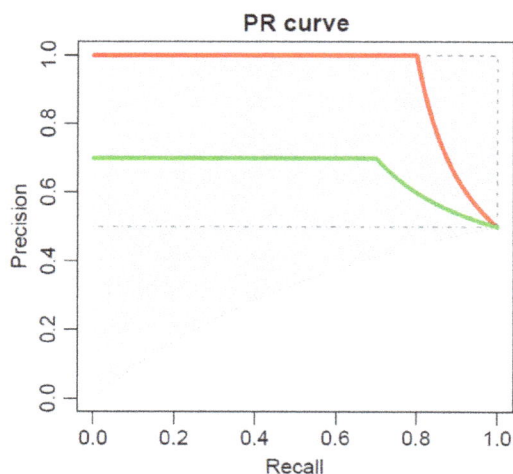

**Fig. 10.** The PR curves after applying the gene selection method for sample size of 10.

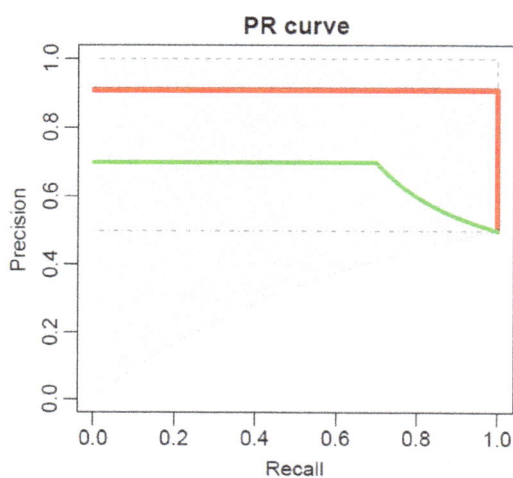

**Fig. 11.** The PR curves after applying the gene selection method for sample size of 20.

## 6. Conclusion and future work

In this research, a gene selection method was proposed in the first stage to eliminate redundant and additional genes in a microarrays dataset. Then in the second stage, using the dataset obtained from the first step, the genes that differ in abundance in different phenotypes are identified to detect the presence of diseases. In the first step, a hybrid of the filter and wrapper gene selection method is introduced. In the filter section, genes are obtained using the ant colony algorithm and the information gain ratio which is calculated by fuzzy rough sets. Then the gene set obtained from the filter phase is evaluated in the wrapper section. Finally, the best subset of genes is collected. The gene set obtained from the first stage is used as the input of the second

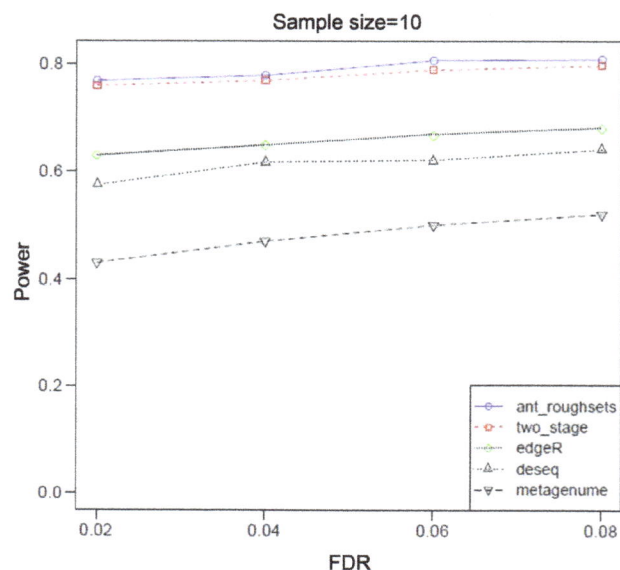

**Fig. 12.** The power in detection of the true differentially abundant genes for the five methods at various levels of FDR for sample size of 10.

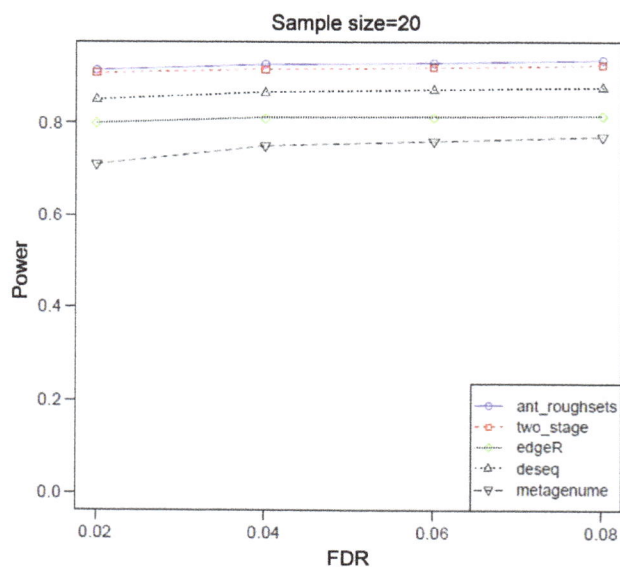

**Fig. 13.** The power in detection of the true differentially abundant genes for the five methods at various levels of FDR for sample size of 20.

stage is used as the input of the second stage. Then the genes that are statistically differentially abundant in two or more phenotypes are identified using negative binomial distribution. The proposed method is implemented by the statistical software R. The results show that the proposed method is highly effective due to high accuracy, $\Psi$, ROC curves, and the increase of the AUC as compared to other existing methods, but this method has low runtime.

# References

[1] V. Bolón-Canedo, N. Sánchez-Maroño, A. Alonso-Betanzos, J.M. Benítez, F. Herrera, A review of microarray datasets and applied feature selection methods, Inform. Sciences, 282 (2014) 111-135.

[2] H. Salem, G. Attiya, N. El-Fishawy, Classification of human cancer diseases by gene expression profiles, Appl. Soft Comput. 50 (2017) 124-134.

[3] P. Agarwalla, S. Mukhopadhyay, Bi-stage hierarchical selection of pathway genes for cancer progression using a swarm based computational approach, Appl. Soft Comput. 62 (2018) 230-250.

[4] H.H. Inbarani, A.T. Azar, G. Gothi, Supervised hybrid feature selection based on PSO and rough sets for medical diagnosis, Comput. Met. Prog. Bio. 113 (2014) 175-185.

[5] Y.Chen, Q.Zhu, H. Xu, Finding rough set reducts with fish swarm algorithm, Knowl-Based Syst. 81 (2015) 22-29.

[6] I.K. Park, G.S. Choi, Rough set approach for clustering categorical data using information-theoretic dependency measure, Inform. Syst. 48 (2015) 289-295.

[7] Z. Pawlak, A. Skowron, Rudiments of rough sets, Inform. Sciences, 177 (2017) 3-27.

[8] L.I. Kuncheva, Fuzzy rough sets: Application to feature selection, Fuzzy Set. Syst. 51 (1992) 147-153.

[9] R. Jensen, Q. Shen, Fuzzy-rough attributes reduction with application to web categorization, Fuzzy Set. Syst. 141 (2004) 469-485.

[10] M. Pradipta, G. Partha, Fuzzy-rough simultaneous attribute selection and feature extraction algorithm, IEEE T. Cybernetics, 43 (2013) 1166-1177.

[11] S. Zhao, E.C.C. Tsang, D. Chen, X. Wang, Building a rule-based classifier-a fuzzy rough set approach, IEEE T. Knowl. Data En. 22 (2010) 624-638.

[12] M. Dorigo, LM. Gambardella, A cooperative learning approach to the traveling salesman problem, IEEE T. Evolut. Comput. 1 (1997) 53-66.

[13] P. Schloss, J. Handelsman, Introducing SONS, a tool for operational taxonomic unit based comparisons of microbial community memberships and structures, Appl. Environ. Microb. 72 (2006) 6773-6779.

[14] B. Rodriguez-Brito, F. Rohwer, R.A. Edwards, An application of statistics to comparative metagenomics, BMC Bioinformatics, 7 (2006) 162.

[15] J. White, N. Nagarajan, M. Pop, Statistical methods for detecting differentially abundant features in clinical metagenomics samples, PLOS Comput. Biol, 5 (2009) e1000352.

[16] D. Huson, D. Richter, S. Mitra, A. Auch, S. Schuster, Methods for comparative metagenomics, BMC Bioinformatics, 10(Suppl 1) (2009) S12.

[17] Kristiansson, E. et al, ShotgunFunctionalizeR: An R-package for functional comparison of metagenomes, Bioinformatics, 25 (2009) 2737-2737.

[18] G.A. Montazer, S. ArabYarmohammadi, Detection of phishing attacks in Iranian e-banking using a fuzzy-rough hybrid system, Appl. Soft Comput. 35 (2015) 482-492.

[19] M. Podsiadło, H. Rybiński, Rough sets in economy and finance, In: Peters J.F., Skowron A. (eds) Transactions on Rough Sets XVII. Lecture Notes in Computer Science, Vol. 8375, pp. 109-173, 2014.

[20] C.H. Xie, Y.J. Liu, J.Y. Chang, Medical image segmentation using rough set and local polynomial regression, Multimed. Tools Appl. 74 (2015) 1885-1914.

[21] V. Prasad, T.S. Rao, M.S. Babu, Thyroid disease diagnosis via hybrid architecture composing rough data sets theory and machine learning algorithms, Soft Comput. 20 (2016) 1179-1189.

[22] M.P. Francisco, J.V. Berna-Martinez, A.F. Oliva, M.A.A. Ortega, Algorithm for the detection of outliers based on the theory of rough sets, Decis. Support Syst. 75 (2015) 63-75.

[23] J. Dai, Q. Xu, Attribute selection based on information gain ratio in fuzzy rough set theory with application to tumor classification, Appl. Soft Comput. 13 (2013) 211-221.

[24] M. Dorigo, L.M. Gambardella, A cooperative learning approach to the traveling salesman problem, IEEE T. Evolut. Comput. 1 (1997) 53-66.

[25] P. Naruekamol, M. Sohn, Q. Li, A two-stage statistical procedure for feature selection and comparison in functional analysis of metagenomes, Bioinformatics, 31 (2014) 157-165.

# 14

# A green approach for the synthesis of silver nanoparticles using *Lithospermum officinale* root extract and evaluation of their antioxidant activity

**Saeed Mollaei[1,*], Biuck Habibi[2], Alireza Amani Ghadim[3], Milad Shakouri[1,2]**

[1] *Phytochemical Laboratory, Department of Chemistry, Faculty of Sciences, Azarbaijan Shahid Madani University, Tabriz, Iran*

[2] *Electroanalytical Chemistry Laboratory, Department of Chemistry, Faculty of Sciences, Azarbaijan Shahid Madani University, Tabriz, Iran*

[3] *Applied Chemistry Laboratory, Department of Chemistry, Faculty of Sciences, Azarbaijan Shahid Madani University, Tabriz, Iran*

## HIGHLIGHTS

- Green method for synthesis of the silver nanoparticles using *Lithospermum officinale*.

- Evaluation of antioxidant activity of synthesized silver nanoparticles.

- Therapeutic potential of the synthesized nanoparticles due to their antioxidant activity.

## GRAPHICAL ABSTRACT

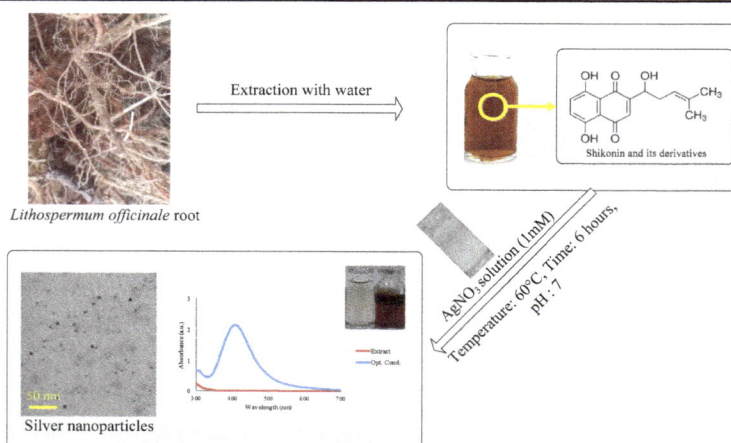

*Lithospermum officinale* root

Extraction with water

Shikonin and its derivatives

AgNO₃ solution (1mM)
Temperature: 60°C, Time: 6 hours, pH : 7

Silver nanoparticles

## ARTICLE INFO

*Keywords:*

*Lithospermum officinale*
Root aqueous extract
Silver nanoparticles
Green method
Antioxidant

## ABSTRACT

Recently, the synthesis of silver nanoparticles has become an important subject in the bionanotechnology field. Many different chemical and physical methods could be used for silver nanoparticles synthesis, but they are limited due to the usage of toxic chemicals and the production of dangerous by-products. However, the usage of plant extract for silver nanoparticles synthesis is a green single-step method without using toxic chemicals. Herein, silver nanoparticles were synthesized using *Lithospermum officinale* root aqueous extract and their antioxidant activity was evaluated in vitro. The results showed that 1 ml of the extract could reduce 9 ml of silver ions (1 mM) to silver nanoparticles by heating the reaction mixture (60 °C) for 6 hours at pH 7.0. The synthesized silver nanoparticles were detected by UV–Vis spectroscopy, TEM, FT-IR, DLS, and XRD. The synthesized silver nanoparticles spectrum had a maximum peak at 390nm, and TEM analysis indicated spherical particles, higher stability (zeta potential: -15.3 mV) and an average size of 7 nm. The antioxidant activity of the synthesized silver nanoparticles was 0.07 mg/ml compared to *L. officinale* root aqueous extract (0.142 mg/ml) which indicated higher antioxidant activity. So, it is concluded that the synthesized silver nanoparticles could be considered a clinical therapeutic potential due to its antioxidant property.

* *Corresponding author:* ; *E-mail address: s.mollaei@azaruniv.edu*

# 1. Introduction

The field of nanotechnology, which is related to manufacturing materials at the range of nanometers, is growing rapidly because of its application in technology and science. Metal nanoparticles, especially silver nanoparticles, are widely used in many fields such as catalysis [1], plasmonics [2], optoelectronics [3], biological sensor [4,5], agricultural [6,7] and the pharmaceutical industry [8-10].

Many physical and chemical methods such as thermal decomposition [11], electrochemical [12], laser ablation [13], microwave, [14] etc. [15,16] have been applied for silver nanoparticles synthesis, but these are various limitations for these mentioned methods to such as high cost, harmful solvent and chemical systems, environmental contamination, hazardous by-products, higher temperature and higher pressure.

In comparison to physicochemical methods, green chemistry methods are cost effective, easily available, eco-friendly, and nontoxic [17,18]. In this method naturally occurring biomaterials, such as the extract of plants, were used for the metal nanoparticles synthesis applied in medicine and pharmaceutics [17,18]. It has been confirmed that the plant metabolites, such as polyphenols, phenolic acids, terpenoids, alkaloids, sugars, and proteins, play an important role as reducing, capping and stabilizing agents [19,20]. Many articles are available for silver nanoparticles synthesis using plants such as *Citrullus lanatus* [21], *Avena sativa* [22], *Prangos ferulaceae* [23], *Gmelina arborea* [24], *Parkia speciosa* [25], and *Azadirachta indica* [26]. Also in many cases, the biological activity of silver nanoparticles synthesized by plant extract has been researched. Elemike et al. [10] synthesized biological active silver nanoparticles using *Costus afer* extract, they found the nanoparticles indicated better antibacterial and antioxidant activity compared to the leaf extract. Jemal et al. [27] demonstrated a green and cost-effective method for silver nanoparticles synthesis from leaf extracts of *Allophylus serratus* and evaluated their antibacterial activity. The silver nanoparticles indicated antimicrobial activity against both gram negative and gram positive bacteria. Similarly, the antimicrobial potential of silver nanoparticles synthesized by the extract of mint plant leaves was researched by Sarkar and Paul [28]. Their results showed that these nanoparticles have good inhibition activity towards *Escherichia coli*

and *Pseudomonas aeruginosa*.

To the best of our knowledge there have been no reports for silver nanoparticles synthesis using *L. officinale* root aqueous extract. The dried root of Lithospermum species is used as an herbal medicine in many countries. So, the aim of this research was to synthesize and characterize the silver nanoparticles using *L. officinale* root aqueous extract as a reducing, capping and stabilizing agent. In addition, the antioxidant property of the synthesized silver nanoparticles was investigated.

# 2. Material and Methods

## 2.1. Synthesis of silver nanoparticles

The silver nanoparticles were synthesized using *L. officinale* roots collected from Azarshahr, (37° 45' N, 45° 57' E) eastern Azarbaijan province, Iran. The roots were cleaned, weighed (20 g) and cut into small pieces. After adding the roots to 200 ml autoclaved double distilled water and shaking for 30 minutes at room temperature, the mixture was centrifuged for 2 minutes and then filtered through a 0.45 μm filter. The experiments were performed at different silver ion concentrations (0.25, 0.50, 1.0, 2.0, and 5 mM), temperatures (25, 40 and 60 °C), pH (from 3 to 13 using HCl or NaOH), and the ratios of extract to silver nitrate (1:9; 3:7, and 5:5 v/v) in order to obtain the optimum condition for synthesis. The primary observation of synthesis indicated a color change from yellow to brown, which was confirmed by UV-Visible spectroscopy. After centrifuging the nanoparticles solution at 15000 rpm, the precipitation was washed with distilled water to eliminate all impurities and dried in a 50 °C oven for 48 h. The obtained powder was put into a refrigerator for further analysis.

## 2.2. Characterization of silver nanoparticles

The synthesis of nanoparticles was initially observed by the color change and later confirmed using UV-Vis spectrum. The samples were monitored as a function of reaction time by a UV-Visible spectrophotometer (PG, UK) between 300-700 nm at a resolution of 10 nm. In order to determine the stability and size distribution of the nanoparticles, zeta potential and DLS were measured using a Zetasizer Nano ZS 3600 (Malvern Instrument Ltd, UK). The synthesized silver nanoparticles were then scanned with an AFM (Nano Surf® AG, Switzerland, Product: BTO 2089, BRO). Also, the morphology of nanoparticles was determined using the TEM technique (JEM-2100F, Jeol Ltd., Japan) operating at 200 kV. The crystalline nature of the silver nanoparticles was characterized by the XRD technique using an X-Ray diffractometer (Bruker AXS D8 ADVANCE). A Perkin Elmer infrared spectrophotometer (model FT-PC-160) was used to determine surface chemistry and functional groups, such as hydroxyls and carbonyls, which attach to the surface during the synthesis of the nanoparticles.

## 2.3. Assay of antioxidant activity

The antioxidant activity of the samples was evaluated using the DPPH method [29]. Briefly, the samples in different concentrations ranging from 50 μg/ml to 300 μg/ml were added to tubes containing 1ml of DPPH solution (0.1 mM in methanol). After vigorous shaking, the mixtures were kept in the dark for 30 min (at room temperature) and then measurement of the absorbance was done at 517 nm against a control (L-Ascorbic acid as the control). The radical scavenging capacity was calculated based on Eq. (1).

$$\text{Inhibition } \% = \frac{A_{control} - A_{sample}}{A_{control}} \times 100 \qquad (1)$$

## 2.4. Total phenolic content

The Folin-Ciocalteau method was used to determine the total phenolic content [30]. In brief, 200 μL of the sample (1 mg/ml) was mixed with 200 μL of Folin-Ciocalteu reagent (10 % v/v). Three minutes later 2 ml of sodium carbonate (20% w/v) was added, and the sample was allowed to stand for 2 hours (in the dark and at room temperature). After measuring the absorbance at λ= 650 nm, the total phenolic contents was calculated using gallic acid as a standard.

## 2.5. Statistical analysis

The data were statistically analyzed by Statistical Analysis System (SAS) software and are the mean ±SD of three replications.

## 3. Results and Discussion

The presence research describes silver nanoparticles synthesis using *L. officinale* root aqueous extract. To obtain silver nanoparticles with antioxidant activity, the reaction time and temperature, pH, silver nitrate concentration, and the ratio of extract to silver nitrate were optimized.

## 3.1. Biosynthesis of silver nanoparticles

### 3.1.1. Effect of reaction time

As shown in Figure 1a, when the reaction time was extended beyond 10 minutes, the solution started to turn darker and the intensity of the SPR peak (at 418 nm) increased significantly during 2 hours, this was due to the continuous formation of the nanoparticles with the passage of time. In other words, an increase of the SPR peak intensity indicated an enhancement in the silver nanoparticles synthesis. The SPR peak increased and shifted somewhat to higher wavelengths with reaction time increment until it remained constant after 6 hours. The red-shift in the peak wavelength showed that the particles size increases as the reaction time increases. These results were in agreement with Verma and Mehata's [31] report, which showed controllable synthesis of silver nanoparticles using Neem leaves. Therefore, it could be concluded the optimum time for the synthesis of nanoparticles was around 6 hours.

### 3.1.2. Effect of reaction temperature

To find out the effect of temperature, the silver nanoparticles synthesis was done at 25, 40, and 60 °C. The results shown in Figure 1b reveal that increasing temperature increased the intensity of SPR peaks and the color of solution changed to yellowish brown, which may be due to a faster reaction rate at higher temperature. Also, as the temperature increased a blue-shift appeared from 460 nm to 405 nm because of a reduction in the silver nanoparticles size. The average kinetic energy of silver ions is related to the temperature reaction. As the temperature increases the silver ions would have a more dispersed range of speeds, this causes more collisions between the silver ions and secondary metabolites present in the extract and increases the consumption of silver ions. So, the possibility of particle size growth decreases and uniform size silver nanoparticles form. These results agree well with the findings from Ibrahim [17], Verma & Mehata [31], and Kasture et al. [32] who

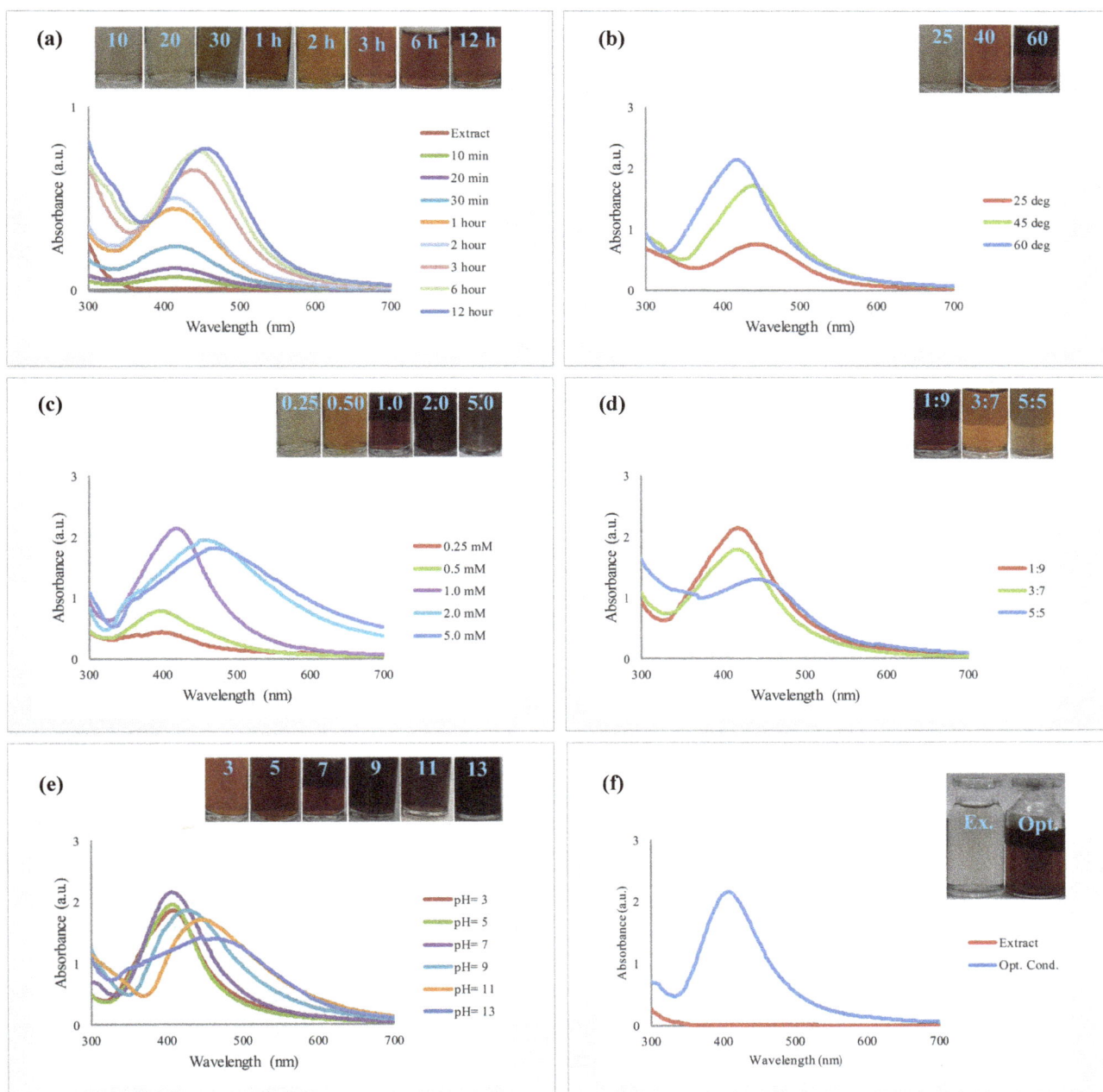

**Fig. 1.** Visual observation and UV-Vis absorption spectra of the silver nanoparticles at (a) different reaction time, (b) different reaction temperature, (c) different silver nitrate concentration, (d) different concentration ratio of the extract and silver nitrate, (e) different pH, and (f) optimum condition including: reaction time, 6 hours; silver nitrate concentration, 1 mM; temperature, 60°C, extract pH, 7; and concentration ratio of the extract and silver nitrate, 1:9 v/v.

all concluded that as the reaction temperature increased the reaction rate also increased, which may result in smaller-sized nanoparticles. Therefore, the optimum temperature for achieving smaller-sized nanoparticles was 60 °C.

### 3.1.3. Effect of silver nitrate concentration

As shown in Figure 1c, when the silver nitrate concentration increased from 0.25 to 1 mM the SPR peak intensity significantly increased, which indicated that the silver nanoparticles synthesis occurred at higher rate. But, further increasing the silver nitrate concentration from 1 to 2 mM and above caused reduction of peak intensity, further broadening of the peak, a red-shift from 405 nm to 490 nm, and settlement of particles. The formation of large-scale micro particles may be due to the presence of large amounts of silver atoms in a small volume of the solution, which causes a high attraction between silver atoms and the agglomeration of nanoparticles. Hence, the rapid settling of particles may be related to their large size [33]. So, 1 mM of silver nitrate was selected to achieve a large amount of silver nanoparticles with low dispersion.

### 3.1.4. Effect of extract and silver nitrate ratio

Optimization of the root extract and silver nitrate ratio was carried out with a change in the volume of root extract and silver nitrate (1 mM). When the volume of the extract to silver salt was in the ratio of 5:5 v/v, a broad peak of less intensity (at 460 nm) was observed (Figure 1d), which might be because of more reduction processes on the surface of the formed nuclei in the presence of higher concentration of reducing agents [34]. A similar result was also observed in the case of gold nanoparticles synthesis using *Cocos nucifera* extract [18]. As the volume of the extract was decreased from 5:5 to 3:7 v/v a narrow peak with a blue shift (from 460 to 407 nm) occurred (Figure 1d), which might be due to a reduction in the particles' size. Also, the absorption intensity increased as the volume of the extract decreases from 3:7 to 1:9 v/v, indicating an enhancement in the silver nanoparticles synthesis. Therefore, the optimum ratio of the extract and silver nitrate for the silver nanoparticles synthesis was evaluated at 1:9 v/v.

### 3.1.5. Effect of pH

Figure 1e shows the effect of pH on the synthesis of silver nanoparticles. The pH of the extract can influence the electrical charges of natural products which might act as capping and stabilizing agents. So, the size and shape of the particles can be changed by the pH of the extract. The results (Figure 1e) show that by increasing the pH values from 3.0 to 7.0, the intensity of the SPR peaks increased and maximum silver nanoparticles production occurred. In addition, the SPR peak became broader and shifted toward the long wavelength region as the pH value increased. Any shift of the SPR peak toward the longer wavelength is accompanied by an increase in the size of the synthesized silver nanoparticles, and the broadening of the SRP peak indicates the presence of a wider range of particle size in the solution. Also, the decrease of the SRP peak intensity showed the decrement in the silver nanoparticles synthesis. This

may happen due to the ionization of natural phenolic compounds in *L. officinale* root extract. So, the results suggested that neutral pH (pH= 7.0) was more suitable to obtain nanoparticles with a small size. Afreen and Vandana [35] synthesized silver nanoparticles at neutral conditions (pH=7) using aqueous extract of Rhizopus stolonier. Similarly, Elemike et al. [36] reported that silver nanoparticles showed maximum stability at neutral pH (6.8-7.0) using an aqueous extract of *Lasienthra africanum*. These findings were in a close agreement with Figure 1f in the present study.

### 3.2. Characterization of silver nanoparticles

The particle size and morphology of the nanoparticles synthesized by *L. officinale* root aqueous extract was determined by TEM (Figure 2a). It is evident by the morphology of the nanoparticles which was almost spherical and the particles diameter was from 2 to 11 nm with an average size of 7 nm. These results agree well with the particle size calculated from XRD analysis. The results obtained from DLS clearly indicated a narrow range of size distribution from 7 to 22 nm with an average mean size of 9 nm (Figure 2b). The size obtained from DLS was usually larger than that from TEM. This may be due to the surface adsorption of natural products on the nanoparticles, the aggregation of nanoparticles, and the surface adsorption of water onto the silver nanoparticles stabilized with *L. officinale* root aqueous extract. All of these reasons may be have a negative effect on the results obtained by the DLS method [37]. Also, the size of silver nanoparticles was confirmed by AFM (Figure 2c) and it was found to be less than 15 nm. The zeta potential of the synthesized silver nanoparticles was measured and showed a sharp peak at -15.66 mV. So, the repulsive forces between negative charges of the nanoparticles caused the stability of the nanoparticles in the suspension.

The XRD analysis was done to determine the crystalline nature of the synthesized nanoparticles by *L. officinale* root aqueous extract. The XRD spectrum (Figure 3) showed four distinct separate peaks at $2\theta=38.08°$, $44.47°$, $64.17°$ and $77.57°$ corresponding to lattice plane values indexed at (111), (200), (220), and (311) planes, respectively, which were in agreement with the database of face centered cubic (FCC) structures from the Joint Committee Powder Diffraction Standards (JCPDS) file No. 04-0783.

(a)

(b)

(c)

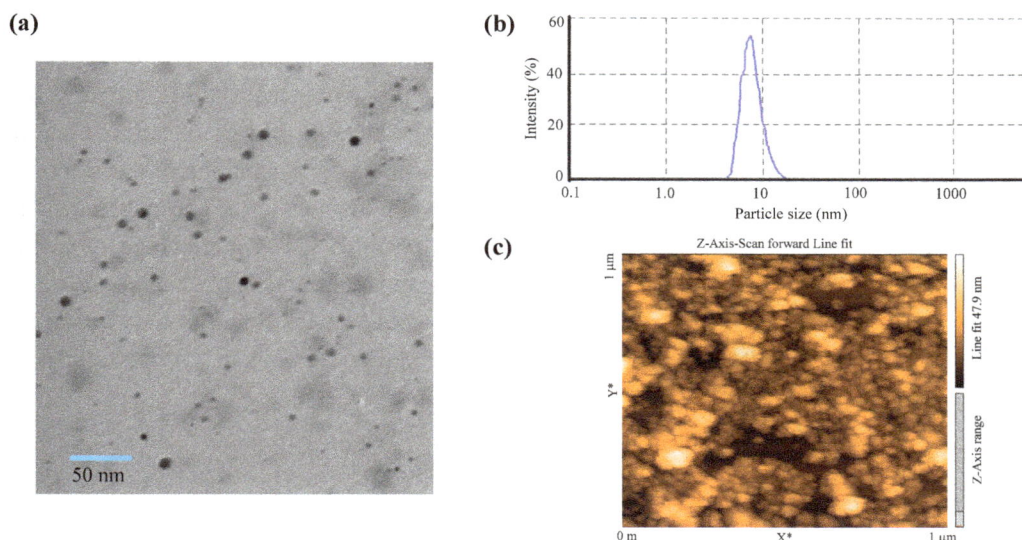

**Fig. 2.** a) TEM micrograph, b) Size distribution, and c) AFM image of the silver nanoparticles synthesized using *L. officinale* root aqueous extract.

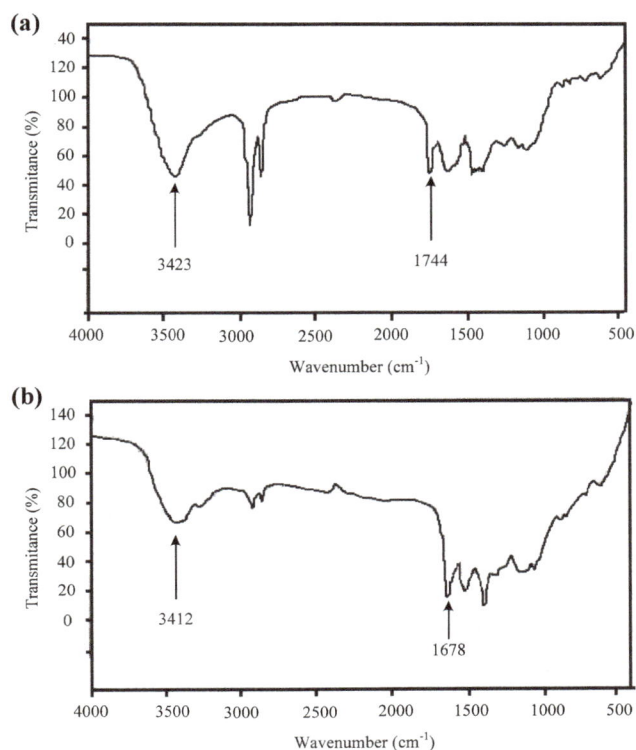

The phytochemical analysis of *Lithospermum* species showed the presence of naphthazalin skeleton metabolites such as shikonin and its derivatives in the outer surface of the roots [38]. The presence of these metabolites may play an important role in silver ion reduction. FT-IR spectroscopy (Figure 4) was used to determine the major functional groups on the *L. officinale* extract which may be responsible for the synthesis and stability of the silver nanoparticles. The –OH and C=O peaks at 3423 cm$^{-1}$ and 1744 cm$^{-1}$, respectively, were due to the presence of shikonin and its derivatives in the extract. Also, the peak at 1625 cm$^{-1}$ was related to the C=C stretching vibration of aromatic rings. So, the FT-IR spectrum of the silver nanoparticles indicated an increase of the C=O peak and a decrease of the –OH peak, which may be related to the oxidation of –OH groups to C=O groups. So, the proposed mechanism

may be due to the presence of shikonin and its derivatives in *L. officinale* extract (Figure 5), which play an important role in the reduction of silver ion to silver nanoparticles. Also, the trace shift in the absorption peaks of the FT-IR spectrum indicated presence of the natural products on the surface of the silver nanoparticles.

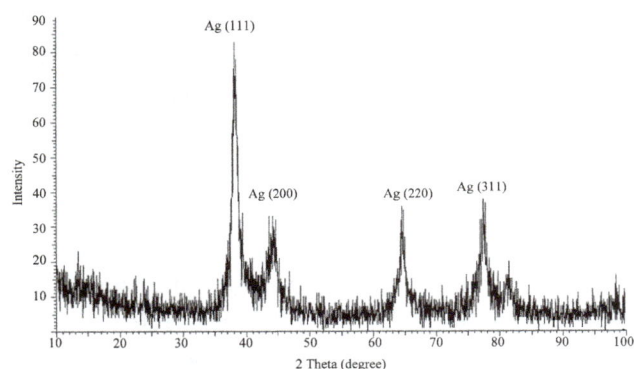

(a)

(b)

**Fig. 4.** FT-IR Spectrum of A) *L. officinale* root aqueous extract and B) biosynthesized silver nanoparticles.

**Fig. 3.** X-ray diffractometer of the silver nanoparticles synthesized using *L. officinale* root aqueous extract.

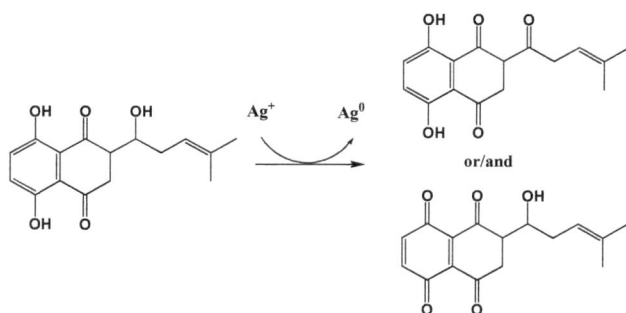

**Fig. 5.** The proposed mechanism for the synthesis of silver nanoparticles, due to the presence of shikonin and its derivatives.

### 3.3. Antioxidant activity of synthesized nanoparticles

The antioxidant potential of the synthesized nanoparticles was assayed using the DPPH method and compared to *L. officinale* root aqueous extract. The results shown in Table 1 indicated that the nanoparticles had more antioxidant activity in comparison with *L. officinale* root extract, and the recorded IC50 values were 79±0.9 and 142±2.1 µg/ml, respectively. The enhanced antioxidant capacity of the synthesized silver nanoparticles may be because of the phenolic compounds present on the surface of the nanoparticles, which could cause antioxidant activities via transferring a hydrogen atom or an electron to the reactive species [39]. Also, the analysis of the total phenolic content (TPC) showed (Table 1) that the TPC of the synthesized silver nanoparticles was higher (76.83±0.15 mg/g) compared to the extract (57.9±0.23 mg/g), which had a positive correlation with the antioxidant capacity. Phull et al. [9] evaluated antioxidant activity of the synthesized silver nanoparticles from crude extract of *Bergenia ciliate* and suggested that the synthesized nanoparticles can be used as an antioxidant. Abdel-Aziz et al. [8] reported the presence of strong antioxidant activity of nanoparticles-containing leaf extract compared to *Chenopodium murale* leaf extract. So, the antioxidant potential of the synthesized nanoparticles could be attributed to phenolic compounds which adhered to them and which originated from the extract, and these nanoparticles could be applied as natural antioxidants to cure degenerative diseases which are related to oxidative stress.

### 4. Conclusion

The synthesis of the silver nanoparticles was done on the base of a green chemistry method, in which *L. officinale* root aqueous extract was used as a reducing and stabilizing agent. In this process, one ml of the extract reduced 9 ml of silver ions (1 mM) into silver nanoparticles during heating of the reaction mixture (60 °C) for 6 hours at pH 7.0. The synthesized nanoparticles had a uniform and spherical shape with an average particles size around 7 nm, which indicated significant antioxidant activity. So, this study provided a better and faster method for silver nanoparticles synthesis with antioxidant activity.

### References

[1] F.U. Khan, Y. Chen, Z.U.H.K. Khan, A. Ahmad, K. Tahir, L. Wang, M.R. Khan, P. Wan, Antioxidant and catalytic applications of silver nanoparticles using *Dimocarpus longan* seed extract as a reducing and stabilizing agent, J. Photoch. Photobio. B. 164 (2016) 344-351.

[2] C.M. Cobley, S.E. Skrabalak, D.J. Campbell, Y. Xia, Shape-controlled synthesis of silver nanoparticles for plasmonic and sensing applications, Plasmonics, 4 (2009) 171-179.

[3] S. Jeong, H. Choi, J.Y. Kim, T. Lee, Silver-based nanoparticles for surface plasmon resonance in organic optoelectronics, Part. Part. Sys. Charact. 32 (2015) 164-175.

[4] K.S. Lee, M.A. El-Sayed, Gold and silver nanoparticles in sensing and imaging: sensitivity of plasmon response to size, shape, and metal composition, J. Phys. Chem. B. 110 (2006) 19220-19225.

[5] V. Thamilselvi, and K. V. Radha, A review on the diverse application of silver nanoparticle, IOSR J. Pharm. 7 (2017) 21-27.

[6] M.Y. Babu, V.J. Devi, C.M. Ramakritinan, R. Umarani, N. Taredahalli, A.K. Kumaraguru, Application of biosynthesized silver nanoparticles in agricultural and marine pest control, Curr. Nanosci. 10 (2014) 1-8.

[7] S.M. Ali, N.M.H. Yousef, N.A. Nafady, Application of biosynthesized silver nanoparticles for the control of land snail *Eobania vermiculata* and some plant pathogenic fungi, J. Nanomater. 2015 (2015) Article ID 218904, doi:10.1155/2015/218904.

[8] M.S. Abdel-Aziz, M.S. Shahee, A.A. El-Nekeety, M.A. Abdel-Wahhab, Antioxidant and antibacterial

activity of silver nanoparticles biosynthesized using *Chenopodium murale* leaf extract, J. Saudi Chem. Soc. 18 (2014) 356-363.

[9] A. Phull, Q. Abbas, A. Ali, H. Raza, S. Jakim, M. Zia, I. Haq, Antioxidant, cytotoxic and antimicrobial activities of green synthesized silver nanoparticles from crude extract of *Bergenia ciliate*, Fut. J. Pharm. Sci. 2 (2016) 31-36.

[10] E.E. Elemike, O.E. Fayemi, A.C. Ekennia, D.C. Onwudiwe, E.E. Ebenso, Silver nanoparticles mediated by *Costus afer* leaf extract: synthesis, antibacterial, antioxidant and electrochemical properties, Molecules, 22 (2017) 701.

[11] T. Togashi, K. Saito, Y. Matsuda, I. Sato, H. Kon, K. Uruma, M. Ishizaki, K. Kanaizuka, M. Sakamoto, N. Ohya, M. Kurihara, Synthesis of water-dispersible silver nanoparticles by thermal decomposition of water-soluble silver oxalate precursors, J. Nanosci. Nanotechno. 14 (2014) 6022-6027.

[12] R.A. Khaydarov, R.R. Khaydarov, O. Gapurova, Y. Estrin, T. Scheper, Electrochemical method for the synthesis of silver nanoparticles, J. Nanopart. Res. 11 (2009) 1193-1200.

[13] S. Machmudah, T. Takayuki Sato, Wahyudiono, M. Sasaki, M. Goto, Silver nanoparticles generated by pulsed laser ablation in supercritical $CO_2$ medium, High Pressure Res. 32 (2012) 1-7.

[14] A. Pal, S. Shah, Microwave-assisted synthesis of silver nanoparticles using ethanol as a reducing agent, Mater. Chem. Phys. 114 (2009) 530-532.

[15] H. Wang, X. Qiao, J. Chen, S. Ding, Preparation of silver nanoparticles by chemical reduction method, Colloid. Surface. A. 256 (2016) 111-115.

[16] K. Gudikandula, S.C. Maringanti, Synthesis of silver nanoparticles by chemical and biological methods and their antimicrobial properties, J. Exp. Nanosci. 11 (2016) 714-721.

[17] H.M.M. Ibrahim, Green synthesis and characterization of silver nanoparticles using banana peel extract and their antimicrobial activity against representative microorganisms, J. Radiat. Res. Appl. Sci. 8 (2015) 265-275.

[18] H. Parab, N. Shenoy, S.A. Kumar, S.D. Kumar, A.V.R. Reddy, One pot spontaneous green synthesis of gold nanoparticles using *cocos nucifera* (coconut palm) coir extract, J. Mater. Envir. Sci. 7 (2016) 2468-2481.

[19] S.A. Aromal, D. Philip, Green synthesis of gold nanoparticles using *Trigonellafoenum-graecum* and its size dependent catalytic activity, Spectrochim. Acta Part A. 97 (2012) 1-7.

[20] S. Ahmed, M. Ahmad, B.L. Swami, S. Ikram, A review on plants extract mediated synthesis of silver nanoparticles for antimicrobial applications: A green expertise, J. Adv. Res., 7 (2016) 17-28.

[21] M. Ndikau, N.M. Noah, D.M. Andala, E. Masika, Green synthesis and characterization of silver nanoparticles using *Citrullus lanatus* fruit rind extract, Int. J. Anal. Chem. 2017 (2017) Article ID 8108504, doi:10.1155/2017/8108504.

[22] N. Amini, G. Amin, Z. Jafari Azar, Green synthesis of silver nanoparticles using *Avena sativa L.* extract, Nanomed. Res. J. 2 (2017) 57-63.

[23] B. Habibi, H. Hadilou, S. Mollaei, A. Yazdinezhad, Green synthesis of Silver nanoparticles using the aqueous extract of *Prangos ferulaceae* leaves, Inter. J. Nano Dim. 8 (2017) 132-141.

[24] J. Saha, A. Begum, A. Mukherjee, S. Kumar, A novel green synthesis of silver nanoparticles and their catalytic action in reduction of Methylene Blue dye, Sust. Envir. Res. 27 (2017) 245-250.

[25] I. Fatimah, Green synthesis of silver nanoparticles using extract of *Parkia speciosa Hassk* pods assisted by microwave irradiation, J. Adv. Res. 7 (2016) 961-969.

[26] S. Ahmed, M. Ahmad, B. L. Swami, S. Kram, Green synthesis of silver nanoparticles using *Azadirachta indica* aqueous leaf extract, J. Radiat. Res. Appl. Sci. 9 (2016) 1-7.

[27] K. Jemal, B.V. Sandeep, S. Pola, Synthesis, characterization, and evaluation of the antibacterial activity of *Allophylus serratus* leaf and leaf derived callus extracts mediated silver nanoparticles, J. Nanomater. 2017 (2017) Article ID 4213275, 11 pages, doi:10.1155/2017/4213275.

[28] D. Sarkar, G. Paul, Green synthesis of silver nanoparticles using *Mentha asiatica* (Mint) extract and evaluation of their antimicrobial potential, Int. J. Curr. Res. Biosci. Plant Biol. 4 (2017) 77-82.

[29] K. Schlesier, M. Harwat, V. Boèhm, R. Bitsch, Assessment of antioxidant activity by using different in vitro methods, Free Rad. Res. 36 (2002) 177-187.

[30] S.A. Baba, S.A. Malik, Determination of total phenolic and flavonoid content, antimicrobial and antioxidant activity of a root extract of *Arisaema jacquemontii* Blume, J. Taibah Uni. Sci. 9 (2015)

449-454.

[31] A. Verma, M.S. Mehata, Controllable synthesis of silver nanoparticles using Neem leaves and their antimicrobial activity, J. Radiat. Res. Appl. Sci. 9 (2016) 109-115.

[32] M.B. Kasture, P. Patel, A.A. Prabhune, C.V. Ramana, A.A. Kulkarni, B.L.V. Prasad, Synthesis of silver nanoparticles by sophoro lipids: Effect of temperature and sophorolipid structure on the size of particles, J. Chem. Sci. 120 (2008) 515-520.

[33] B.S. Maria, A. Devadiga, V.S. Kodialbail, M.B. Saidutta, Synthesis of silver nanoparticles using medicinal *Zizyphus xylopyrus* bark extract, Appl. Nanosci. 5 (2015) 755-762.

[34] A. Gangula, R. Podila, M. Ramakrishna, L. Karanam, C. Janardhana, A.M. Rao, Catalytic reduction of 4-nitrophenol using biogenic gold and silver nanoparticles derived from *Breynia rhamnoides*, Langmuir, 27 (2011) 15268-15274.

[35] B. Afreen, A. Vandana, Synthesis and characterization of silver nanoparticles by *Rhizopus stolonier*, Int. J. Biomed. Adv. Res. 2 (2011) 148-158.

[36] E.E. Elemike, D.C. Onwudiwe, O. Arijeh, H.U. Nwankwo, Plant-mediated biosynthesis of silver nanoparticles by leaf extracts of *Lasienthra africanum* and a study of the influence of kinetic parameters, B. Mater. Sci. 40 (2017) 129-137.

[37] S. Das, P. Roy, S. Mondal, T. Bera, A. Mukherjee, One pot synthesis of gold nanoparticles and application in chemotherapy of wild and resistant type *visceral leishmaniasis*, Colloid. Surface. B. 107 (2013) 27-34.

[38] K. Tatsumi, M. Yano, K. Kaminade, A. Sugiyama, M. Sato, K. Toyooka, T. Aoyama, F. Sato, K. Yazaki, Characterization of shikonin derivative secretion in *Lithospermum erythrorhizon* hairy roots as a model of lipid-soluble metabolite secretion from plants, Front. Plant Sci. 7 (2016) 1066, doi: 10.3389/fpls.2016.01066.

[39] V. Goodarzi, H. Zamani, L. Bajuli, A. Moradshahi, Evaluation of antioxidant potential and reduction capacity of some plant extracts in silver nanoparticles' synthesis, Mol. Biol. Res. Commun. 3 (2014) 165-174.

# 15

# Maltodextrine nanoparticles loaded with polyphenolic extract from apple industrial waste: Preparation, optimization and characterization

Shohreh Saffarzadeh-Matin*, Majid Shahbazi

*Department of Chemical Technologies, Iranian Research Organization for Science and Technology (IROST), Tehran, Iran*

## HIGHLIGHTS

- Maltodextrine with low hydrolytic conversion (DE: 8) was utilized as the wall material.

- RSM was utilized for the modeling and optimization of three independent variables.

- The surfactant concentration and the mixture intermixing were significant factors.

- Spherical shaped NPs with a size of 52 nm with a loading efficiency of 98% were produced.

- Hydrogen bonding is the main mode of interaction between the core and the shell.

## GRAPHICAL ABSTRACT

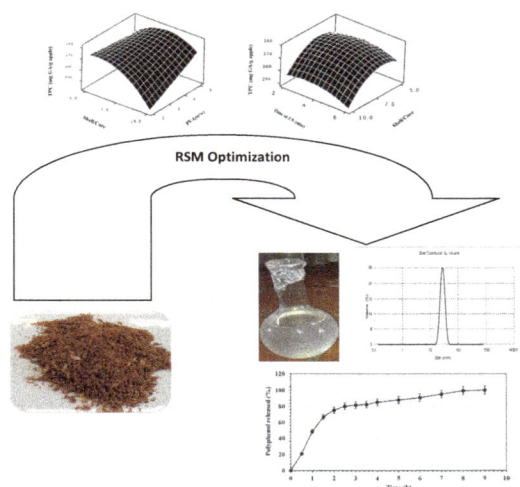

## ARTICLE INFO

*Keywords:*

Process optimization
Natural polymers
Nanoencapsulation
Apple pomace polyphenolic extract
Nanoprecipitation

## ABSTRACT

The main aim of this study was to prepare apple pomace polyphenolic extract (APPE-referred to as a core) loaded into biodegradable and commercially available natural polymer such as maltodextrin (MD-referred to as a shell). The polymer coating potentially improves its low stability and bioavailability and also directs the control release of the encapsulated material. The MD-nanoparticles (NPs) loaded with the APPE were prepared by a modified nanoprecipitation method. An experimental central composite design was utilized for the modeling, optimization and to assess the influence (and interactions) of the shell to core ratio, surfactant concentration, and sonication time (as the independent variables) on the NPs preparation to maximize the level of polyphenols loading and the NPs formation yield (referred to as dependant variables). The adopted models were verified statistically and experimentally. The results showed that amongst the independent variables, the shell to core ratio and the surfactant concentration were statistically significant in the experimentally selected ranges. By adopting the optimal process conditions, the spherical shaped NPs were prepared with a mean average size of 52 nm (confirmed by the Dynamic Light Scattering and FE-SEM techniques) and polyphenols loading efficiency of 98%. FT-IR spectroscopy confirmed the successful entrapment of the core in the shell of NPs. Hydrogen bonding is one of the modes of interactions between the hydrophilic moieties of polyphenols and MD. The in vitro polyphenols release of the NPs through simulating cancerous tumor acidity conditions represented a sustainable release, indicating potential anticancer application of the NPs.

* Corresponding author: ;  E-mail address: saffarzadeh@irost.ir

## 1. Introduction

Polyphenols with natural origins e.g., apple pomace polyphenols could be potentially used as food supplements, pharma- and cosmeceuticals due to their antimicrobial, antiradical and antioxidant activities [1,2]. The significance of these natural antioxidants in preventative medicine, particularly for reducing cancer risk, is well known and has been the subject of recent reviews [3]. The high activity of natural polyphenols makes them prone to chemical instability against environmental conditions such as oxygen, moisture, etc. Moreover, most natural polyphenols show low in vivo bioavailability and solubility, which may weaken or even suppress their full beneficial health effects [4]. In this regard, nanoencapsulation of polyphenols, i.e., their incorporation into biodegradable natural polymers, not only enhances their chemical stability but also provides passive and active targeting and controlled release [5]. Additionally, their low in vivo bioavailability and solubility will be improved considerably [6], which affects their absorption by the gastrointestinal system [7]. For example, the recent substantial research conducted on the nanoencapsulation of curcumin [8-10] and quercetin [11-13] as model polyphenols are indicative of the importance of this new emerging technology in preventative medicine. By adopting a biomimicry approach it is believed that polyphenolic compounds, e.g., green tea polyphenols, act synergistically together as their nature's pattern. Therefore their mixed extraction, purification, and bioavailability have been the subject of intensive research over recent years [14,15].

The main objective of this study was to identify the most suitable operating conditions for the green production of natural polymer based-nanoparticles loaded with polyphenol extract produced from Iranian industrial apple pomace.

Apple is one of the most frequently consumed fruit in the Iranian diet. According to the latest FAOSTAT report (2012), the total worldwide production of apple is approximately 60-70 million metric tons (MMT)/ annually, and the Iranian contribution is about 5% (above 3 MMT) of this production, making it the fourth largest world producer. In Iran apples are mostly consumed fresh and roughly 0.57 MMT are processed into juice, jams, and syrup [16]. Therefore, during the processing of apple products a considerable amount of industrial apple waste is produced annually in Iran. This waste is mostly used as animal feed. The recovery of value added natural by-products from industrial wastes e. g., apple pomace, is a global research trend to reduce environment pollution in addition to increase profitability [3,17]. According to a study conducted on the polyphenol profile of the mixture of flesh and peel of the main Iranian apple cultivars [18], the main constituents were (-)-epicatechin, cyanidin-3-galactoside, quercetin-3-galactoside followed by chlorogenic acid and phloridzin dehydrate (Figure 1).

In the encapsulation process varieties of natural and/or synthetic polymers could be used as a matrix and/or coating to retard the diffusion of oxygen and/ or small organic molecules, which further leads to oxidative stability [19]. Polysaccharides such as maltodextrin, chitosan and cellulose are examples of such wall compounds [20], e.g., chitosan was used as wall material for encapsulation of a natural antioxidant extracted from yerba mate [21].

However, use of the MD (high DE: 18-20) is restricted as an appropriate shell material due to its low glass transition temperature (Tg). Low Tg may prompt crystals formation under increasing temperature, for example during the storage period and also the applications [22]. Therefore, the structural integrity of the particles will be disrupted and premature release of the encapsulated active components will occur [23]. In this study, MD with low hydrolytic conversion (DE: 8) was utilized as the wall material. We propose that a lower DE value increases the MD polymer solubility in organic polar solvents and reduces its water solubility. These will presumably result in a more rapid precipitation, greater encapsulation efficiency, and finally potential higher stabilization during the storage period.

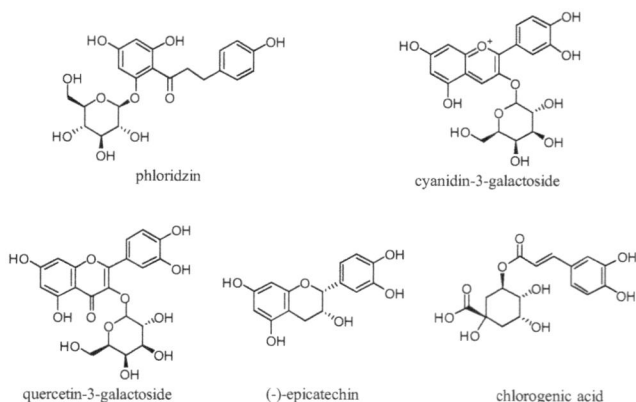

**Fig. 1.** Chemical structure of main polyphenolic constituents of industrial apple pomace.

In this study the nanoprecipitation method was utilized for the fabrication of the novel MD- APPE composite NPs. Nanoprecipitation is a straightforward process which generates a narrow unimodal size distribution of NPs [24] and is mostly suitable for hydrophobic compounds [25]. However, the amphiphilic character of various mixed polyphenolic compounds (due to the presence of both aromatic rings and hydroxyl groups) make the suitability of the nanoprecipitation method questionable as the technique of choice for the encapsulation. Therefore, high molecular weight polyvinyl alcohol (dissolved into water) was used as surfactant to reduce (or even zeroing) the polyphenolics leakage towards the aqueous phase.

Low power sonication homogenization was used to improve the solvent-nonsolvent intermixing due to its increasing the aqueous phase viscosity. The operating conditions were expressed in terms of the wall to the core ratio, the surfactant concentration (w/w %) and the sonication homogenization time (as independent variables) to maximize the entrapment of the apple pomace polyphenol extract and the NPs formation yield (as response variables).

In this study a 33-full factorial design and RSM were employed to investigate the interactions as well as the optimization of the variables. RSM has been recently adopted and used in the optimization of encapsulation process of various natural polyphenols [26,27].

## 2. Experimental

### 2.1. Materials

Maltodextrin (DE: 8, Mw: 15,000 g mol-1) was supplied by Arian glucose CO (Iran). Ethanol (96%) was obtained from the Zakaria Jahrom Institute (Iran). Folin Ciocalteu's phenol reagent, Gallic acid, and polyvinyl alcohol (Mw: 72000 g mol-1) were obtained from Merck Chemical Co (Germany). All other reagents were of analytical grade. Deionized water was used in all experiments.

Apple pomace including the peel and seeds was provided immediately after processing in October, 2015 by the Sanich Co. (Iran). The waste was transferred within a few hours to a laboratory, dried in an air flow cabinet oven at 30 °C for 48 h, and powdered in a hammer mill crusher. The ground material was consecutively passed through the sieves of various mesh sizes and the

fraction between 35 and 60 mesh sizes (a mean particle size of 0.25-0.50 mm) was collected for further processing. The sized material was placed in an opaque plastic bag and stored at room temperature (20-25 °C) in a dry ventilated area until used.

### 2.2. Methods

#### 2.2.1. Determination of total phenolic compounds (TPC)

The method used to determine the TPC employed Folin-Ciocalteu (FC) reagent; this was done according to a procedure described in the literature [28]. Six different concentrations of Gallic acid solutions (25–500 mg/l) were used for the calibration plot. The estimation of phenolic compounds was carried out in triplicate, and the results were averaged (standard deviations <5%). The results were expressed in Gallic Acid (GA) equivalents (mg GAE/g waste). The measurements were carried out by ultraviolet absorption at their maximum wavelength on a MESU LAB UV/V-3000 spectrophotometer, Hong Kong.

#### 2.2.2. Extraction of phenolic compounds

40 g of the powdered apple waste was placed into individual cotton bags in the five cells of a multistage counter current extraction system. For extraction ethanol concentration of 45%, solvent to dry waste ratio of 10, extraction temperature of 65 °C and extracting time of 6 h were selected. Multi stage counter current extraction was performed according to a procedure previously reported [29]; then the extracts produced from each stage were mixed and filtered. The filtrate was further concentrated via evaporation under reduced pressure and further subjected to spray drying [30]. 10 mg of the dry extract was dissolved in 25 ml of hydroethanol of 50% and was analyzed for TPC, which was 288 mg/l GA.

#### 2.2.3. Preparation of the NPs containing apple pomace polyphenols

Accurately weighed amounts of the MD (selected from experimental arrangements designed using RSM (Table 1) and dried APPE (20 mg) were magnetically stirred (Model EM3300T, Labotech Inc., Germany) overnight in ethanol 96% (8 ml) at room temperature.

The colloidal solution was then added to 15 ml of water containing accurately weighed amounts of polyvinyl alcohol as surfactant (selected from Table 1, magnetically stirred and dissolved at 70 °C temperature). Upon dilution with water, a dispersion of NPs was generated instantaneously. The latter was immediately sonicated with 150 W power by using a prob sonicator for the defined times, according to the experimental arrangements designed using RSM (Table 1).

The produced NPs were further collected by ultracentrifugation 3 times (Model 29318, Sigma, Laborzentrifugen GmbH, Germany) at 13500 rpm for 45 min, the supernatants were mixed and analyzed for the TPC. Therefore, the difference between the total amount used to prepare the nanopaticles and the amount that was found in the supernatant was attributed to the amount of polyphenols loaded within the NPs. The collected particles were washed twice with 5 ml of hydro-ethanol 50% (v/v) and subjected to drying using a vacuum oven (Model 3737, Precision Scientific, INC,) at 40 °C and weighed. The filtrates were analyzed for TPC. There was no detectable polyphenols in the washing filtrates as evidenced by TPC analysis, suggesting that polyphenols did not exist on the NPs surfaces. The vacuum dried particles were stored at 4 °C until used for other characterizations.

## 2.3. Ultrasonic generator

The equipment employed in this search was a 20 kHz, 600 W ultrasonic generator, MISONIX Ultrasonic liquid Processor, Model S-4000 (Qsonica, LLC, USA) and a titanium microtip 419 BR as the probe.

## 2.4. Loading efficiency, polyphenol loading content and NPs' yield

The percentages of loading efficiency (% L.E),

polyphenol loading content (% LC) and the NPs yield (% N.Y) were calculated according to equations Eqs. (1), (2) and (3), respectively. The N.Ys' were obtained gravimetrically.

$$L.E = (\text{TPC of the NPs/TPC of the feeding sample}) \times 100 \tag{1}$$

$$L.C = (\text{weight of polyphenol in the NPs/the weight of NPs}) \times 100 \tag{2}$$

$$N.Y = (\text{the weight of NPs / the weight of feeding polymer and sample}) \times 100 \tag{3}$$

The polyphenols loaded in nanospheres expressed as mg GAE/L were also quantified by the Folin-Ciocalteu method after dissolving 50 mg of NPs in 20 ml acetone.

## 2.5. Experimental design

The Minitab Version 16 software was used to conduct the experimental design and the statistical analysis. A three factor and three levels second order regression for the central composite design (CCD) in RSM that consisted of 20 experimental runs was employed. This was not only used to evaluate the interaction of targeted variables, but also to optimize the TPC and the NPs yield (% N.Y) as response variables. The coded values, levels, and real values are listed in Table 1. Regression analysis was performed to establish an empirical second-order polynomial model through Eq. (4).

$$Y = \beta_0 + \sum_{i=1}^{3} \beta_i X_i + \sum_{i=1}^{3} \beta_{ii} X_i^2 + \sum_{i=1}^{3} \sum_{j=i+1}^{3} \beta_{ij} X_i X_j \tag{4}$$

where $Y$ is the predicted response variable, $\beta_0$ is defined as the constant, $\beta_i$ is the linear coefficient, $\beta_{ii}$ is the square coefficient, and $\beta_{ij}$ is the cross-product coefficient. $X_i$ and $X_j$ are two independent variables.

## 2.6. Characterization of the MD- APPE NPs: Size distribution and morphology of the nancapsules

The particle size distribution of NPs was determined by dynamic light scattering (DLS) measurement using Zetasizer nano ZS instrument (Malvern Instruments Ltd, United Kingdom). The appropriate concentration

**Table 2.** Independent variables and coded values employed for optimization of the nanoparticle preparation procedure.

| Independent variables | Factor units | level | | |
|---|---|---|---|---|
| | | -1 | 0 | 1 |
| Shell to core | $X_1$ | 5 | 7.5 | 10 |
| Surfactant (w/w %) | $X_2$ | 2 | 3.5 | 5 |
| Time of ultrasonic homogenizaton (min) | $X_3$ | 2 | 4 | 6 |

of the sample was prepared with the deionized water, and then was filtered with a 0.45 mm Millipore filter, before analysis.

The shape and the surface morphology of NPs were observed by TESCAN WEGA3-SB scanning electron microscopy (FE-SEM).

## 2.7. FTIR spectrum analysis

Each powdered sample was mixed with KBr salt, using a mortar and pestle, and compressed into a thin pellet. Fourier transform infrared (FTIR) spectroscopy experiments were performed on a Perkin-Elmer (model: Spectrum-1) over a frequency window from 4000-400 cm$^{-1}$. Interferograms (64) with a resolution of 4 cm$^{-1}$ were co-added and Fourier transformed for each sample and background.

## 2.8. In vitro polyphenols release of polyphenol loaded NPs

In vitro release studies of polyphenols from the NPs were performed by diffusion technique. Nanospheres (50 mg) inside a cellulose dialysis bag (10 cm, dialysis tubing, molecular weight (Mw) cut off 12,000 Da, Sigma-Aldrich) were suspended inside a beaker containing 100 ml of a release medium consisting of buffer (pH 4.5 and 6.8) at 37 °C and 125 rpm (Brunswick INNOVA 4430 incubator shaker, GMI.inc, USA). The NPs suspensions were aliquoted in 1.5 ml centrifuge tubes in time intervals of half an hour for the first 4 h, an hour for the next 6 h and finally two hours for the third 8 h, centrifuged at 15,000 rpm for 10 min, then decanted and the supernatant was utilized to quantify the in vitro release profile of polyphenols in slight acidic environment models , i.e. pH 4.5 (Acetate Buffer Saline) and pH 6.8 (Phosphate Buffer Saline) using the Folin-Ciocalteu method as described previously. All experiments were carried out in triplicate.

# 3. Results and discussion

## 3.1. Fitting the response surface models and models verification for the responses

Independent variables and coded values employed for optimization of the nanoparticle preparation procedure are presented in Table 1. The design arrangement and

experimental results of the nanoparticle formation are shown in Table 2. The multiple regression coefficients were calculated using Minitab 16 software for both responses. By applying the coefficients into the generalized model (Eq. (4)), the second order polynomial equations for the acquired TPC (Eq. (5)) and the NPs yield (Eq. (6)) were obtained in terms of coded values of shell to core ratio ($X_1$), surfactant weight percentage ($X_2$), and time of the sonication homogenization ($X_3$).

$$Y_1 = 179.83 + 19.67X_1 + 2.58X_2 + 14.02X_3 - 1.93X_{12} - 0.98X_{22} - 1.53X_{32} \tag{5}$$

$$Y_2 = 49.40 + 4.79X_1 - 0.83X_2 + 4.28X_3 - 0.48X_{12} - 0.06X_{22} - 0.36X_{32} - 0.52X_1X_2 - 0.04X_1X_3 - 0.41X_2X_3 \tag{6}$$

**Table 2.** The arrangement and responses of the three-factor, three-level second-order regression for central composite design.

| Run order | Shell to core ($X_1$) | Surfactant (% w/w) | Time of sonication (min) | TPC (mg GAE/l) | N.E. (%) |
|---|---|---|---|---|---|
| 1 | 7.5 (0) | 3.5 (0) | 4 (0) | 280.56 | 73.62 |
| 2 | 7.5 (0) | 3.5 (0) | 6 (1) | 270.47 | 70.98 |
| 3 | 10 (1) | 3.5 (0) | 4 (0) | 265.22 | 69.60 |
| 4 | 5 (-1) | 2 (-1) | 6 (1) | 267.67 | 70.76 |
| 5 | 7.5 (0) | 3.5 (0) | 4 (0) | 272.44 | 71.50 |
| 6 | 7.5 (0) | 3.5 (0) | 2 (-1) | 275.11 | 72.19 |
| 7 | 10 (1) | 5 (1) | 6 (1) | 256.89 | 67.41 |
| 8 | 10 (1) | 2 (-1) | 6 (1) | 238.67 | 62.63 |
| 9 | 10 (1) | 2 (-1) | 2 (-1) | 234.67 | 61.58 |
| 10 | 7.5 (0) | 3.5 (0) | 4 (0) | 280.113 | 71.64 |
| 11 | 7.5 (0) | 3.5 (0) | 4 (0) | 278.11 | 72.98 |
| 12 | 7.5 (0) | 2 (-1) | 4 (0) | 276.56 | 73.10 |
| 13 | 5 (-1) | 3.5 (0) | 4 (0) | 268.56 | 70.47 |
| 14 | 10 (1) | 5 (1) | 2 (-1) | 265.78 | 72.37 |
| 15 | 7.5 (0) | 3.5 (0) | 4 (0) | 279.56 | 66.54 |
| 16 | 5 (-1) | 2 (-1) | 2 (-1) | 266.33 | 69.89 |
| 17 | 5 (-1) | 5 (1) | 6 (1) | 261.89 | 68.72 |
| 18 | 5 (-1) | 5 (1) | 2 (-1) | 273.56 | 71.78 |
| 19 | 7.5 (0) | 5 (1) | 4 (0) | 276.89 | 72.66 |
| 20 | 7.5 (0) | 3.5 (0) | 4 (0) | 279.67 | 73.39 |

The estimation of phenolic compounds was carried out in triplicate, and the results were averaged (standard deviations <5%).

The analysis of variance (ANOVA) result (summary tables for TPC and N.Y %) used to check the adequacy of the developed models is presented in Table 2. From the ANOVA table, it can be seen that the models were significant and adequate at a 95% confidence level. The similarity between the experimental values and the predicted ones (using the models) for both response variables were another indication of satisfactory models. According to the summary table of ANOVA (Table 3), the calculated F-value of the lack of fit for both TPC (3.31) and N.Y % (3.33) did not exceed the tabulated values of the F-distribution (3.5) found from the standard table at the 95% confidence level. This implies that for both models, the lack of fit is not significant (p>0.05) relative to the pure errors; therefore, both models were statistically significant and the responses were optimized.

On the other hand, the F-values from the regression models (16.11 for TPC and 12.22 for N.Y %), which are the calculated values using the adjusted mean square of regression models divided by the adjusted mean square of the residual errors, are more than 1. This indicated that there were significant differences between the models data and the mean values, which were not due to sampling or experimental error (rejecting the null hypothesis). So the experimental values of TPC ($Y_1$) and N.Y % ($Y_2$), as dependent variables, were fitted to the second-order polynomial equations using RSM as shown in Eqs. (3) and (4). It is worth mentioning that the coefficients of the determinations ($R^2$) of TPC and N.Y % were 0.935 and 0.917, respectively, which were another indication that both models adequately fit the chosen parameters.

The absolute values indicated that the shell to core ($X_1$) followed by time of the sonication homogenization ($X_3$) and the weight percentage of surfactant ($X_2$) are the significant factors affecting both the acquired TPC loaded within the MD and nanoparticle formation yield.

### 3.2. Analysis of response surfaces for the responses

The three-dimensional response surfaces of four independent variables were obtained by keeping two of the variables constant. The constants were equal to the natural value of the zero level. The response surfaces for TPC and N.Y % are shown in Figures 2(a-c) and (a'-c'), respectively.

The most important results are outlined as:

1-The combined effect of the surfactant concentration and the shell/core ratio variables on the acquired TPC loaded within the NPs, (Figure 2a), indicates in the range of the surfactant concentration as low as 2-3.5% w/w and the core/shell ratio from 5-8 that the polyphenol loading value remains relatively constant and high (275-280 mg GAE/l), and then starts to decrease as the core/shell ratio increases up to 10. The highest level of loaded polyphenols (maximum TPC value) was obtained in the shell/core ratio ranging of 6-7.5 and the surfactant concentration more than 4% w/w.

Under controlled conditions, after the addition of the organic solvent to the nonsolvent, a dispersion of NPs is generated instantaneously by spontaneous diffusion of the solvent in the aqueous phase. Moreover, as the polymer concentration in the organic phase increases, a high viscosity of the polymeric solution prevents the appropriate organic phase diffusion towards the aqueous phase, so the coprecipitation of the maltodextrin/ polyphenol solution into water was restricted and the level of TPC loading was decreased. This was consistent with the results gained by other studies [31]. This result also shows that in this critical point, due to the presence of the surfactant in the aqueous phase, the interfacial

**Table 3.** Summary table of analysis of variance for TPC and NY %.

| Source | DF | TPC | | | | N.Y % | | | |
| | | Adj. SS | Adj. MS | F | P | Adj. SS | Adj. MS | F | P |
|---|---|---|---|---|---|---|---|---|---|
| Regression | 9 | 2881.55 | 320.173 | 16.11 | 0.00 | 188.63 | 20.96 | 12.22 | 0.00 |
| Residual Error | 10 | 198.72 | 19.87 | | | 17.15 | 1.71 | | |
| Total | 19 | 3080.27 | | | | 205.79 | | | |
| Lack of fit | 5 | 152.65 | 30.53 | 3.31 | 0.11 | 13.187 | 2.64 | 3.33 | 0.11 |
| R-Sq | | 93.55 | | | | 91.67 | | | |
| R-Sq (adj) | | 87.74 | | | | 84.17 | | | |

**(a)**

**(a')**

**(b)**

**(b')**

**(c)**

**(c')**

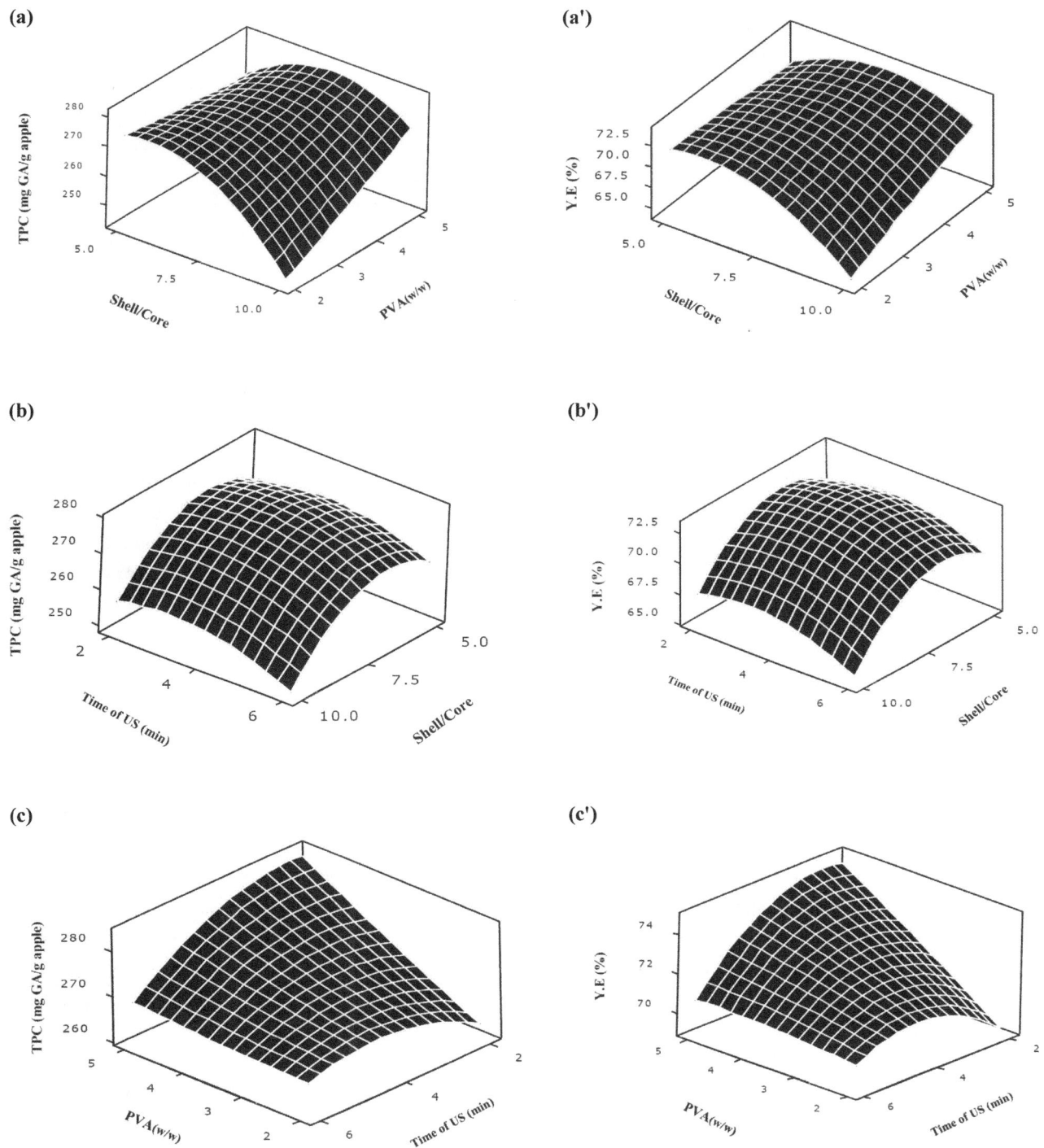

**Fig. 2.** Response surface plots of the combined effect of shell-to-core ratio and surfactant (PVA) concentration (a), shell-to-core ratio and time of sonication (b), PVA concentration and time of sonication (c) on the TPC of loaded NPs, shell-to-core ratio and PVA concentration (a'), shell-to-core ratio and time of ultrasonication (b'), PVA concentration and time of sonication (c') on the yield of NPs preparation.

tension between two liquids is appropriately lowered and the organic phase diffusion towards the aqueous phase is improved, leading to the highest level of TPC loading. However, the high viscosity of the aqueous phase would also hamper the diffusion of the organic solvent; therefore, the surfactant/ polymer ratio must

be carefully optimized [32]. This mechanism is more prominent at medium and high polyvinyl alcohol concentrations (more than 7%). The critical influence of the type and the concentration of various surfactants, e.g. polyvinyl alcohol, on nanoparticles loaded with silymarin as a hydrophobic compound using the nanoprecipitation method has been investigated [33].

2- Figure 2b shows the combined effect of the sonication time and the shell/core ratio variables on the acquired TPC. It indicates as the shell/core increases, during nearly the entire range of the time of sonication, the polyphenol loading decreases. This clearly indicates the negative effect of increasing the polymer concentration, attributed to a shell/core ratio more than 7, in preventing the appropriate organic phase diffusion into the aqueous phase despite the sonication/ enforced (helped) stirring. However, a maximum peak is observed in the shell/core 6.6-6.8 and the sonication time of 3.7-3.9 min, which is attributed to a polyphenol loading level more than 95%.

This result indicates the positive effects of moderate homogenization time in polyphenols loadings via the nanopercipitation method. The positive influence of the solvent-nonsolvent intermixing to reduce the mean NPs diameter during the nanopercipitation process has already been investigated [34,35].

The interaction effect of the surfactant concentration and the sonication time variables on the acquired TPC, (Figure 2c), indicates as the surfactant concentration increase in the range of the sonication time between 2-5 min, the polyphenol loading increases, until a maximum peak is observed in the moderate surfactant concentration of 4-5% w/w and the sonication time of 2-4.2 min.

This observation might be attributed to a less favorable mixing efficiency during the sonication/homogenization process resulting from a higher viscosity of the aqueous phase, above a critical concentration of surfactant. The less favorable mixing efficiency in a higher viscosity of the aqueous phase has been already reported [32]; however, it has neither been optimized nor its interaction with other variables been studied.

Interestingly the relatively same patterns (in terms of rises and falls) were observed with the nanoparticle formation yield (Figures 2(a'-c')), meaning that both dependant (response) variables were directly correlated and followed the same trends upon the programmed variations in the independent variables. This correlation has not been previously reported.

Our study showed that in a critical surfactant concentration an interfacial tension lowering effect is observed. This finding also showed the constructive influence of the solution-nonsolvent intermixing in the particle size distribution of NPs, the level of loading

efficiency, and the formation yield of the NPs. On the other hand, the sonication homogenization transforms the maltodextrin solution into small droplets immediately after entering the water, leading to acceleration of the solvent-nonsolvent intermixing process. Moreover the possible formation of large aggregates due to flocculation of particles is overcome, which leads to NPs with unimodal size distributions and low polydispersity. Furthermore, the sonication homogenization as well as the surfactant application lowered the NPs growth in the initial steps; leading to finer NPs. Our observation is in contrast with research that reported no substantial improvement in the diffusion rate of organic solvent towards aqueous solvent in the surfactant presence. These results are not clearly in agreements with some of the principles of the common nanoprecipitation models based on the so-called "Marangoni effect" in which interfacial tension and mechanical turbulence were not considered as the driving forces during the course of the NPs formation [36].

### 3.3. Optimization of the NPs formation process

To achieve the maximum level of TPC and NPs formation yield, the optimal level of extraction parameters were generated based on both two separate single response variables and in combination. Multiple graphical and numerical optimizations were run to determine the optimum level of the independent variables, with desirable response targets, and further verified experimentally. The optimal conditions were expressed as the wall to core ratio of 7.5, the percentage of surfactant of 4% w/w, and the time of sonication homogenization of 2 min. The NPs were prepared in triplicate according to the optimal conditions to assess the experimental reproducibility and the models verification. The response surface models were verified by similarities between the observed and the predicted values. For the NPs obtained experimentally in optimal process conditions, the L.E of 98% , the L.C of 8.62 % and the N.Y of 75% were quantified according to the Eqs. (1-3), respectively.

The DLS result on these NPs showed a mean particle size of 52 nm with the 40 nm distribution width (Figure 3a), which was composed of just one population, corresponding to 100% of all particles. This is less than the previously reported studies on PLA-grape extract NPs, which was 351.9 and 291.6 nm for the seed and skin grape extracts, respectively [37]; and also (PLGA-PEG) NPs loaded with pomegranate extract, which was in the range of 120-200 nm [38]. Both NPs were prepared by the modified emulsion-solvent evaporation method. It seems that presumably at the initial stage of the nanoprecipitation process, sonication inhibited the crystal growth. Then it was followed by adsorption of the polyvinyl alcohol on to the NPs, leading to further inhibition growth and producing finer NPs. This would certainly influence their final bioavailability due to the higher surface to volume ratio.

The FE-SEM micrographs of NPs' suspension (Figure 3b) showed a spherical shape of the dispersed particles with smooth surfaces, and the NPs sizes were in agreement with the results governed by DLS measurements.

### 3.4. Fourier Transform Infra-Red (FT-IR) characterization of MD-APPE NPs

The MD, the APPE, and the loaded NPs were evaluated by FT-IR spectroscopy to identify the functional groups of the active components and also any mode of interactions between the polymeric wall and the polyphenolic core. The resulting spectrums are shown in Figure 4. For the APPE, the major peaks were assigned to the stretching vibration of hydroxyl groups (a broad band around 3390 cm$^{-1}$), asymmetrical stretching vibration of $CH_2$ groups ($v(CH_2)$ 2928 cm$^{-1}$), accompanied by the corresponding $\delta(CH_2)$ bending vibration (1350, 1226, and 632 cm$^{-1}$), $-CO$ stretching (1727 cm$^{-1}$), and aromatic bending and stretching ($v(C-C)$ conjugated with $C=C$ (1350 cm$^{-1}$) and $v(C=C)$ (1615 cm$^{-1}$). Additionally, two peaks at lower wave numbers were assigned to $C-H$ and $C-C$ out of plane bending vibrations at 875 and 776 cm$^{-1}$, respectively, which are associated with 1, 4-disubstituted benzene molecules [39]. A very strong broad peak at 1051 cm$^{-1}$ corresponds to the $v(C-O-C)$, which is associated with the glycosydic bond.

As it is clearly shown in Figure 4 and Table 4, the wave numbers of the major peaks of the APPE (c) and MD (b) were shifted in the NPs (a) and also their intensities were changed. A very strong broad peak correspondent to the glycosydic bond at 1051 cm$^{-1}$

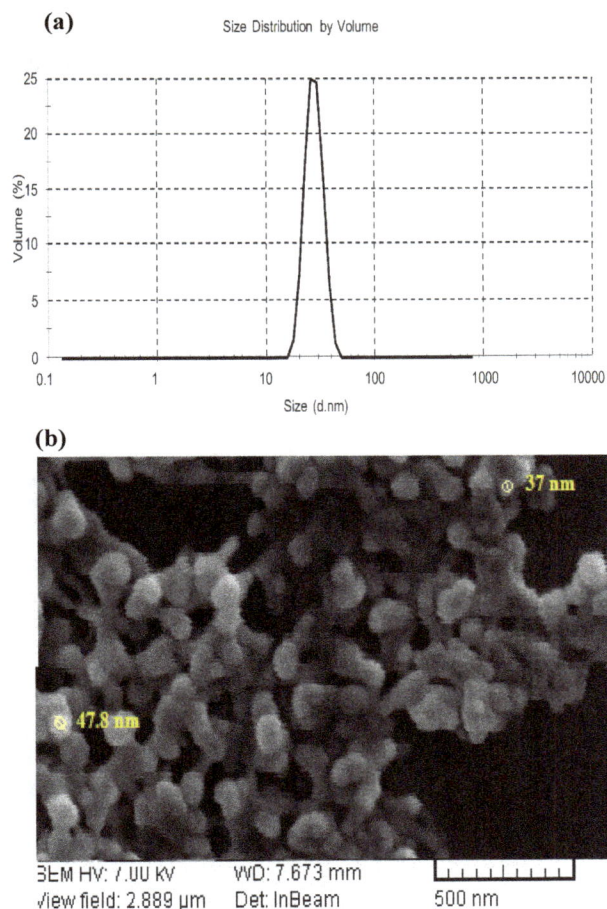

**Fig. 3.** Particle size spectrum of MD-APPE NPs, DLS result (a), FE-SEM image (b).

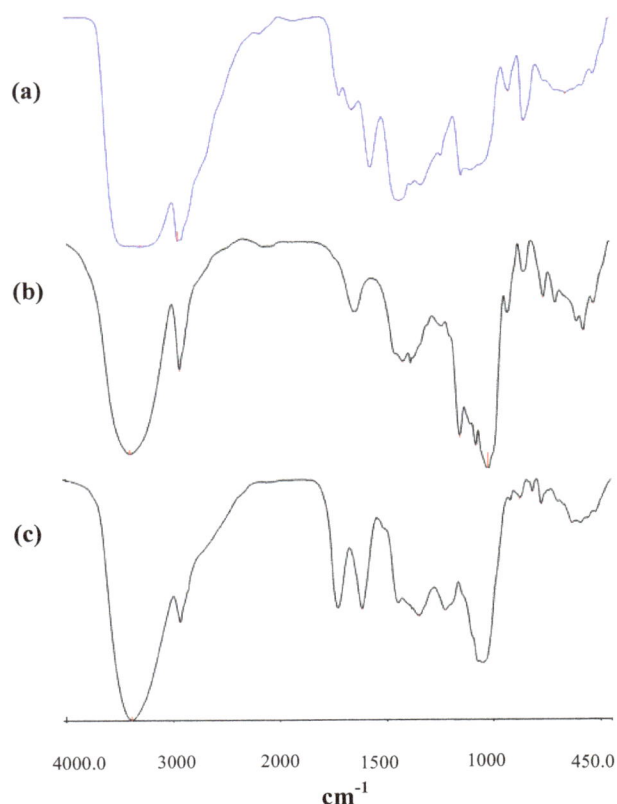

**Fig. 4.** FT-IR spectra of MD-APPE NPs (a), MD (b) and APPE (c).

was absent in the NPs. These observations indicate the successful entrapment and also an interaction between the wall and the core. Hydrogen bonding is suggested as one of the mode of interactions between the hydrophilic moieties of polyphenols and MD. The hydrogen bonds can be formed between the hydroxyl groups of APPE and MD (intra- and inter-chain bonds) or between the hydroxyl groups of polymer chains and surfactant molecules [13].

### 3.5. In-vitro release of MD-APPE NPs, simulating cancerous tumor acidity conditions

Figure 5 presents the release profile of MD-APPE NPs under mild acidic conditions (pH 4.5, 6.8). Both

**Table 3.** FT-IR major peaks assignments of the APPE, the MD and the loaded NPs.

| Assignment | APPE | MD | NPs |
|---|---|---|---|
| $\nu$(O–H$\cdots$O) | 3390 (s,b) | 3398 (s,b) | 3291 (s,b) |
| $\nu_a$(CH$_2$) | 2928 (s) | 2925 (s) | 2940 (s) |
| $\nu_s$(CH$_2$) | - | - | - |
| $\nu$(C=O) ester | | 1642 (s) | |
| $\nu$(C=O$\cdots$H) ester | | | |
| $\nu$(C=O$\cdots$H weak) acid | 1727 (vs) | | |
| $\nu$(C=O$\cdots$H strong) acid | | | |
| $\nu$(C=C) phenolic acid | 1615 (vs) | | |
| $\nu$(C–C) aromatic | | | |
| $\nu$(C–C) aromatic | | | |
| $\nu$(C–C) aromatic (conjugated with C=C) | 1500 (vw,sh) | | |
| $\delta$(CH$_2$) scissoring | | 1420 (m,b) | 1435 (vs,b) |
| $\nu$(C–C) aromatic (conjugated with C=C) | 1424 (vw,sh) | | |
| $\delta$(CH$_2$) wagging and twisting | 1350 (m,b) | 1383 (w) | 1380 (vw) |
| $\delta$(OH) | 1227 (s,b) | 1153 (s) | 1228 (vw) |
| $\nu_a$(C–O–C), ester | | | |
| $\nu_s$(C–O–C), ester | | | |
| $\nu$(C–O–C), glycosydic bond | 1051 (vs,b) | – | – |
| $\nu$(CvO) | 902 (vw) | 857 (m) | |
| $\gamma$(C–H) aromatic | 875 (w) | | |
| $\delta$(CH$_2$) rocking | 632(s,b) | 762 (m) | 655 (m,b) |

$\nu$: stretching; $\delta$: bending; $\gamma$: out of plane; a: asymmetric; s: symmetric
s: strong; m: medium; w: weak; vs: very strong; vw: very weak; b: broad; sh: shoulder

simulations represented a sustainable release of MD-APPE NPs and also the higher release rate of phenolics in pH 4.5 than pH 6.8, prolonged to 9 and 15 h, respectively. This shows an initial burst release on average of 50-65% and 70-80% of APPE within 2-3 h at pH 6.8 and 4.5, respectively. An initial burst release of the nanoencapsulated polyphenols were also reported by other studies, which was attributed to the fraction of active component placed near the surface of the NPs [37,40].

The MD-APPE nanoparticles with the sustainable release ability produced in this study could be considered as a potential candidate to target certain cancerous tumors, e.g. colon cancer. Application of the pH-sensitive polymeric NPs loaded with an anticancer drug to target solid tumors with slightly acidic extracellular pH (pH 4.5-6.8) environment has been reviewed [41]. However, more specific in vitro and in vivo experiments are necessary to shed more light into the feasibility of this potential application.

### 4. Conclusion

A modified nanoprecipitation method was successfully implemented for the preparation of the MD-APPE NPs and RSM was utilized for the process conditions modeling and optimization. The results indicated that

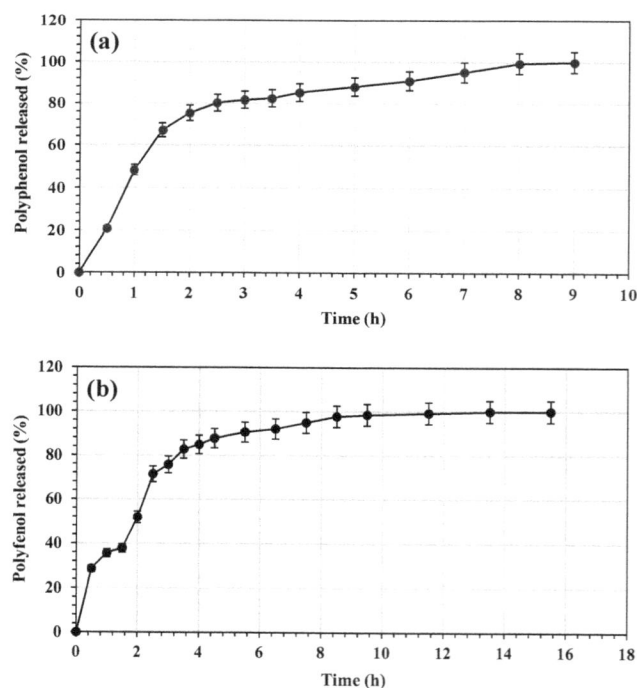

**Fig. 5.** In vitro release profile of MD-APPE NPs at pH 4.5 (a) and 6.8 (b).

the all the independent variables; i.e. the shell to core, the time of sonication homogenization, and the weight percentage of surfactant, were statistically significant factors affecting both the acquired TPC loading efficiency within the MD and nanoparticle formation yield. In the optimal condition, NPs with a mean average size of 52 nm with the distribution width of 40 nm and high loading efficiency of 98% were produced. The scientific basis of our hypothesis was strengthened by the use of FT-IR spectrums, DLS measurement and FE-SEM image and in vitro release studies.

Future studies include the stability and shelf life of the final product during food processing operating conditions, such as pH fluctuations in different finished products and thermal treatment during pasteurization and sterilization.

## Acknowledgment

The authors wish to thank the Iran Nanotechnology Initiative Council (INIC) for partial financial support and the staff of the Department of Chemical Technologies at the Iranian Research Organization for Science and Technology (IROST), and mostly Dr. Bashiri Sadr for their co-operation and support. We also appreciate the efforts of Sanich Co. for providing the apple industrial wastes.

## References

[1] J. Boyer, R.H. Liu, Apple phytochemicals and their health benefits, Nutr. J. 3 (2004) 1-15.

[2] B. Suárez, Á.L. Álvarez, Y.D. Garc'ia, G. del Barrio, A.P. Lobo, F. Parra, Phenolic profiles, antioxidant activity and in vitro antiviral properties of apple pomace, Food Chem. 120 (2010) 339-342.

[3] G.S. Dhillon, S. Kaur, S.K. Brar, Perspective of apple processing wastes as low-cost substrates for bioproduction of high value products: A review, Renew. Sust. Energ. Rev. 27 (2013) 789-805.

[4] Z. Fang, B. Bhandari, Encapsulation of polyphenols-a review, Trends Food Sci. Tech. 21 (2010) 510-523.

[5] O.I. Parisi, F. Puoci, D. Restuccia, G. Farina, F. Iemma, N. Picci, Polyphenols and their formulations: Different strategies to overcome the drawbacks associated with their poor stability and bioavailability, in: Polyphenols in Human Health and Disease, Elsevier Academic Press, USA, 2014, pp. 29-45.

[6] G. Spigno, F. Donsì, D. Amendola, M. Sessa, G. Ferrari, D.M. De Faveri, Nanoencapsulation systems to improve solubility and antioxidant efficiency of a grape marc extract into hazelnut paste, J. Food Eng. 114 (2013) 207-214.

[7] H.B. Nair, B. Sung, V.R. Yadav, R. Kannappan, M.M. Chaturvedi, B.B. Aggarwal, Delivery of antiinflammatory nutraceuticals by nanoparticles for the prevention and treatment of cancer, Biochem. Pharmacol. 80 (2010) 1833-1843.

[8] H. Souguir, F. Salaün, P. Douillet, I. Vroman, S. Chatterjee, Nanoencapsulation of curcumin in polyurethane and polyurea shells by an emulsion diffusion method, Chem. Eng. J. 221 (2013) 133-145.

[9] A. Altunbas, S.J. Lee, S.A. Rajasekaran, J.P. Schneider, D.J. Pochan, Encapsulation of curcumin in self-assembling peptide hydrogels as injectable drug delivery vehicles, Biomaterials, 32 (2011) 5906-5914.

[10] P. Salehi, O.V. Akinpelu, S. Waissbluth, E. Peleva, B. Meehan, J. Rak, S.J. Daniel, Attenuation of cisplatin ototoxicity by otoprotective effects of nanoencapsulated curcumin and dexamethasone in a guinea pig model, Otol. Neurotol. 35 (2014) 1131-1139.

[11] S. Ghosh, S.R. Dungdung, S.T. Chowdhury, A.K. Mandal, S. Sarkar, D. Ghosh, N. Das, Encapsulation of the flavonoid quercetin with an arsenic chelator into nanocapsules enables the simultaneous delivery of hydrophobic and hydrophilic drugs with a synergistic effect against chronic arsenic accumulation and oxidative stress, Free Radical Bio. Med. 51 (2011) 1893-1902.

[12] L. Dian, E. Yu, X. Chen, X. Wen, Z. Zhang, L. Qin, Q. Wang, G. Li, C. Wu, Enhancing oral bioavailability of quercetin using novel soluplus polymeric micelles, Nanoscale Res. Lett. 9 (2014) 684, 11 pages.

[13] A.R. Patel, P.C.M. Heussen, J. Hazekamp, E. Drost, K.P. Velikov, Quercetin loaded biopolymeric colloidal particles prepared by simultaneous precipitation of quercetin with hydrophobic protein in aqueous medium, Food Chem. 133 (2012) 423-429.

[14] C.F. Rodrigues, K. Ascencao, F. A.M. Silva, B. Sarmento, M. B.P.P. Oliveira, J.C. Andrade, Drug-delivery systems of green tea catechins for improved stability and bioavailability, Curr. Med. Chem. 20 (2013) 4744-4757.

[15] S.M. Henning, Y. Niu, Y. Liu, N.H. Lee, Y. Hara, G.D. Thames, R.R. Minutti, C.L. Carpenter, H. Wang, D. Heber, Bioavailability and antioxidant effect of epigallocatechin gallate administered in purified form versus as green tea extract in healthy individuals, J. Nutr. Biochem. 16 (2005) 610-616.

[16] Anonymous, Agricultural statistics Volume III - horticultural crops, Tehran, 2014.

[17] C.M. Galanakis, Recovery of high added-value components from food wastes: Conventional, emerging technologies and commercialized applications, Trends Food Sci. Tech. 26 (2012) 68-87.

[18] S. Faramarzi, A. Yadollahi, M. Barzegar, K. Sadraei, S. Pacifico, T. Jemric, Comparison of Phenolic compounds' content and antioxidant activity between Some Native Iranian apples and standard cultivar "Gala", J. Agr. Sci. Tech.-Iran, 16 (2014) 1601-1611.

[19] J. Ubbink, J. Krüger, Physical approaches for the delivery of active ingredients in foods, Trends Food Sci. Tech. 17 (2006) 244-254.

[20] C.E. Mora-huertas, H. Fessi, A. Elaissari, Polymer-based nanocapsules for drug delivery, Int. J. Pharm. 385 (2010) 113-142.

[21] R. Harris, E. Lecumberri, I. Mateos-Aparicio, M. Mengíbar, A. Heras, Chitosan nanoparticles and microspheres for the encapsulation of natural antioxidants extracted from *Ilex paraguariensis*, Carbohyd. Polym. 84 (2011) 803-806.

[22] F. Avaltroni, P.P.E. Bouquerand, V. Normand, Maltodextrin molecular weight distribution influence on the glass transition temperature and viscosity in aqueous solutions, Carbohyd. Polym. 58 (2004) 323-334.

[23] E.K. Bae, S.J. Lee, Microencapsulation of avocado oil by spray drying using whey protein and maltodextrin, J. Microencapsul. 25 (2008) 549-560.

[24] H. Fessi, F. Puisieux, J.P. Devissaguet, N. Ammoury, S. Benita, Nanocapsule formation by interfacial polymer deposition following solvent displacement, Int. J. Pharm. 55 (1989) 1-4.

[25] S. Khoee, M. Yaghoobian, An investigation into the role of surfactants in controlling particle size of polymeric nanocapsules containing penicillin-G in double emulsion, Eur. J. Med. Chem. 44 (2009) 2392-2399.

[26] S. Saikia, N.K. Mahnot, C.L. Mahanta, Optimisation of phenolic extraction from *Averrhoa carambola* pomace by response surface methodology

and its microencapsulation by spray and freeze drying, Food Chem. 171 (2015) 144-152.

[27] G.B. Celli, A. Ghanem, M.S.-L. Brooks, Optimized encapsulation of anthocyanin-rich extract from haskap berries (*Lonicera caerulea* L.) in calcium-alginate microparticles, J. Berry Res. 6 (2016) 1-11.

[28] M. Pinelo, M. Rubilar, M. Jerez, J. Sineiro, M.J. Núñez, Effect of solvent, temperature, and solvent-to-solid ratio on the total phenolic content and antiradical activity of extracts from different components of grape pomace, J. Agr. Food Chem. 53 (2005) 2111-2117.

[29] Q. Wang, S. Ma, B. Fu, F.S.C. Lee, X. Wang, Development of multi-stage countercurrent extraction technology for the extraction of glycyrrhizic acid (GA) from licorice (*Glycyrrhiza uralensis* Fisch), Biochem. Eng. J. 21 (2004) 285-292.

[30] M. Valipour, Process conditions optimization in the polyphenolic extraction (one- and multi-counter current) from Iranian industrial apple pomace, Chemistry MSc thesis, Iranian Research Organization for Science and Technology (IROST), 2016.

[31] U. Bilati, E. Allémann, E. Doelker, Development of a nanoprecipitation method intended for the entrapment of hydrophilic drugs into nanoparticles, Eur. J. Pharm. Sci. 24 (2005) 67-75.

[32] S. Galindo-Rodriguez, E. Allémann, H. Fessi, E. Doelker, Physicochemical parameters associated with nanoparticle formation in the salting-out, emulsification-diffusion, and nanoprecipitation methods, Pharm. Res. 21 (2004) 1428-1439.

[33] S.A. Guhagarkar, V.C. Malshe, P. V Devarajan, Nanoparticles of polyethylene sebacate: A new biodegradable polymer, AAPS PharmSciTech. 10 (2009) 935-942.

[34] M.E. Matteucci, M.A. Hotze, K.P. Johnston, R.O. Williams, Drug nanoparticles by antisolvent precipitation: Mixing energy versus surfactant stabilization, Langmuir, 22 (2006) 8951-8959.

[35] M.R. Kulterer, M. Reischl, V.E. Reichel, S. Hribernik, M. Wu, S. Köstler, R. Kargl, V. Ribitsch, Nanoprecipitation of cellulose acetate using solvent/nonsolvent mixtures as dispersive media, Colloid. Surface. A 375 (2011) 23-29.

[36] E. Lepeltier, C. Bourgaux, P. Couvreur, Nanoprecipitation and the "Ouzo effect": Application to drug delivery devices, Adv. Drug Deliver. Rev. 71 (2014) 86-97.

[37] K. Fernández, J. Aburto, C. von Plessing, M. Rockel, E. Aspé, Factorial design optimization and characterization of poly-lactic acid (PLA) nanoparticle formation for the delivery of grape extracts, Food Chem. 207 (2016) 75-85.

[38] A.B. Shirode, D.J. Bharali, S. Nallanthighal, J.K. Coon, S.A. Mousa, and R. Reliene, Nanoencapsulation of pomegranate bioactive compounds for breast cancer chemoprevention, Int. J. Nanomed. 10 (2015) 475-484.

[39] J.A. Heredia-Guerrero, J.J. Benìtez, E. Domìnguez, I.S. Bayer, R. Cingolani, A. Athanassiou, A. Heredia, Infrared and Raman spectroscopic features of plant cuticles: A review, Front. Plant Sci. 5 (2014) 305.

[40] S.K. Pandey, D.K. Patel, R. Thakur, D.P. Mishra, P. Maiti, C. Haldar, Anti-cancer evaluation of quercetin embedded PLA nanoparticles synthesized by emulsified nanoprecipitation, Int. J. Biol. Macromol. 75 (2015) 521-529.

[41] E.S. Lee, Z. Gao, Y.H. Bae, Recent progress in tumor pH targeting nanotechnology, J. Control. Release, 132 (2008) 164-170.

# 16

# Modified CNTs/Nafion composite: The role of sulfonate groups on the performance of prepared proton exchange methanol fuel cell's membrane

**Kamran Janghorban[1], Payam Molla-Abbasi[2,*]**

[1] Department of Chemical Engineering, Farahan Branch, Islamic Azad University, Farahan, Iran

[2] Department of Chemical Engineering, Faculty of Engineering, University of Isfahan, Isfahan, Iran

## HIGHLIGHTS

- Functionalization of CNTs by a silica layer and sulfonated groups

- Improvement of proton conductivity and selectivity of the prepared composite membrane

- Decreasing the methanol permeability by decreasing the size of the nanochannel

- Enhancement of water uptake and ion exchange capacity by introducing the sulfonated groups

## GRAPHICAL ABSTRACT

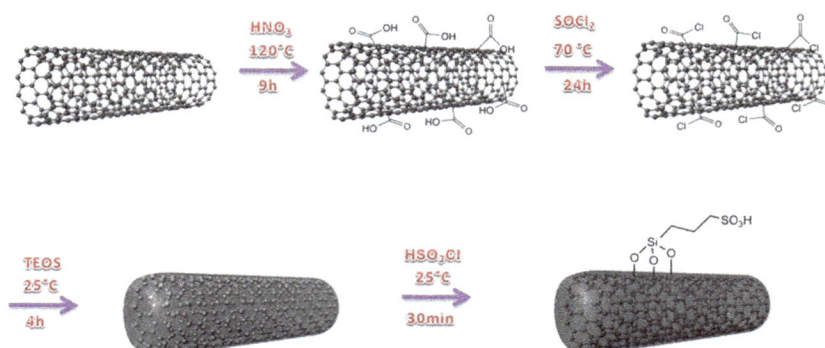

## ARTICLE INFO

*Keywords:*

Fuel cells
Nafion
Sulfonate modified CNTs
Proton conductivity
Membrane permeability

## ABSTRACT

A novel Nafion®-based nanocomposite membrane was synthesized to be applied as direct methanol fuel cells (DMFCs). Carbon nanotubes (CNTs) were coated with a layer of silica and then reacted by chlorosulfonic acid to produce sulfonate-functionalized silicon dioxide coated carbon nanotubes (CNT@$SiO_2$-$SO_3H$). The functionalized CNTs were then introduced to Nafion®, and subsequently, methanol permeability, proton conductivity, ion exchange capacity (*IEC*) and water uptake properties of the prepared membranes were investigated. The experimental results showed that the water uptake and *IEC* of the Nafion®/CNT@$SiO_2$-$SO_3H$ (1 wt%) membrane increased in comparison with the recast Nafion®. *IEC* was enhanced from 0.9 meq/g for the recast Nafion® to 0.946 meq/g for Nafion®/CNT@$SiO_2$-$SO_3H$, which could be attributed to the presence of sulfonate groups on the surface of CNTs. In addition, the proton conductivity of the sulfonate modified CNT/Nafion® composite was enhanced in a wide range of temperatures. Selectivity of the fabricated membrane was found to be more than 8-fold higher than that of recast Nafion® 117, demonstrating the promising potential of the produced membranes for DMFC applications.

* *Corresponding author:* ; *E-mail address: p.abbasi@eng.ui.ac.ir*

# 1. Introduction

With the ongoing development of industry and the progressively increasing need for consumption of energy, fuel cell technology is becoming considered as an important alternative to fossil fuel-based energy converters. Among the different types of fuel cells, direct methanol fuel cells (DMFCs) are more attractive for using as a portable power source because of their advantages such as high energy density, low molecular weight and liquid form of methanol [1].

Performance of methanol fuel cells is strongly dependent on polymer polyelectrolyte, which plays the most important role in the fuel cell. This section of the fuel cell is formed by a polymer membrane with a particular structure capable of proton exchange. The polyelectrolyte layer is located between the anode and cathode creating a preferred path for the proton transfer [2].

In addition to having proton exchange ability, this membrane should also have a minimum permeability to methanol. Moreover, the mechanical and chemical stability, flexibility and durability of polyelectrolyte membranes are very important for use in fuel cells.

Perfluorosulfic acid (PFSA) based membranes, such as Nafion®, are the most common polyelectrolytes for fuel cells because of their excellent proton conductivity and also high mechanical, chemical and thermal resistance. Nafion is a semi-crystalline polymer and its structure is comprised of a hydrophobic poly-tetrafluoroethylene body with short perfluorovinyl ether side-chains which are terminated by hydrophilic sulfonate ionic groups [3]. This exceptional two-phase structure has brought a lot of attention to Nafion in recent years. Due to their hydrophilic nature of absorption of water molecules into the membrane and swelling to facilitate proton transfer, sulfonic ionic groups are leading to the formation of nano-channels. Moreover, their hydrophobic Teflon matrix creates good mechanical strength for the membrane. However, Nafion also has major disadvantages such as high methanol crossover, low proton conductivity at low humidity or high temperatures, loss of mechanical and thermal stability at elevated temperatures and restricted operating temperatures [4,5].

Modification of the Nafion membrane by incorporation of fillers is one of the best routes to improve its performance. By incorporating organic or inorganic nano-scale additives like silica, metal oxides, clay, graphene, zeolite, and others into the Nafion polyelectrolyte matrices, the methanol permeability can be significantly reduced, but unfortunately once nanoparticles are located in the route of nano channels the proton transfer is decreased. This problem can be solved by using functionalized nanoparticles which provide new proton conduction sites on their surfaces [6].

Among the mentioned additives, carbon nanotubes (CNTs) have recently been considered as one of the best options due to their high aspect ratios, nanometer scale diameter, high specific surface area, substantial structural and their excellent mechanical and chemical properties. However, the problem of insolubility of the carbon nanotubes and the inherently poor compatibility between the carbon nanotubes and the Nafion matrix in obtaining a uniform dispersion of carbon nanotubes in the polymer matrix has limited the use of CNTs [7]. Within the past few years, several efforts have been to modify the surface of carbon nanotubes to counterbalance the mentioned issue. An effective way to ameliorate the proton conductivity and the hydration properties of Nafion at elevated temperatures is to fabricate the composite membrane with inorganic additives that have been surface modified with compounds such as hygroscopic oxides (e.g. $SiO_2$, $TiO_2$, $ZrO_2$) to improve the water retention at operating temperatures above 90 °C [8]. The incorporation of hydrophilic inorganic hygroscopic oxide particles produced promising results due to their propensity to embed water in their interlayer regions which makes them more hydrophilic or more permeable to water [9]. Amjadi et al. [10] have studied the Nafion® 117 membranes doped with $SiO_2$ particles by in-situ sol-gel reactions. The results showed that the membranes modified with $SiO_2$ increased the fuel cell performance at 110 °C and in low humid conditions. Also, the membrane including 5-7 wt% $SiO_2$ content exhibited higher water uptake than the pure Nafion. Jung et al. [11] explored the proton conductivity of Nafion with different contents of $SiO_2$ and concluded that the proton conductivity was increased for the silicon oxide content of 12.4% at 125 °C. In another work, Adjemian et al. [12] prepared Nafion/$SiO_2$ composite membranes by the sol-gel technique, and reported high water retention characteristics above 100 °C.

Additionally, the inclusion of sulfonate into the Nafion membrane has also shown encouraging results. For example, the addition of sulfonated graphene oxide to Nafion enhances the water retention characteristics of the membranes at higher temperatures. Chang et al. [13] studied the structural and functional properties of Nafion® 117 membranes filled with sulfonated graphene oxide (SGO). The composite membrane which can be implemented in direct methanol fuel cells (DMFCs) showed lower methanol permeability and water uptakes, a reduced swelling ratio, enhanced proton conductivity in low relative humidity and extremely high methanol selectivity. Moreover, Jae-Hong et al. [14] prepared Nafion nanocomposite membranes using sulfonated $SiO_2$ nanoparticles, and their results indicated that the addition of sulfonated $SiO_2$ nanoparticles into the Nafion matrix is very effective in improving the membrane performance including ion exchange capacity (*IEC*), proton conductivity, methanol permeability and mechanical strength.

The principal aim of this paper is to fabricate nanocomposite membranes based on Nafion by incorporating the sulfonate-functionalized silicon dioxide coated carbon nanotubes (CNT@$SiO_2$-$SO_3$H) to investigate the performance of the prepared membrane. The most important aim of this research is to investigate the functionalization of CNTs and presence of sulfonate groups on the performance of the prepared membrane. To this end, the properties of the prepared membranes, such as water retention, ion exchanger, proton conductivity, methanol permeability and selectivity, were investigated and compared with pristine Nafion.

## 2. Experimental

### 2.1. Materials

Nafion® 117 was acquired from the E. I. DuPont de Nemours Company (USA) and multi-walled carbon nanotubes (MWNTs, outer diameter: 20-30 nm, inside diameter: 5-10 nm, average length: 10-30 μm and surface area: 110-130 $m^2$/g) were purchased from Nano Amor (Houston, TX, USA). Nitric acid ($HNO_3$, 68%) was used for oxidation of MWNTs (Merck, Germany), N,N-dimethylformamide (DMF) (Merck, Germany), thionyl chloride (Sin Chem, Indian), tetrahydrofuran (THF) (Merck, Germany), tetraethoxysilane (TEOS) (Sigma Aldrich, Germany) were used for synthesis of the silica coated MWNTs. Also, Chlorosulfonic acid (Sigma Aldrich, Germany) was used for surface modification of MWNTs and ethanol (99.7%) was used.

### 2.2. Fabrication of nanoparticle

First, CNTs were treated with nitric acid (68 wt%). For this purpose, a mixture of CNTs and nitric acid was magnetically stirred on a heater stirrer equipped with reflux condenser at 120 °C for 9 h [15,16]. In this step, COOH groups were introduced to the surface of CNTs. The resulting CNT-COOH was dispersed in thionyl chloride in a sonication bath for 1 h, and then an adequate amount of DMF (5 ml) was added to the mixture followed by stirring for 24 h at 70 °C under reflux conditions to attain the functionalized CNTs with chlorocarbonyl groups (CNT-COCl). [6].

To add the silane groups into the CNTs surfaces, CNT-COCl was dispersed into ethanol, and then stirred for 15 min. After that, $NH_3$ (4 ml, 30 wt%) and distilled water were added to the mixture of ethanol and CNT-COCl. The TEOS was quickly injected into the flask, and the reaction vessel was stirred at 25 °C for 4 h. The prepared mixture was centrifuged at 4000 rpm for 30 min to gain silicon dioxide coated CNTs (CNT@$SiO_2$) [16,17]. In the following procedure to modify the surface of the sulfonate groups, a suction flask charged with the nanoparticles (80 mg) and equipped with a constant-pressure dropping funnel and a gas inlet tube for conducting HCl gas over an adsorbing solution (i.e., water) was used. Chlorosulfonic acid was added dropwise over a period of 30 min at room temperature. HCl gas immediately evolved from the reaction vessel. After the addition was completed, the mixture was shaken for 30 min. A black solid of sulfonate-functionalized silicon dioxide coated carbon nanotubes (CNT@$SiO_2$-$SO_3$H) was obtained [18]. The schematic procedure of CNT functionalization is illustrated in Figure 1.

### 2.3. Nanocomposite membranes fabrication

First, certain amounts of sulfonate-functionalized silicon dioxide coated carbon nanotubes (CNT@$SiO_2$-$SO_3$H) powder were suspended in deionized water at 25 °C, stirred for 1.5 h, and sonicated for 30 min. Then, the suspensions were added to Nafion® 5 wt% (diluted by water and isopropyl alcohol) at room temperature and stirred for 3 h to gain 1 wt% of CNT@$SiO_2$-$SO_3$H polymer composite. The mixtures were sonicated for two sequential 30 min periods. The resultant solutions were cast on petri dishes then incubated in ambient

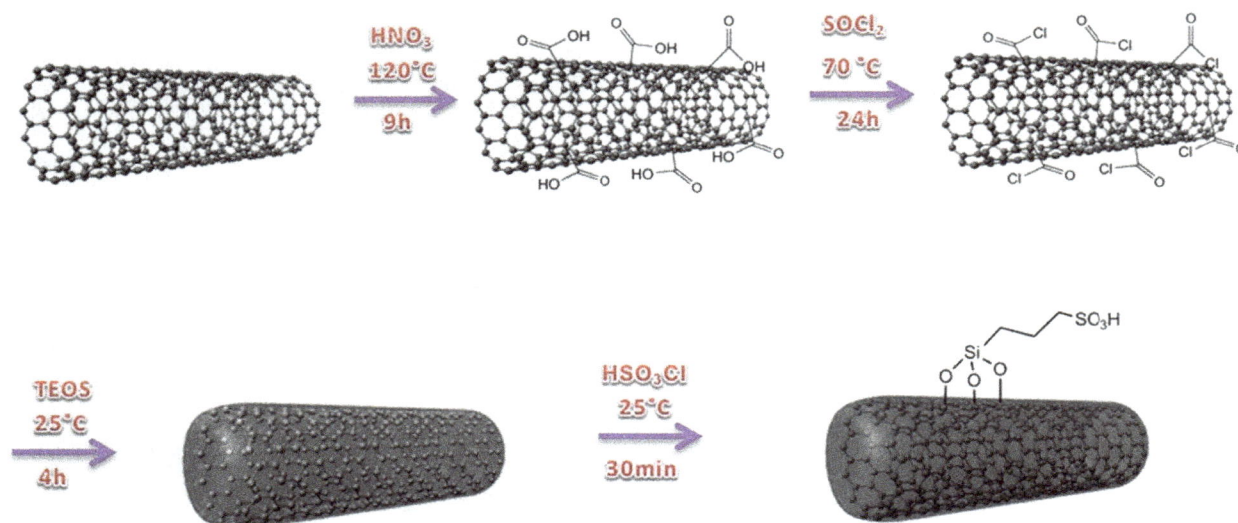

**Fig. 1**. Schematic of the CNTs functionalization process.

temperature overnight and further dried in 70 °C for 12 h. Lastly, the fabricated membranes were annealed at 120 °C overnight to stabilize the microstructure of Nafion. The same procedure was used to fabricate the recast Nafion.

## 2.4. Characterization

### 2.4.1. Field Emission Scanning Electron Microscope (FESEM)

A FESEM (HITACHI S-4160, Japan) is used to visualize very small topographic details on the surface or cross-section area of the fractured objects in liquid nitrogen. Prior to the analysis, the samples were coated with a thin layer of gold.

### 2.4.2. Water uptake

A vital property of any membrane is its water uptake rate (WUR), which provides information about the water retention capacity of the membrane. Water uptake can be calculated from the difference in weights between the wet and dry samples. To attain precise results, this process was repeated several times [16]. The percentage of water uptake was then calculated by the following equation:

$$WU\% = (\frac{m_{wet} - m_{dry}}{m_{wet}}) \times 100 \tag{1}$$

where $m_{wet}$ and $m_{dry}$ are the weights of wet and dry membranes, respectively.

### 2.4.3. Ion exchange capacity (IEC)

The ion exchange capacity (IEC) in meq/g is presented as the ratio of moles of sulfonic acid groups per gram of dried membrane ($W_{dry}$). The titration method was employed to measure the value of ion exchange capacity (IEC). The IEC is measured using Eq. (2).

$$IEC(meq / g) = (\frac{V_{NaOH} \times N_{NaOH}}{W_{dry}}) \times 100 \tag{2}$$

where $V_{NaOH}$ and $N_{NaOH}$ are the volumes of NaOH used up in titration and the concentration of NaOH, respectively.

### 2.4.4. Proton Conductivity

Ionic conductivity is usually measured based on proton conducting membrane resistance at room temperature by AC impedance spectroscopy. A four-point probe apparatus, manufactured by Bekktech Company, was employed to measure in-plane bulk conductivity. The proton conductivity was calculated by Eq. (3).

$$\sigma = L / (R \times A) \tag{3}$$

where $L$ is the membrane thickness, $A$ is the membrane cross-sectional area, and $R$ is the resistance.

### 2.4.5. Methanol permeability

A system including two tanks connected by a thin polymer film was used to calculate the methanol permeability. One of the tanks was filled with methanol (cell A) and the other tank was filled with distilled water (cell B). These tanks were stirred constantly, and the concentration of methanol in the water was measured by a gas chromatography apparatus (model 890, Hewlett-Packard Company). The methanol permeability was then calculated using the Eq. (4).

$$C_{B(t)} = \frac{A}{V_B} \times \frac{DK}{L} \times C(\Delta t) \qquad (4)$$

where $C_{B(t)}$ is methanol concentration in cell B (mol/l), the product of $DK$ is the methanol permeability of the membrane (cm²/s), $C$ is the concentration of methanol in Cell A (mol/l), $A$ is the cross-section area of membrane (cm²), $L$ is the thickness of membrane (cm), $V_B$ is the volume of each tank, and $\Delta t$ is the time of each methanol crossover measurement.

## 3. Results and discussion

Based on the chemistry of the sulfonate group and the presence of O⁻ in its structure, sulfonate groups have been identified as inorganic groups which are capable of proton exchange. Accordingly, they have been investigated as proton conductors for applications in fuel cells, hydrogen sensors, and hydrogen separation membranes. Sulfonated molecules exhibit high water solubility, and consequently, protons are incorporated into the materials as charge-compensating defects when the sulfonate are doped with cations [19].

FESEM was used to investigate the surface modification of CNTs and observe the construction of the silica layer on the surface of CNTs. Figure 2 illustrates the obtained images from CNT and CNT@SiO₂. Comparing the FESEM images seen in Figure 2 (a) and (b), the CNTs have been successfully coated by a thin layer of silica. As can be seen in Figure 2, the CNTs' diameter was increased by the addition of silica, which could be used as evidence for the successful coating of a silica layer on the surface of the CNTs. This layer is a substrate for future functionalization of CNTs.

The main reason for using only 1 wt% of CNT@SiO₂-SO₃H in preparation of Nafion membranes is due to the fact that an increase in concentration of CNT in membranes can lead to an increase in the probability of charge transferring across the prepared membranes. In order to prevent such phenomenon and minimize the risk of forming a short-circuit in the final membranes,

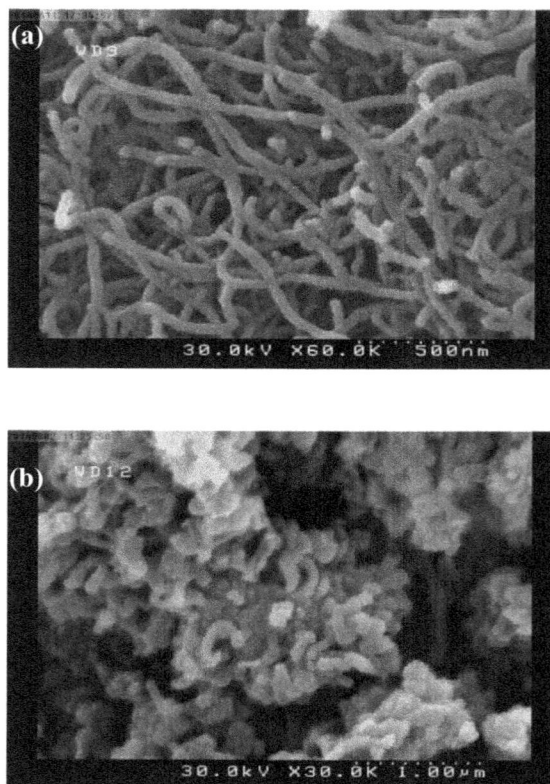

**Fig. 2.** Field emission Scanning electron microscope (FESEM) images of (a) CNT and (b) CNT@SiO₂.

the concentration of CNTs used in PEMs must be less than the amount of the percolation threshold. The permissible limit in a polymeric membrane for percolation threshold has been reported at about 2 wt% [20].

In Figure 3, the value of water uptake (*WU*), ion exchange capacity (*IEC*), $\lambda$ of the recast Nafion, and the prepared composite membrane containing 1 wt% of nanoparticles are depicted. The ratio of mole number of water molecules to the fixed-charged sulfonate groups, denoted as lambda ($\lambda$), was calculated from the following equation:

$$\lambda = \frac{WU}{IEC \times M_{water}} \qquad (5)$$

where *WU*, *IEC* and $M_{water}$ are the water uptake, ion exchange capacity and molecular weight of water (18 g/mol), respectively. In fact, the average number of water molecules per ionic group $\lambda$ shows how many water molecules can be bound to the ionic groups of polyelectrolytes. Hydration of the polyelectrolytes and proton conduction across the membranes are often explained with $\lambda$ [4].

As can be seen, by adding CNT@SiO₂-SO₃H to the

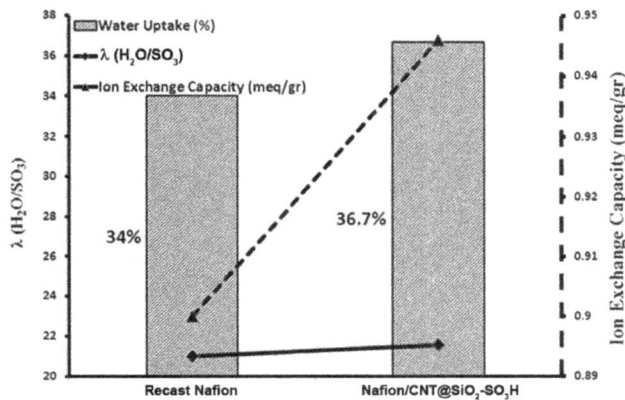

**Fig. 3.** Comparison of water uptake, ion exchange capacity, and $\lambda$ in recast Nafion and Nafion/CNT@SiO$_2$-SO$_3$H (1 wt%).

Nafion membrane, the water uptake increased from 34 wt% for the recast Nafion to 36.7 wt% for the prepared nanocomposite membrane. Moreover, *IEC* also increased and $\lambda$ was enhanced from 0.900 to 0.945.

As a conclusion, the value of ion exchange capacity and water uptake of the membrane containing CNT@SiO$_2$-SO$_3$H nanoparticle is higher than those of the recast Nafion sample. Moreover, by using the sulfonate groups in surface modification of nanoparticles, in addition to be considered as an acidic proton exchanger group the water absorption in the membrane containing this nanoparticle is higher than other membranes because of a good interaction between O$^-$ groups on the surface of CNTs and H$_2$O molecules. So, by adding the sulfonate decorated CNTs to the Nafion membrane the *WU* and *IEC* of the prepared membrane were increased.

This analysis was applied to investigate the effect of surface modified nanoparticles, especially CNT@SiO$_2$-SO$_3$H, on the proton conductivity of composite membrane. Figure 4 illustrates the effect of the modified CNTs on the proton conductivity of Nafion nanocomposite membranes in the range of 25-90 °C. With regard to the results, the proton conductivity of recast Nafion and Nafion/CNT@SiO$_2$-SO$_3$H (1 wt%) were calculated at room temperature as 0.086 and 0.0913 S/cm, respectively. Phosphonated and sulfonated polymers are known to conduct protons at low water content [21]. An increase in the proton conductivity with the addition of CNT@SiO$_2$-SO$_3$H nanoparticle is related to the enhanced *IEC* as a result of introducing sulfonic acid groups on the surface of the nanoparticles. Moreover, CNT@SiO$_2$-SO$_3$H nanoparticles might absorb water on their surfaces through a strong interaction with surface –SO$_3$H groups and the formation

of hydrogen bonds, leading to enhanced water retention and proton conductivity.

The Arrhenius equation was used to prove the dependency of proton conductivity to temperature in different membranes. According to the following equation, $E_a$, $\sigma$ and $\sigma_0$ are the activation energy of proton conduction, the proton conductivity, and pre-exponential factor in this equation, respectively. $T$ is the absolute temperature and $R$ is the universal gas constant (8.314 J.mol$^{-1}$.K$^{-1}$).

$$\sigma = \sigma_0 e^{-E_a/(RT)} \qquad (6)$$

Arrhenius plots of conductivity (logarithm of conductivity v.s $1/T$) for the recast Nafion® and modified Nafion® membranes at different temperatures are shown in Figure 4 (b). Activation energy was calculated using the Arrhenius plot and Eq. (6). The slope of these plots

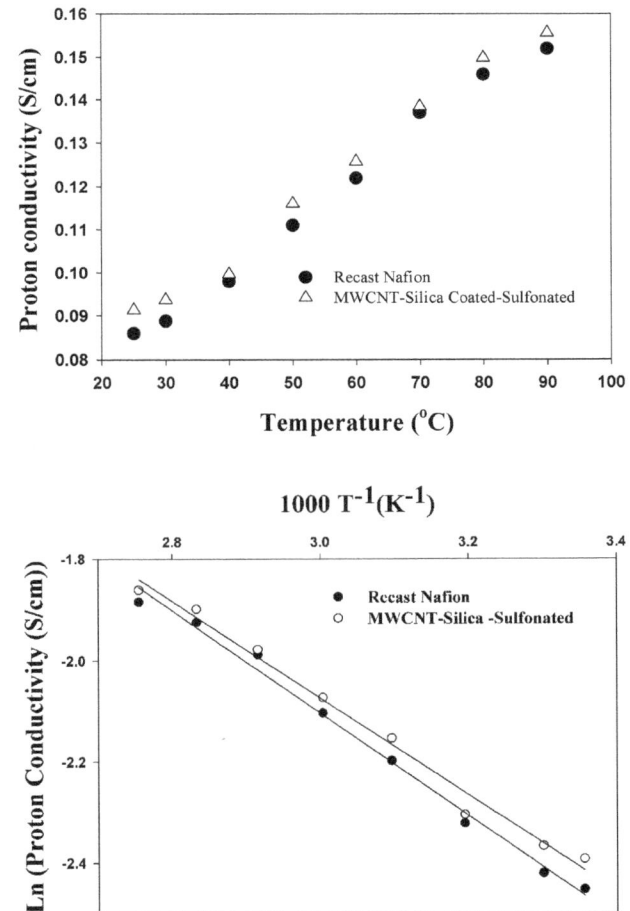

**Fig. 4.** (a) Proton conductivity of Nafion/CNT@SiO$_2$-SO$_3$H in the temperature range of 25-90 °C in comparison with recast Nafion. (b) Demonstrates the Arrhenius amount of proton conductivity for Nafion/CNT@SiO$_2$-SO$_3$H in comparison with recast Nafion.

denotes the activation energy ($E_a$) of the proton exchange at different temperatures.

According to the results from the Arrhenius plots, the activation energies of recast Nafion and Nafion/CNT@SiO$_2$-SO$_3$H are calculated as 0.1140 and 0.1066 (kJ/kmol), respectively. As can be seen, the activation energy of the composite membrane, which includes sulfonate groups, is lower than that of the recast Nafion membrane. Sulfonate groups reduced the activation energy of O–H bond by exposing the ionic groups locating on the chains of Nafion facilitating the movement of protons through the Grotthus mechanisms.

Figure 5 compares the amount of methanol permeability for the prepared membranes at room temperature. The values of methanol permeability are equal to $2.25 \times 10^{-6}$ and $2.91 \times 10^{-7}$ cm$^2$/S for Nafion and Nafion/CNT@SiO$_2$-SO$_3$H, respectively. It is seen that the methanol permeability of the membrane including Nafion/CNT@SiO$_2$-SO$_3$H has been noticeably reduced in comparison with the recast Nafion, this is mostly due to shrinking and stretching out of the route of methanol diffusion. The greatest reduction in methanol permeability occurs in the membrane including CNT@SiO$_2$-SO$_3$H nanoparticles and could be attributed to the reduced size of nano channels and obstruction of methanol's permeation route in the Nafion matrix by the presence of CNTs and also to the barrier role of these particles against methanol transfer [14].

With regard to membranes' modification, two factors, namely, proton conductivity and methanol permeability are simultaneously concerned. In fact, potential of the new composite membranes for direct methanol fuel cell (DMFC) is evaluated using the ratio of proton conductivity to methanol permeability, called selectivity

($\emptyset$) (S.s/cm$^3$). Generally, membranes exhibiting higher selectivity than Nafion® 117 indicate a higher potential to improve electrochemical cell performance. This is impossible unless high performance membrane electrode assemblies (MEA) can be manufactured using novel composite membranes which include organic or inorganic materials with higher selectivity. The selectivity of membrane was calculated by the following equation:

$$Selectivity (\phi) = \frac{Value \ of \ Proton \ Conductivity}{Value \ of \ Methanol \ Permeability} \qquad (7)$$

The value of $\emptyset$ for the recast Nafion and Nafion/CNT@SiO$_2$-SO$_3$H (1 wt%) are calculated at 38222 and 313745 S.s/cm$^3$, respectively, at room temperature. As can be noted in Figure 5, the membrane modified by CNT@SiO$_2$-SO$_3$H has higher selectivity in comparison to the recast Nafion.

As mentioned above, the addition of modified CNTs to the Nafion membrane can cause the methanol permeability to be decreased because of the barrier role of inorganic added CNTs in the route of methanol transfer. On the other hand, the presence of sulfonated groups on the surface of modified CNTs improved the interaction of H$^+$ and nano channels of the Nafion structure as well as increased the proton conductivity of the prepared modified membranes. Decreasing the methanol permeability and increasing the proton conductivity in a wide range of temperatures caused the selectivity of the prepared membrane, one of the most important performance parameters of PEMs, to be improved.

## 4. Conclusion

The sulfonate-functionalized silicon dioxide coated carbon nanotube (CNT@SiO$_2$-SO$_3$H) nanoparticles were successfully synthesized from tetraethoxysilane (TEOS) and chlorosulfonic acid. According to the results attained from proton conductivity, methanol permeability, ion exchange capacity and other analyzes, the nanocomposite membranes prepared with Nafion and (CNT@SiO$_2$-SO$_3$H) nanoparticles exhibited a higher level of proton conductivity, higher water retention and lower methanol permeability as compared to the recast Nafion membrane. Proton conductivity was higher in the modified-CNTs/Nafion membrane than the Nafion membrane in a wide range of temperature. Methanol permeability decreased from $2.25 \times 10^{-6}$ cm$^2$/S for the Nafion membrane to $2.91 \times 10^{-7}$ cm$^2$/S for the modified composite membrane.

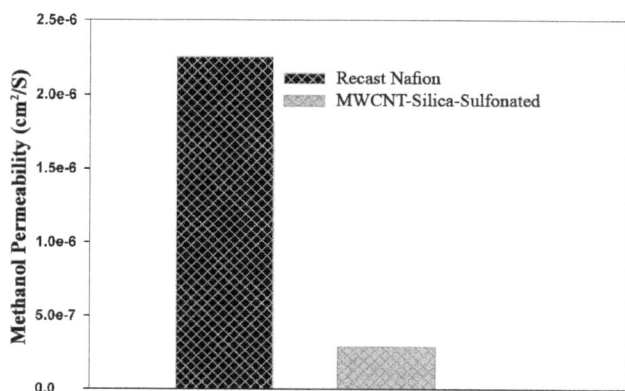

**Fig. 5.** Methanol permeability of Nafion/CNT@SiO$_2$-SO$_3$H in comparison with recast Nafion selectivity (S.s/cm$^3$).

Moreover, the selectivity of this fabricated membrane base on Nafion and CNT@SiO$_2$-SO$_3$H nanoparticles is more than 8-fold higher than that of the recast Nafion® 117, demonstrating its promising potential for DMFC applications.

## References

[1] X. Li, A. Faghri, Review and advances of direct methanol fuel cells (DMFCs) part I: design, fabrication, and testing with high concentration methanol solutions. J. Power Sources, 226 (2013) 223-240.

[2] S. Peighambardoust, S. Rowshanzamir, M. Amjadi, Review of the proton exchange membranes for fuel cell applications, Int. J. Hydrogen Energ. 35 (2010) 9349-9384.

[3] K.A. Mauritz. R.B. Moore, State of understanding of Nafion, Chem. Rev. 104 (2004) 4535-4586.

[4] M.M. Hasani-Sadrabadi, E. Dashtimoghadam, F.S. Majedi, H. Moaddel, A. Bertsch, P. Renaud, Superacid-doped polybenzimidazole-decorated carbon nanotubes: a novel high-performance proton exchange nanocomposite membrane, Nanoscale, 5 (2013) 11710-11717.

[5] A. Hacquard, Improving and understanding direct methanol fuel cell (DMFC) performance, MSc Thesis, Worcester Polytechnic Institute, 2005.

[6] M.S. Asgari, M. Nikazar, P. Molla-Abbasi, M.M. Hasani-Sadrabadi, Nafion®/histidine functionalized carbon nanotube: High-performance fuel cell membranes, Int. J. Hydrogen Energ. 38 (2013) 5894-5902.

[7] C-S. Wu and H-T. Liao, Study on the preparation and characterization of biodegradable polylactide/ multi-walled carbon nanotubes nanocomposites, Polymer, 48 (2007) 4449-4458.

[8] N.H. Jalani, K. Dunn, R. Datta, Synthesis and characterization of Nafion®-MO$_2$ (M= Zr, Si, Ti) nano-composite membranes for higher temperature PEM fuel cells, Electrochim. Acta. 51 (2005) 553-560.

[9] Z-G. Shao, H. Xu, M. Li, I-M. Hsing, Hybrid Nafion-inorganic oxides membrane doped with heteropolyacids for high temperature operation of proton exchange membrane fuel cell, Solid State Ionics, 177 (2006) 779-85.

[10] M. Amjadi, S. Rowshanzamir, S. Peighambardoust, S. Sedghi, Preparation, characterization and cell performance of durable Nafion/SiO$_2$ hybrid membrane for high temperature polymeric fuel cells, J. Power Sources, 210 (2012) 350-357.

[11] D. Jung, S. Cho, D. Peck, D. Shin, J. Kim, Performance evaluation of a Nafion/silicon oxide hybrid membrane for direct methanol fuel cell, J. Power Sources, 106 (2002) 173-177.

[12] K. Adjemian, S. Lee, S. Srinivasan, J. Benziger, A. Bocarsly, Silicon oxide Nafion composite membranes for proton-exchange membrane fuel cell operation at 80-140 °C, J. Electrochem. Soc. 149 (2002) A256-A261.

[13] H-C. Chien, L-D. Tsai, C-P. Huang, C-y. Kang, J-N. Lin, F-C. Chang, Sulfonated graphene oxide/ Nafion composite membranes for high-performance direct methanol fuel cells, Int. J. Hydrogen Energ. 38 (2013) 13792-13801.

[14] J-H. Kim, S-K. Kim, K. Nam, D-W. Kim, Composite proton conducting membranes based on Nafion and sulfonated SiO$_2$ nanoparticles, J. Membrane Sci. 415 (2012) 696-701.

[15] I.D. Rosca, F. Watari, M. Uo, T. Akasaka, Oxidation of multiwalled carbon nanotubes by nitric acid, Carbon, 43 (2005) 3124-3131.

[16] P. Molla-Abbasi, K. Janghorban, M.S. Asgari, A novel heteropolyacid-doped carbon nanotubes /Nafion nanocomposite membrane for high performance proton-exchange methanol fuel cell applications, Iran. Polym. J. 27 (2018) 77-86.

[17] W. Stöber, A. Fink, E. Bohn, Controlled growth of monodisperse silica spheres in the micron size range, J. Colloid Interf. Sci. 26 (1968) 62-69.

[18] Y. Xiong, Z. Zhang, X. Wang, B. Liu, J. Lin, Hydrolysis of cellulose in ionic liquids catalyzed by a magnetically-recoverable solid acid catalyst, Chem. Eng. J. 235 (2014) 349-355.

[19] A. Singhvi, S. Gomathy, P. Gopalan, A. Kulkarni, Effect of aliovalent cation doping on the electrical conductivity of Na$_2$SO$_4$: Role of charge and size of the dopant, J. Solid State Chem. 138 (1998) 183-192.

[20] J-M. Thomassin, J. Kollar, G. Caldarella, A. Germain, R. Jérôme, C. Detrembleur, Beneficial effect of carbon nanotubes on the performances of Nafion membranes in fuel cell applications, J. Membrane Sci. 303 (2007) 252-257.

[21] J. Sun, X. Jiang, A. Siegmund, M.D. Connolly, K.H. Downing, N.P. Balsara, Morphology and proton transport in humidified phosphonated peptoid block copolymers, Macromolecules, 49 (2016) 3083-3090.

# 17

# Electrophoretic deposition and corrosion behavior study of aluminum coating on AZ91D substrate

**Hossein Aghajani***, **Maryam Pouzesh**

*Department of Materials Engineering, University of Tabriz, Tabriz, Iran*

## HIGHLIGHTS

- In this study, aluminum powder coating was developed on AZ91D magnesium alloy substrate by electrophoretic deposition.

- To determine the optimal condition of deposition, the effects of $AlCl_3.6H_2O$ concentration, applied voltage, and deposition time were investigated.

- A well-stabilized suspension and a uniform deposition were obtained at the $AlCl_3.6H_2O$ concentration of 0.6 mM, applied voltage of 70 V and deposition time of 18 min.

## GRAPHICAL ABSTRACT

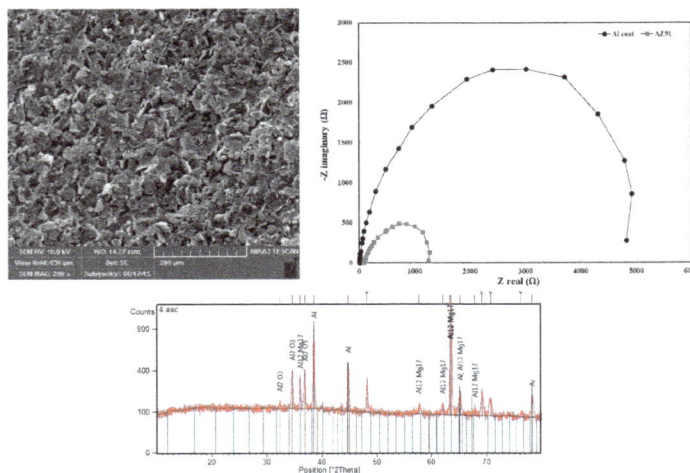

## ARTICLE INFO

*Keywords:*

Aluminum coating
Electrophoretic deposition
AZ91D magnesium alloy
Coating morphology
Suspension stability

## ABSTRACT

Aluminum coating was prepared on AZ91D magnesium alloy substrate using the electrophoretic deposition (EPD) method in absolute ethanol solvent. In order to determine the optimal concentration of $AlCl_3.6H_2O$ additive, the zeta potential and size of particles in the suspension were measured in the presence of different concentrations of $AlCl_3.6H_2O$. The results showed that an appropriate coating is obtainable in the presence of 0.6 mM $AlCl_3.6H_2O$ as an additive. The effects of applied voltage, deposition time, and additive concentration on deposition weight, deposition thickness, and coating morphology were also studied. A uniform coating with smaller pores and higher density was obtained at the additive concentration of 0.6 mM, deposition time of 18 min, and applied voltage of 70 V. The thickness of this coating was measured at about 256.91 µm. According to the results of corrosion behavior studies, the corrosion current density was measured at 29.16 and 12.85 µA/cm² for uncoated and aluminum-coated AZ91D alloy, respectively.

*\* Corresponding author: ; E-mail address: h_aghajani@tabrizu.ac.ir*

# 1. Introduction

Magnesium alloys possess excellent mechanical and physical properties such as high specific strength and stiffness, low density [1], high strength-to-weight ratio [2], good electromagnetic shielding [3], great damping capability [4], and satisfactory thermal and electrical conductivity [5,6]. These alloys are widely used in automotive, aerospace, military, electronic [4,7] and ceramic industries [8]. Another application of EPD was reported by [9] to fabricate a $YSZ/Al_2O_3$ nanostructured composite coating on an iron-nickel based superalloy. Additionally, these alloys suffer from high flammability, low melting point, high chemical activity, and low corrosion resistance, resulting in limited industrial applications [10].

Generally, the corrosion resistance of magnesium alloys can be improved using heat treatment and coating processes [1]. A series of coating methods and surface treatments has been developed to improve the corrosion, wear, and heat resistance of these alloys. Among these methods the electrophoretic deposition (EPD) technique is well-considered with a variety of new applications in coating technology [11-13]. This is not only due to its versatility and ability to combine with various materials, but also because of the simple accessories required for this technique [14,5]. During EPD, charged powder particles dispersed in a liquid medium are moved and deposited on a conductive substrate with the opposite charge by applying a DC electric field [14]. Aluminum has many advantages such as good corrosion resistance, thermal and electrical conductivity, and excellent mechanical properties [15,16].

In the present work, the electrophoretic deposition of aluminum on AZ91D magnesium alloy substrate was studied. The dispersion of the suspensions was investigated in the presence of different concentrations of $AlCl_3 \cdot 6H_2O$. In addition, the effects of applied voltage and deposition time on coating morphology were thoroughly examined. Finally, heat treatment and corrosion studies were performed.

# 2. Factors affecting EPD

It should be noted that the kinetics of electrophoretic deposition and the deposition quality depend on a large number of parameters which are related to the suspension and its process. The parameters related to the suspension are the particle size, dielectric constant of liquid, conductivity of suspension, viscosity of suspension, zeta potential and stability of suspension. Also, the process related parameters are the concentration of solid in suspension, conductivity of substrate, applied voltage and deposition time.

Some of these parameters are inter-related to one another. It is noted that the quality of electrophoretic deposition depends heavily on the suspension conditions [17]. In general, a stable suspension can provide a better deposition during the EPD process. The stability of suspension can be measured by zeta potential. Generally, its higher absolute value shows a better dispersion of the particles in the suspension. The electrical conductivity of the suspension has an important role in the process during EPD [18]. Experiments have shown that as the ionic concentration in the suspension increases, the conductivity of the suspension increases rapidly [17]. Also, the dielectric constant of the suspending medium directly affects the conductivity of suspension and it increases as the dielectric constant increases [14]. After fixing the suspension parameters, the process parameters can be chosen to have a desired deposition. Normally the amount of deposit increases as the applied voltage increases. Similarly, a higher deposition rate is expected with increasing particle concentration and deposition time [14].

# 3. Experimental procedure

Aluminum powder (Sigma-Aldrich, 99.5%) with a flake shape and a mean particle size of <5 μm was used as the raw material. Aluminum chloride hexahydrate $AlCl_3 \cdot 6H_2O$ (Beijing Guohua Chemical Factory, China) was employed as the additive, and absolute ethanol (99.6%) was used as the solvent. The Al powder (10 g/l) was dispersed in ethanol and different amounts of $AlCl_3 \cdot 6H_2O$ (0.1-5 mM) were added to the suspensions. The suspensions were magnetically stirred for 24 hours and then ultrasonically deflocculated for 180 min to prepare a well-dispersed stable suspension.

An AZ91D magnesium alloy with a thickness of 2 mm and working area of 1.44 cm² was utilized as the substrate (cathode). In addition, a low-carbon 316 stainless steel with the same working area and a thickness of 0.1 mm was used as the anode. The distance between the two parallel electrodes was fixed at 1.2 cm during deposition.

The EPD of Al particles was performed using additive concentrations in the range of 0.1 to 5 mM, applied voltage in the range of 10 to 80 V, and deposition time in the range of 2 to 18 min. The zeta potential of the suspensions with different concentrations of $AlCl_3 \cdot 6H_2O$ was measured using a zeta potential analyzer (Malvern-HSA3000), and the deposit weight was measured by weighing the cathode before and after deposition (RADWAG ± 0.0001 g). The surface morphology and thickness of coatings were studied by scanning electron microscopy (FE-SEM, TESCAN-MIRA3 FEG). The size of particles in the suspensions was determined using the dynamic light scattering (DLS) system (Microtrac Nanotrac Wave). Finally, the corrosion behavior of the coating was examined using the electrochemical impedance spectrum.

## 4. Results and discussion

The kinetics of electrophoretic deposition and coating quality is highly dependent on a number of parameters, e.g. applied voltage, additive concentration, deposition rate, and substrate conductivity. Hence, a proper control mechanism should be considered on individual parameters in the process of electrophoretic deposition [14,19].

### 4.1. Effect of $AlCl_3 \cdot 6H_2O$ concentration on zeta potential

Yang et al. studied the influence of $AlCl_3 \cdot 6H_2O$ additive concentration on the stability of aluminum suspensions and rate of deposition were studied [20]. The results showed that positively charged particles of aluminum should be deposited on the cathodic substrate during the deposition process. While the deposition rate is directly related to zeta potential, zeta potential is much lower in the alkaline range than the acidic range [14]. Zeta potential increases and acidity decreases with an increase in additive concentration [21]. Zeta potential has more impact on the stability of the suspension and electrophoretic mobility and can be changed by adding additives such as $AlCl_3 \cdot 6H_2O$. The effect of additive concentration on the zeta potential of ethanol-contained suspension of aluminum particles is depicted in Figure 1.

It should be noted that the amount and type of additive has a great influence on the charging of the particles present in the suspension. All zeta potentials are positive including the suspension without $AlCl_3 \cdot 6H_2O$. This indicates that positively charged Al particles should be deposited on the cathode substrate. For a suspension without $AlCl_3 \cdot 6H_2O$, a reasonable mechanism to adjust the charge is to produce $H^+$ ions from small amounts of existing $H_2O$ in the commercial alcohol by electrolytic dissociation, and then absorb these onto the aluminum particles to make them electrified. When metal ions are introduced into the suspension through the addition of $AlCl_3 \cdot 6H_2O$, the resulting aluminum alkoxide and aluminum hydroxide ions are absorbed on the surface of aluminum particles and make a surface charge density [20].

Based on Figure 1, zeta potential increases as the $AlCl_3 \cdot 6H_2O$ concentration is increased up to 1 mM, but then decreases with further increases in the additive concentration. It can be concluded that the increase in zeta potential is due to the enhancement of metal ions in the suspension as a result of additive concentration enhancement. Thus, the density of particle surface charge increases with the absorption of aluminum hydroxide and aluminum alkoxide on the aluminum particles in the suspension. In addition, by increasing the surface charge the electrostatic repulsion force increases between the particles leading to zeta potential enhancement.

Increasing the metal ions and their attraction on the surface of aluminum particles may reduce the thickness of the electrical double layer. This may lead to a reduction in repulsive forces between particles; and consequently, a reduction in zeta potential. Therefore, the reduction of zeta potential in 1.5 mM additive concentration can be attributed to the reduced thickness of the electrical double layer. Nevertheless, higher zeta potential values may be undesirable during the deposition process since they may lead to an excessively conductive suspension and a reduction of electrophoretic mobility.

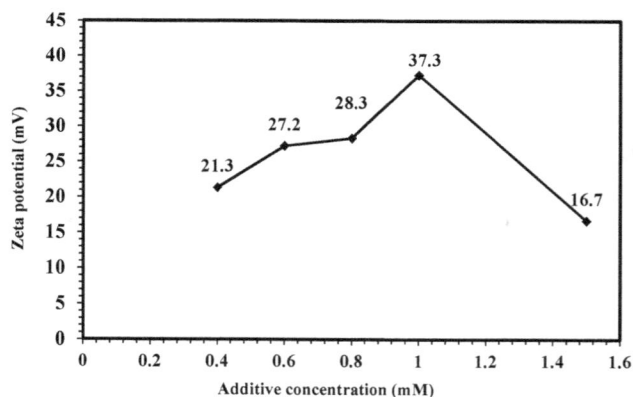

**Fig. 1.** Zeta potential as a function of additive concentration for Al particles in ethanol.

## 4.2. Effect of AlCl₃·6H₂O concentration on the size of particles

The effect of additive concentration on the size of particles is demonstrated in Figure 2. Although there is no general rule for determining the size of particles for electrophoretic deposition, a suitable deposition has been reported in the range of 1 to 20 μm [14]. According to Figure 2, the size of particles in suspensions in the presence of 0.4 to 1.5 mM additive is in the range of 1 to 5 μm. At low additive concentrations, there are few free ions such as aluminum hydroxide and aluminum in the suspension, proving little surface charge on the particles. Therefore, the electrostatic repulsion force necessary for separating the particles is not provided, which can lead to particle agglomeration and increased size.

By increasing the additive concentration up to 0.6 mM, the amount of free ions increased, leading to the enhancement of surface charge on the surface of aluminum particles. Then, the electrostatic repulsion force between particles increased and prevented particle agglomeration. By further increasing the additive concentration up to 1.5 mM, the amount of metal ions and conductivity of the suspension increased. However, the excessive amount of metal ions causes a reduction in the thickness of the electrical double layer. In this case, the particles agglomerate and the size of particles increases. Results revealed that the suspension with the additive concentration of 0.6 mM was a well-stabilized suspension because it had fine particles and an acceptable zeta potential value.

## 4.3. Effect of applied voltage on deposition rate and coating morphology

The surface morphologies of aluminum coatings in

the presence of 0.6 mM AlCl₃·6H₂O at various applied voltages are illustrated in Figure 3. In this figure, white areas indicate coated aluminum particles and gray and black areas represent the porosities or less-coated surface of the substrate. A comparison between the results shows that the coating deposited at the applied voltage of 70 V (Figure 3e) is denser than the others.

The surface morphology of the coatings in the presence of 1 mM AlCl₃·6H₂O and different applied voltages are demonstrated in Figure 4. It is clear that the coating deposited at the applied voltage of 30 V has a lower porosity. By further increasing the applied voltage up to 40 V, the porosities increase and a non-uniform coating forms due to the high velocity of particles.

The deposition weight as a function of applied voltage for the deposition time of 3 min is depicted in Figure 5. According to this figure, by increasing the applied voltage up to 70 V, the deposition weight increases and then decreases for both 0.6 mM and 1mM additive concentrations. At applied voltages higher than 70 V turbulent currents are created which may damage the coating and affect its quality, leading to a reduction in deposition weight.

Figure 5 shows that in the case of 1mM additive concentration the weight of deposition increases slowly at applied voltages lower than 50 V and causes the creation of a uniform coating. However at applied voltages higher than 50 V, the rate of deposition is high leading to the agglomeration of particles and creation of a non-uniform deposition. Therefore, in the presence of 1 mM additive concentration, an applied voltage of lower than 50 V is required to deposit a uniform coating. According to Figures 3 and 5, the applied voltage of 30 V was chosen as the appropriate voltage to have a deposition with a lower porosity and higher homogeneity. In addition, in the case of 0.6 mM additive concentration, voltage of 70 V was selected as the optimum value due to the dense structure and high coating weight shown in Figures 4 and 5, respectively.

## 4.4. Effect of AlCl₃·6H₂O concentration on coating morphology

The surface morphology of coatings for 0.6 and 1 mM additive concentrations are illustrated in Figure 6. Additive concentration affects the viscosity of the suspension. The relationship between viscosity and zeta potential is presented in Eq. (1), which shows that

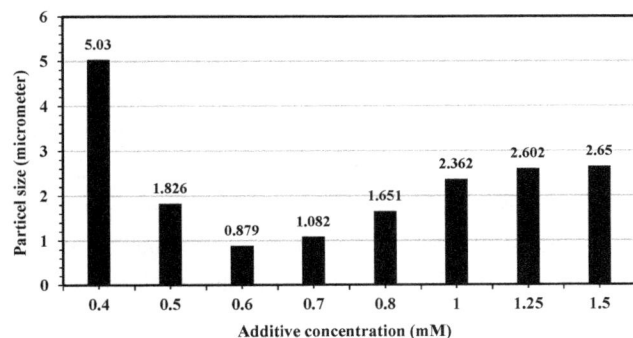

**Fig. 2.** Size of particles as a function of additive concentration for Al particles in ethanol.

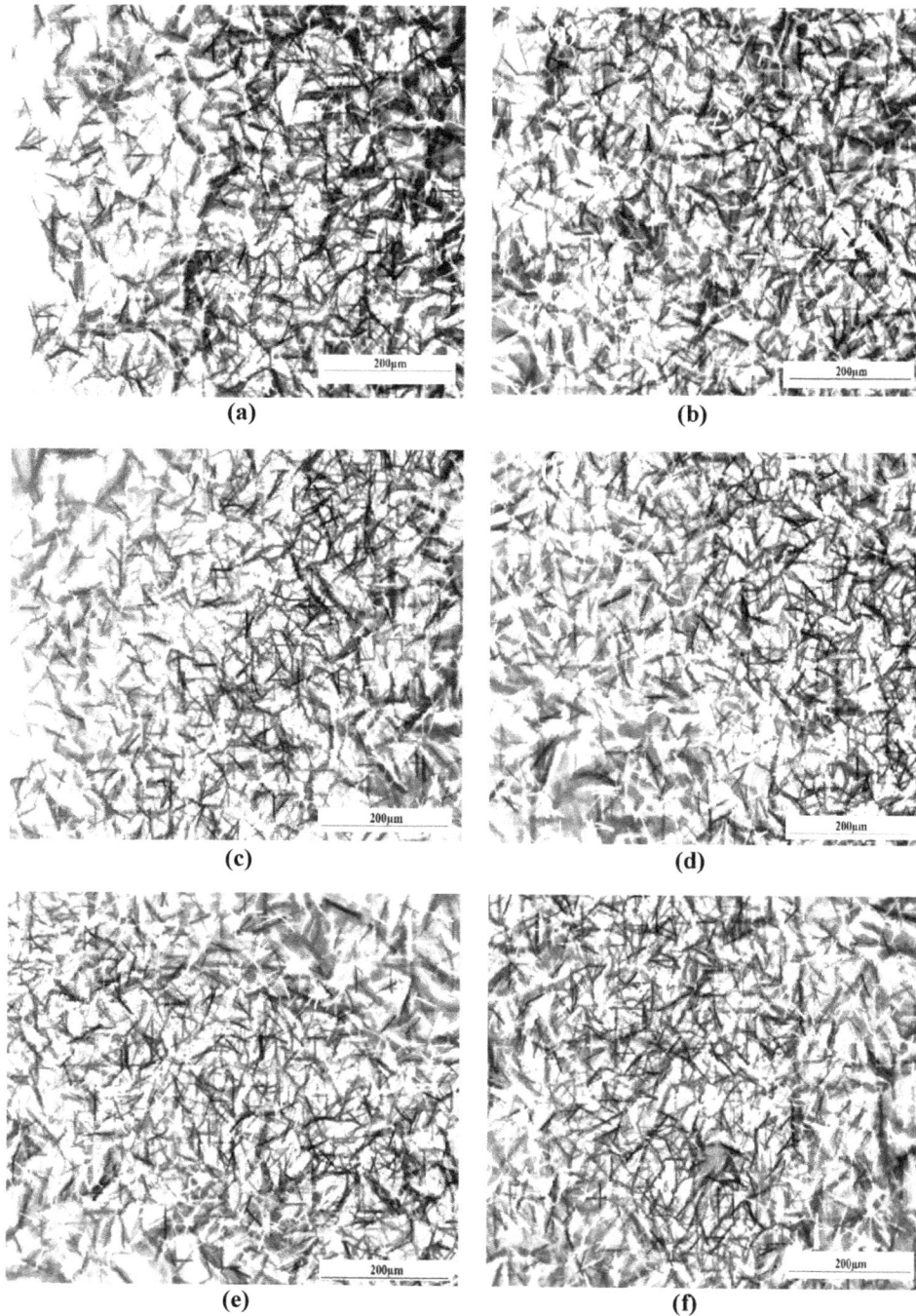

**Fig. 3.** Optical microscope images of aluminum coatings at different applied voltages for 0.6 mM additive concentration and 5 min deposition time, (a) 30 V, (b) 40 V, (c) 50 V, (d) 60 V, (e) 70 and (f) 80 V.

that appropriate conditions for homogeneous deposition can be provided with maximum zeta potential and minimum viscosity [14].

$$\mu = \frac{\varepsilon\varepsilon_0\xi}{\eta} \tag{1}$$

where $\xi$, $\eta$, $\varepsilon_0$, $\varepsilon$ and $\mu$ denote the zeta potential of particles, viscosity of the solvent, vacuum permittivity coefficient, relative permittivity coefficient of the solvent, and electrophoretic mobility, respectively.

As mentioned before, the zeta potential and surface charge of particles increase by increasing additive concentration. In addition, the number of metal ions in the suspension increases. In this case, the majority of the current is carried by free ions in the suspension and only a small portion of it is assigned for the movement of charged particles. Thus, it results in the agglomeration of particles and reduction of electrophoretic mobility.

(a)

(b)

(a)

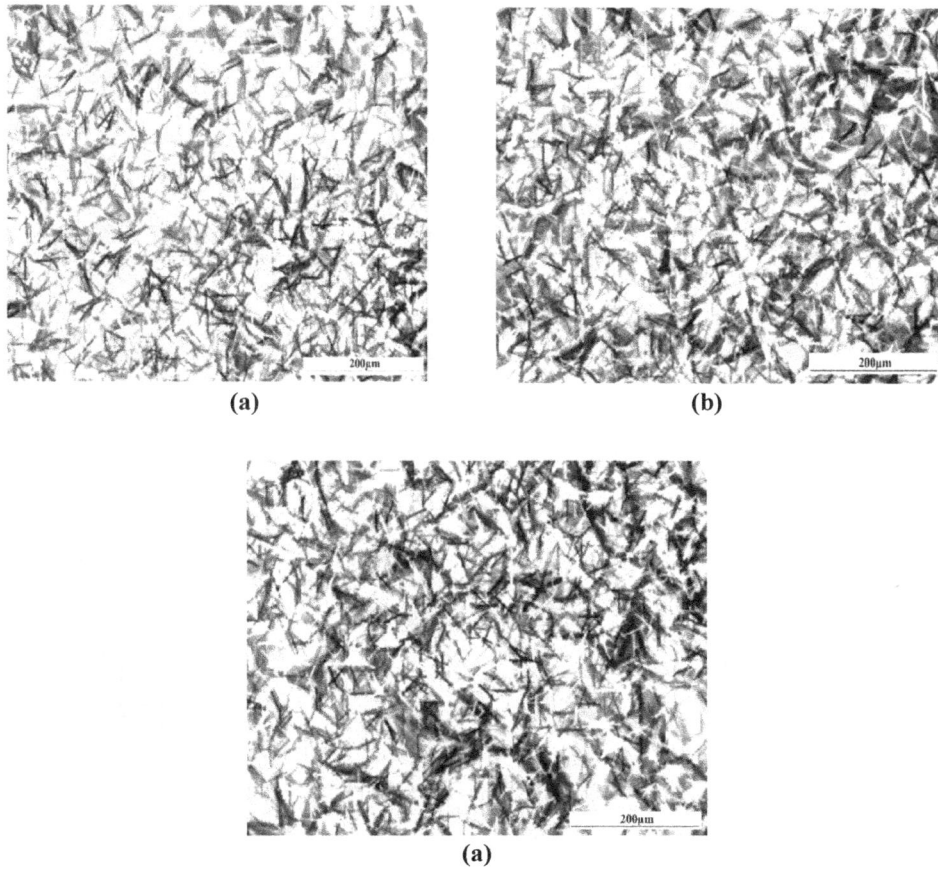

**Fig. 4.** Optical microscope images of aluminum coatings for 5 min deposition time and 1 mM additive concentration. (a) 30 V, (b) 35 V, and (c) 40 V.

charged particles. Thus, it results in the agglomeration of particles and reduction of electrophoretic mobility.

Figure 6 indicates that uniform deposits can be achieved in the presence of 0.6 mM additive concentration. Moreover, the agglomeration of particles in this suspension; and subsequently, the pores in the coating microstructure are less than those of the 1 mM suspension. As a result, the surface morphology of the

**Fig. 5.** The coating weight as a function of the applied voltage at 3 min deposition time.

coating deposited from the 1 mM suspension is coarse and large porosities are observable in this coating.

## 4.5. Effect of deposition time on deposition rate and coating morphology

In order to evaluate the effect of deposition time on deposition weight and its morphology, the structures of coatings deposited at a constant voltage are compared. The surface morphology of aluminum coatings at different deposition times and constant applied voltage of 70 V for 0.6 mM additive concentration is presented in Figure 7. At the initial times of deposition the concentration of particles in the suspension was high, but the time was not adequate for deposition of particles on the surface of the substrate. Therefore, the surface is not completely covered by Al particles and the coatings are not uniform and have a low density.

According to Figure 7, it is clear that a uniform deposition with minimum porosity and high density is obtainable at the deposition time of 18 min. By further increasing the deposition time up to 20 min, the

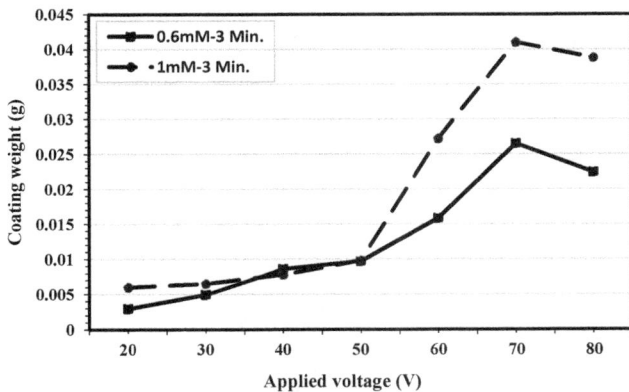

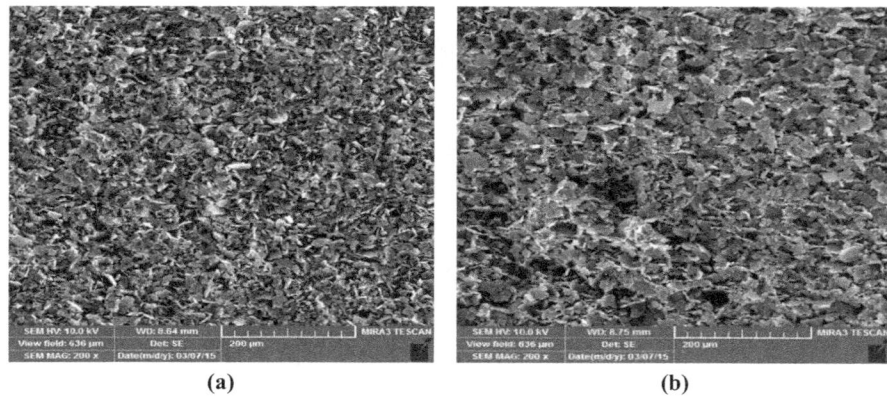

(a)                                                        (b)

**Fig. 6.** The surface morphology of aluminum coatings for (a) 0.6 mM and (b) 1 mM concentrations of $AlCl_3.6H_2O$ (deposition time: 18 min, applied voltage for 0.6 and 1 mM: 70V and 30V, respectively).

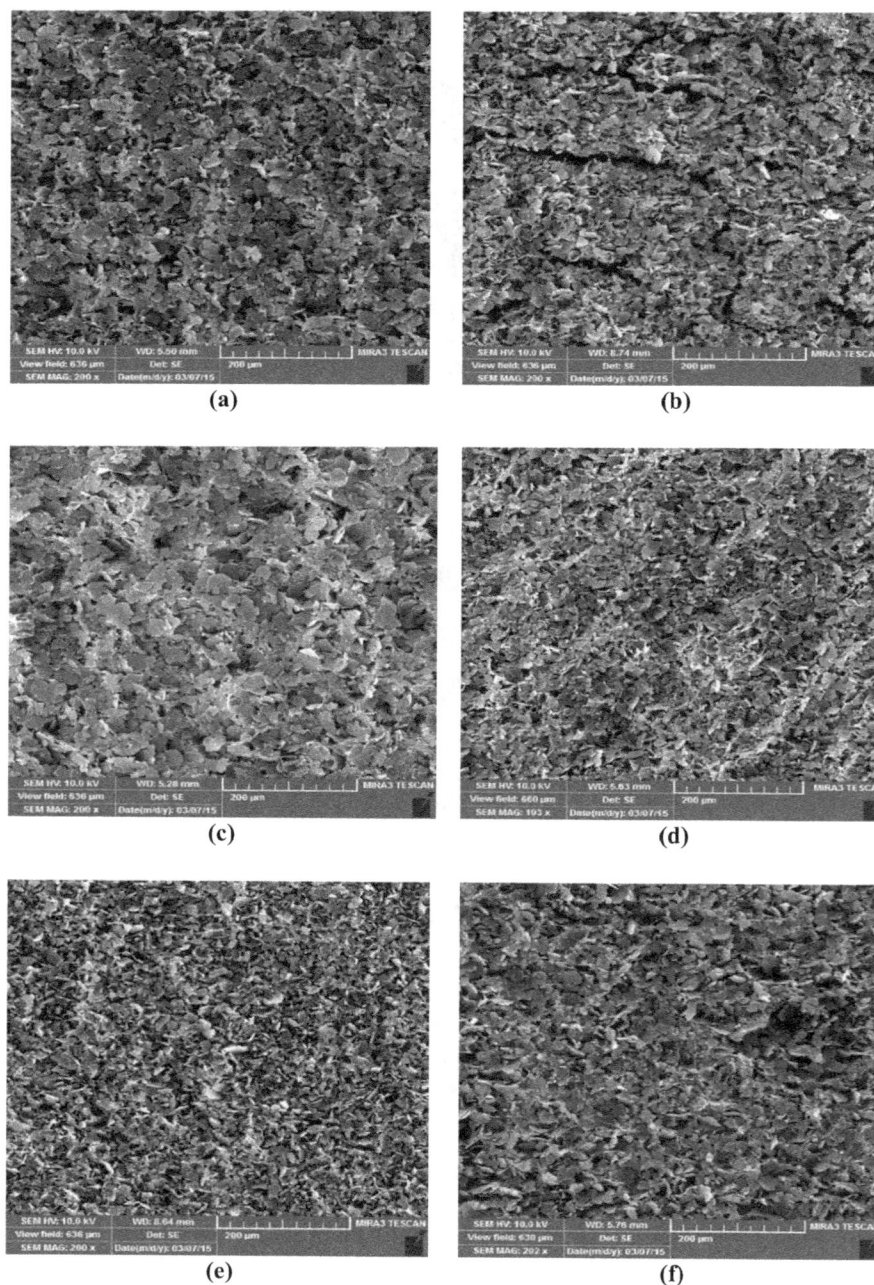

(a)                                                        (b)

(c)                                                        (d)

(e)                                                        (f)

**Fig. 7.** SEM images of surface of the coatings for 0.6 mM additive concentration, coated for (a) 2, (b) 6, (c) 10, (d) 12, (e) 18, and (f) 20 min.

concentration of particles in the suspension decreased and the electrical resistance of the substrate increased. This phenomenon causes the detachment of particles from the coating and reduction of the deposition weight.

The surface morphology of aluminum coatings at the constant applied voltage of 30 V, additive concentration of 1 mM, and various deposition times is shown in Figure 8. It is clear that at the initial times of deposition the substrate surface is not completely covered with coating and contains a low-density layer. Although the density of the coating increases by increasing the deposition time, several large pores are seen in the coating structure (Figure 8d) due to the high concentration of metal ions, this leads to a reduction in the electrophoretic mobility and enhancement in viscosity.

The variation of coating weight versus the deposition time is depicted in Figure 9 for two different additive concentration values. It is obvious that at initial deposition times the weight of deposition increases with time and reaches the highest value in 18 min. By further increasing the deposition time up to 20 min, the deposition weight decreases in both concentrations

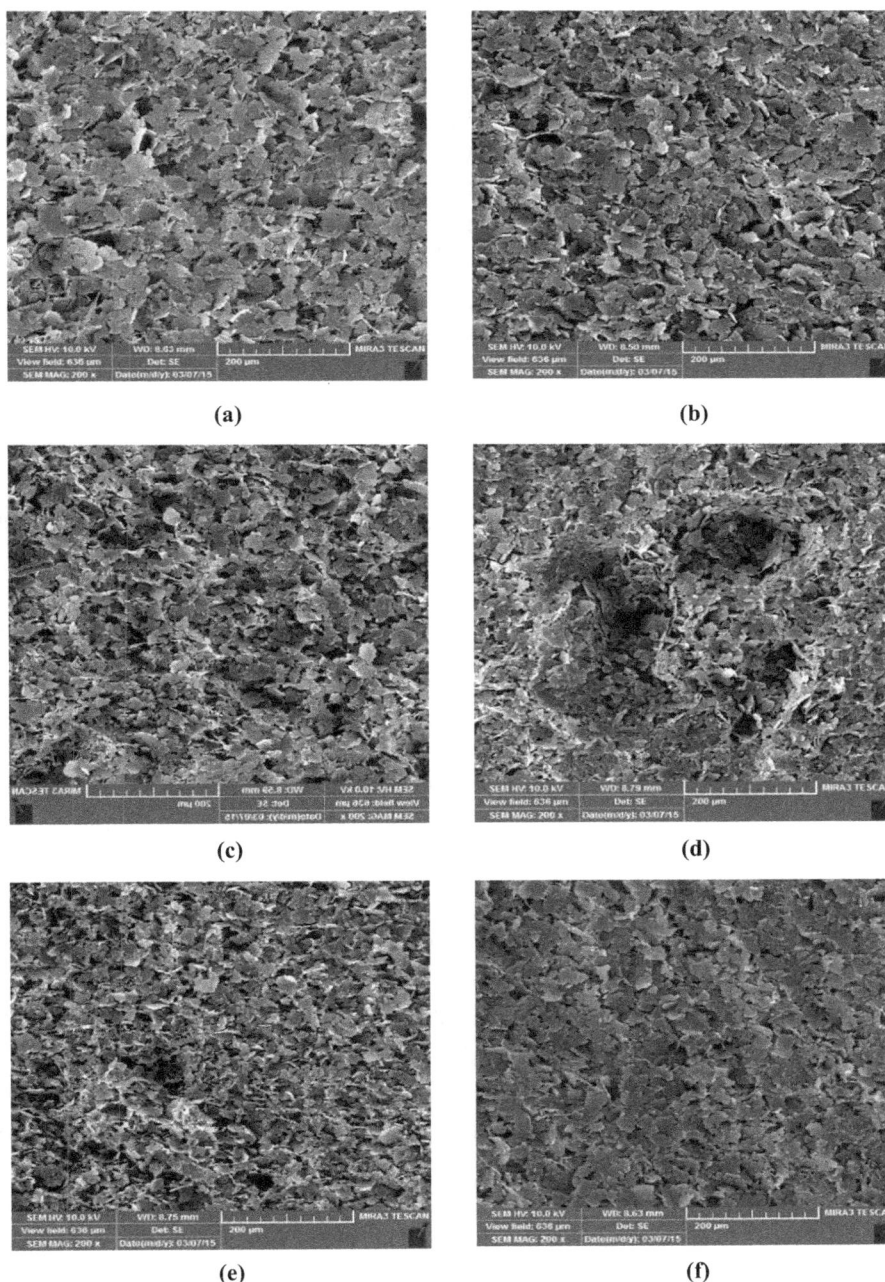

Fig. 8. SEM images of surface of the coating for 1 mM additive concentration, coated for (a) 2, (b) 6, (c) 10, (d) 12, (e) 18 and (f) 20 min.

due to the reduction of suspension concentration and enhancement of the electrical resistance of the substrate. From Figure 9, it is clear that the coating weight for 0.6 mM additive concentration is greater than that of the 1 mM one. This behavior is because of the higher stability of the suspension with 0.6 mM additive concentration. As a result, the particles retain their stability even during prolonged times of deposition leading to less settlement and enhancement of deposition weight.

## 4.6. Thickness analysis of depositions

Scanning electron microscopy (SEM) was used in order to determine the thickness of coatings. The cross-section images of the coatings for both 0.6 and 1mM additive concentrations at two different deposition times are demonstrated in Figure 10. Coating thickness increased as the deposition time increased. The comparison between the cross-sectional views of the coatings show that the thickness of deposition at 1 mM additive concentration is greater than the 0.6 mM one, while the obtained coating weight is higher in 0.6 mM than 1 mM. This is due to the formation of a dense and uniform coating during deposition in the presence of 0.6 mM additive concentration.

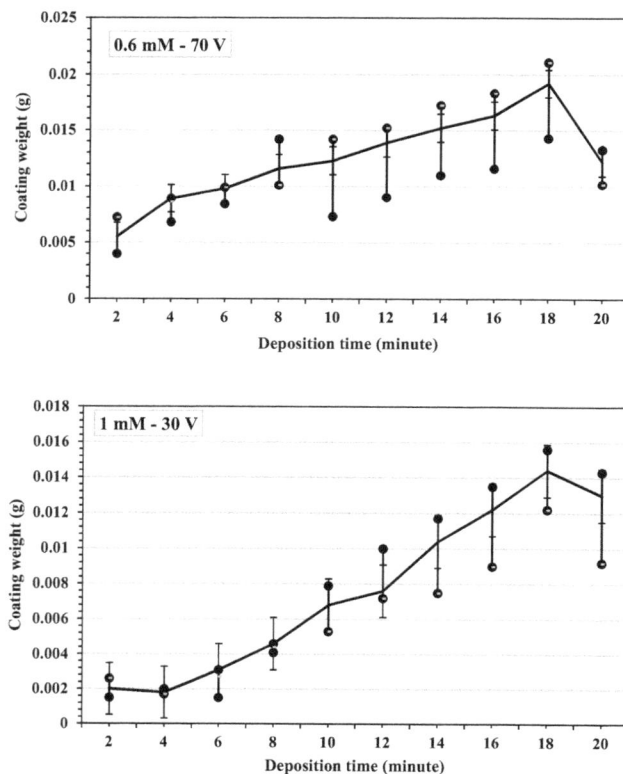

**Fig. 9.** Coating weight versus deposition time.

According to the obtained results, the coating with the additive concentration of 0.6 mM, deposition time of 18 min, and applied voltage of 70 V, is a suitable coating and has appropriate thickness (256.91 μm), weight (0.019 g), and density of deposition.

## 4.7. Heat treatment and microstructural studies

In order to improve the adhesion and density of the green coating, the heat treatment of the optimum coating was carried out at 400 °C for 1 hour. After heat treatment, in order to decrease the probability of crack formation, the sample was slowly cooled in the furnace chamber. To determine the density and adhesion of the coating to substrate, the surface morphology was studied using SEM analysis and the results are provided in Figure 11.

It is clear that heat treatment reduced the porosities and increased the density of the coating. This may be due to the expansion of aluminum particles, their oxidization with increasing temperature, and the formation of an alumina phase. On the other hand, the coating has a high adhesion strength due to the formation of an intermetallic $\beta$ phase ($Mg_{17}Al_{12}$) at the interface of substrate and coating.

In order to determine the resulted phases in the obtained coating, X-ray diffraction analysis (XRD) and energy dispersive spectroscopy (EDS) were utilized. The results of XRD analysis after heat treatment are shown in Figure 12. The results indicated that, beside the metallic FCC aluminum phase, $Al_2O_3$ and $Mg_{17}Al_{12}$ phases are detected in the coating structure. The $Al_2O_3$ phase is formed due to the reaction of aluminum with oxygen in the air, and the intermetallic phase of $\beta$ ($Mg_{17}Al_{12}$) is formed due to the melting of the substrate surface during the heat treatment, which can improve the coating adhesion to the substrate.

Al and $Al_2O_3$ peaks gradually increase and the peaks of $Mg_{17}Al_{12}$ decrease as the coating thickness increases. The alumina ($Al_2O_3$) phase formed during heat treatment plays the role of sintering aid and compensates for the volume shrinkage caused by sintering. This phenomenon leads to the enhancement of the density and adhesion of the coating. The formation of a $\beta$ ($Mg_{17}Al_{12}$) phase at the interface of the coating and substrate improves the adhesion of the coating to the substrate.

The results of EDS analysis of the samples before and after heat treatment are presented in Figure 13. The amounts of elements present in the coatings before and

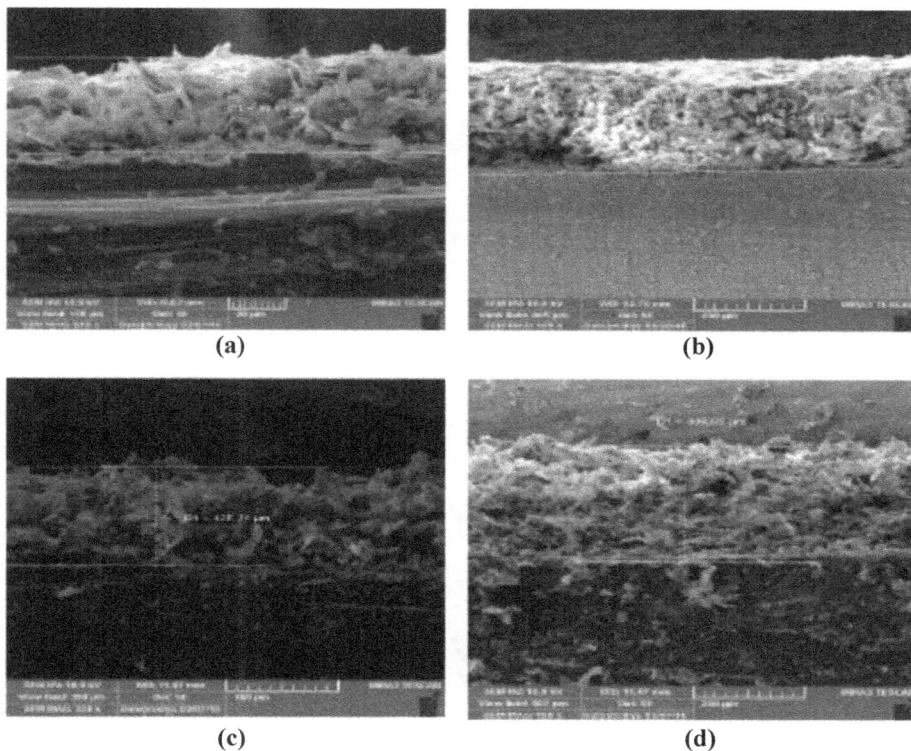

**Fig. 10.** SEM images of the cross-section of coated samples (a) 2 min, 70 V, 0.6 mM; (b) 18 min, 70 V, 0.6 mM; (c) 2 min, 30 V, 1 mM and (d) 20 min, 30 V, 1 mM.

after heat treatment are given in Figure 14. It is obvious that the weight percentage of Mg and O elements increased and the weight percentage of Al decreased after heat treatment.

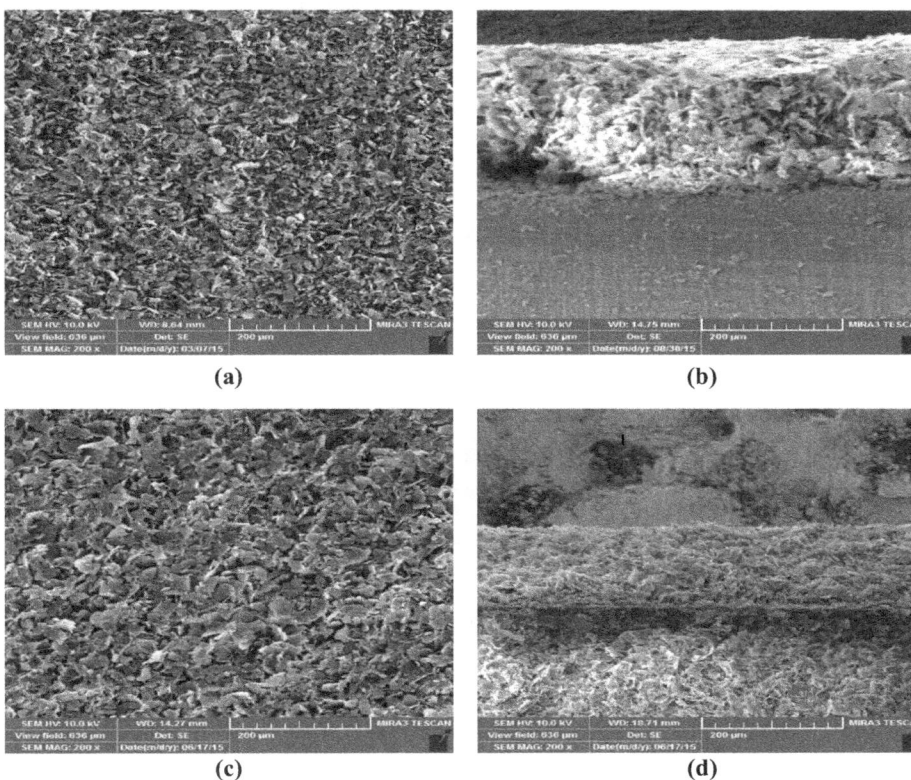

**Fig. 11.** SEM images of surface and cross-section of the coatings, (a) surface and (b) cross-section of the sample before heat treatment, (c) surface and (d) cross-section of the sample after heat treatment at 400 ° C for 1 hour.

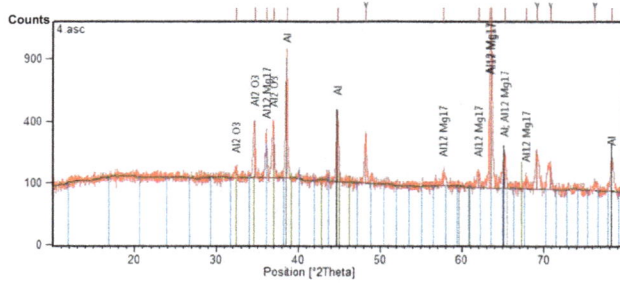

**Fig. 12.** Results of XRD analysis after heat treatment.

**Fig. 14.** Percentage of element weights before and after heat treatment.

## 4.8. Corrosion resistance analysis

The corrosion resistance study of the uncoated and aluminum-coated AZ91D magnesium alloy was performed by polarization and impedance tests. To this end, a solution of 3.5% NaCl was used. Figure 15 illustrates the potentiodynamic polarization curves of both samples.

Polarization resistance, which shows the resistance to transmission time, is calculated using the polarization test results as follows [22]:

$$R_p = \frac{\beta_a \times \beta_c}{2.303 \times i_{corr} \times (\beta_a + \beta_c)} \qquad (2)$$

where $\beta_a$ and $\beta_c$ are slopes of the Tafel anode and cathode (V/decade), respectively. The $i_{corr}$ denotes the corrosion current (amperes). Also, the corrosion rate (CR) in terms

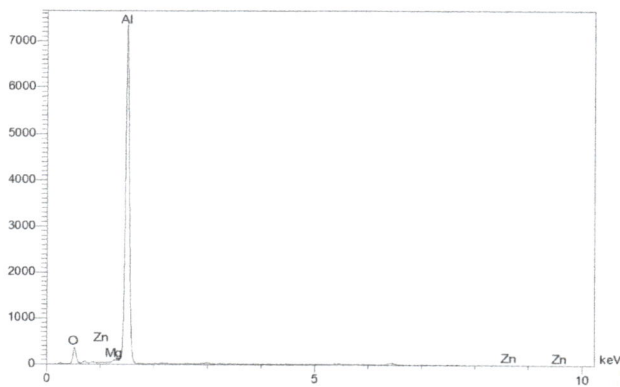

of milligrams per year (mpy) can be calculated as [22]:

$$mpy = \frac{K \times M \times i_{corr}}{Z \times D \times F} \qquad (3)$$

in which M is the base metal atomic mass (in g), D shows the density in g/cm³, F represents the Faraday constant, Z refers to the atomicity, and K=0.129. Table 1 shows the parameters of dynamic polarization test for uncoated and coated samples.

Corrosion current density is associated with the resistance to corrosion. Based on Figure 15, the corrosion current density for the coated sample is less than that of the AZ91D alloy, reflecting the higher corrosion resistance of the coated sample compared to the uncoated one. Moreover, both coated and uncoated samples have an active behavior in the 3.5% NaCl solution and the passive area was not found.

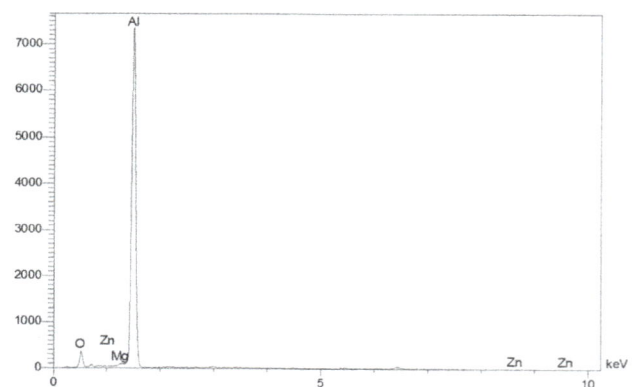

| Element | Line | Intensity | Weight % | Atomic % |
|---------|------|-----------|----------|----------|
| O | Ka | 0.0 | 0.00 | 0.00 |
| Mg | Ka | 0.0 | 0.00 | 0.00 |
| Al | Ka | 4423.5 | 97.04 | 98.75 |
| Zn | Ka | 8.2 | 2.96 | 1.25 |
| | | | 100.00 | 100.00 |

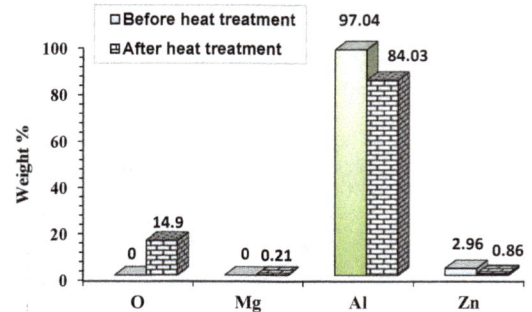

| Element | Line | Intensity | Weight % | Atomic % |
|---------|------|-----------|----------|----------|
| O | Ka | 206.3 | 14.90 | 22.90 |
| Mg | Ka | 15.6 | 0.21 | 0.21 |
| Al | Ka | 6170.8 | 84.03 | 76.56 |
| Zn | Ka | 3.9 | 0.86 | 0.32 |
| | | | 100.00 | 100.00 |

**Fig. 13.** EDS analysis of cross-section (a) before heat treatment and (b) after heat treatment.

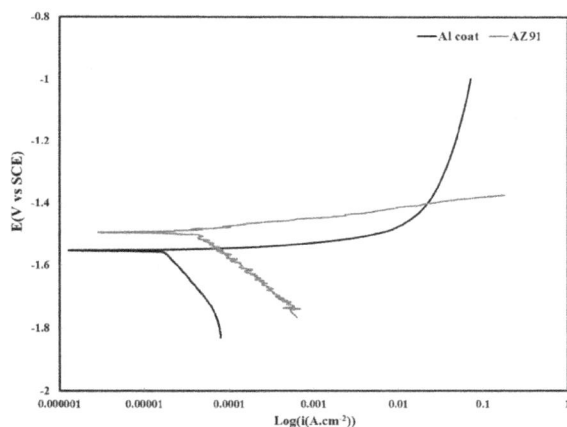

**Fig. 15.** Polarization curve of uncoated and aluminum coated AZ91D magnesium in salt solution NaCl 3.5%.

According to Table 1, the coated sample has a lower corrosion and higher corrosion resistance in comparison with the uncoated AZ91D magnesium, which is due to the higher cathodic potential of aluminum in comparison with AZ91D magnesium alloy. This fact leads to the protection of the substrate from corrosive solution in the presence of aluminum coating.

The presence of cracks and open porosities in the coating can provide a path for the corrosive solution to reach to the substrate, which subsequently leads to galvanic corrosion between the cathodic aluminum coating and anode magnesium substrate. Hence, the thickness and density of the coating must be modified in such a way to minimize galvanic corrosion. The passive layer of alumina and aluminum hydroxide can be a barrier against the penetration of corrosive solution, and corrosion resistance can be improved. The Nyquist curves of uncoated and aluminum-coated AZ91D samples in the 3.5 wt% NaCl solution are shown in

Figure 16. The capacitive ring diameter in the Nyquist curve represents the polarization resistance (resistance to corrosion) of the electrode. A higher polarization resistance represents a lower corrosion rate. According to Figure 16, the capacitive ring for the coated AZ91D alloy is much bigger than that of the uncoated one, showing that the corrosion rate of the coated sample is lower than that of the uncoated sample. The results obtained from the electrochemical impedance test confirmed the results obtained from the polarization test. Both methods emphasized enhancing the corrosion resistance of AZ91D alloy in the presence of aluminum coating.

The results of electrochemical impedance spectroscopy can be simulated using an appropriate electrical equivalent circuit such as the one illustrated in Figure 17 [23]. The equivalent impedance analysis describes the behavior of the corrosion resistance of the coating. In Figure 17, $R_s$ denotes the uncompensated solution resistance, $R_{ct}$ indicates the charge transfer resistance or corrosion resistance on the metal interface and dual layer, Rcoat shows the coating resistance, $CPE_{dl}$ represents the electric double layer capacitor, and the $CPE_{coat}$ refers to the capacitor of coating.

The equivalent impedance of the circuits in Figure 17 can be written as the following, respectively:

$$Z_t = Z(R_s) + Z(R_{ct} \parallel CPE_{dl}) = R_s + \frac{R_{ct}}{R_{ct} \, jCPE_{dl}\omega+1} = Z' + j\,Z'' \quad (4)$$

$$Z_t = Z(R_s) + \frac{(Z(R_{ct}\parallel CPE_{dl}) + Z(R_{coat})) \times Z_{CPE_{coat}}}{(Z(R_{ct}\parallel CPE_{dl}) + Z(R_{coat})) + Z_{CPE_{coat}}} \quad (5)$$

$$Z_t = R_s + \frac{R_{coat} + R_{ct} - R_{coat} R_{ct} CPE_{coat} CPE_{dl}\,\omega^2}{(1 - R_{coat} R_{ct} CPE_{coat} CPE_{dl}\,\omega^2) + j(R_{coat}(CPE_{coat} + CPE_{dl}) + R_{ct}\,CPE_{coat})\omega}$$

The parameters of equivalent circuit in Figure 17 are provided in Table 2.

**Table 1.** Results of dynamic polarization test in 3.5 wt% NaCl solution.

| Samples | Corrosion current density ($\mu$A.cm$^{-2}$) | Corrosion potential (mV vs SCE) | $\beta_a$ (mV/decade) | $\beta_c$ (mV/decade) | $R_p$ ($\Omega$) | C.R. (mpy) |
|---|---|---|---|---|---|---|
| AZ91D | 29.16 | -1490.96 | 39.48 | 229.79 | 501.69 | 26.3 |
| Aluminum coated AZ91D | 12.85 | -1553.5 | 43.46 | 203.33 | 1209.95 | 11.59 |

**Table 2.** Equivalent circuit parameters.

| Samples | $R_S$ ($\Omega$.cm$^2$) | CPET$_{dl}$ ($\mu$F.cm$^{-2}$) | CPEP$_{dl}$ | $R_{ct}$ ($\Omega$.cm$^2$) | CPET$_{coat}$ ($\mu$F.cm$^{-2}$) | CPEP$_{coat}$ | Rcoat ($\Omega$.cm$^2$) |
|---|---|---|---|---|---|---|---|
| Uncoated AZ91D | 12.1 | 13.9 | 0.8 | 1615.7 | --- | --- | --- |
| Aluminum coated AZ91D | 14.2 | 6.75 | 0.9 | 1729.2 | 2.7 | 0.9 | 5637.6 |

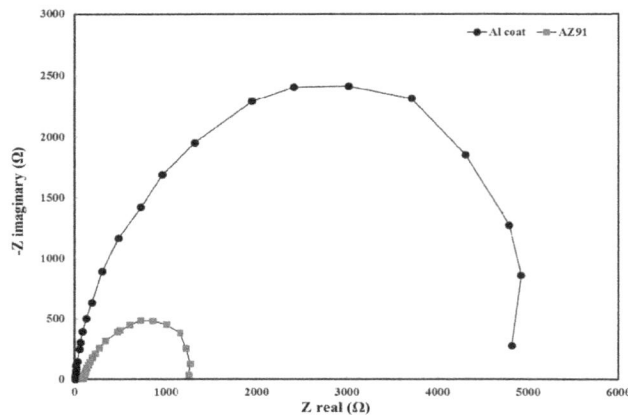

**Fig. 16.** AC electrochemical impedance spectrum of coated and uncoated AZ91D sample in the solution of NaCl 3.5 wt%.

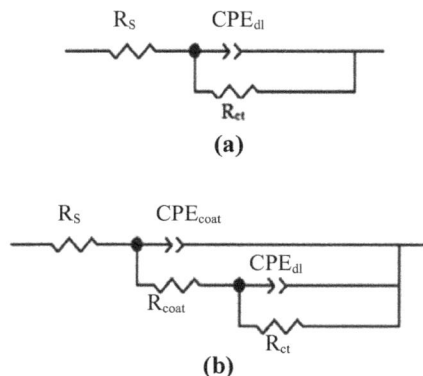

**(a)**

**(b)**

**Fig. 17.** The equivalent circuit for the impedance spectrum analysis, (a) uncoated AZ91D alloy and (b) AZ91D alloy coated with aluminum [23].

## 5. Conclusion

In this paper, aluminum coating was developed on an AZ91D magnesium alloy substrate by electrophoretic deposition. To determine the optimal condition of deposition the effect of $AlCl_3.6H_2O$ concentration, applied voltage, and deposition time was investigated. The DLS analysis showed that the size of the particles in the suspension varies from 1 to 5 μm for the additive concentration in the range of 0.4 to 1.5 mM. A well-stabilized suspension and a uniform deposition were obtained at the $AlCl_3.6H_2O$ concentration of 0.6 mM. The zeta potential value and mean size of particles for this suspension were measured at 27.2 mV and 0.879 μm, respectively. Surface morphology studies showed that a uniform and low-pore coating from this suspension is obtainable at the applied voltage of 70 V and deposition time of 18 min.

## References

[1] F. Czerwinski, Magnesium Alloys: Corrosion and Surface Treatments, InTech Publisher, London, 2011, pp. 195.

[2] G.L. Song, "Electroless" deposition of a pre-film of electrophoresis coating and its corrosion resistance on a Mg alloy, Electrochim. Acta, 55 (2010) 2258-2268.

[3] W. Feng, W. Yue, M.Ping-li, Y.U. Bao-yi, G. Quan-ying, Effects of combined addition of Y and Ca on microstructure and mechanical properties of die casting AZ91 alloy, T. Nonferr. Metal. Soc. 20 (2010) s311-s317.

[4] Y. Zhan, Z. Hong-yang, H. Xiao-dong, J. Dong-ying, Effect of elements Zn, Sn and In on microstructures and performances of AZ91 alloy, T. Nonferr. Metal. Soc. 20 (2010) s318-s323.

[5] A. Nold, J. Zeiner, T. Assion, R. Clasen, Electrophoretic deposition as rapid prototyping method, J. Eur. Ceram. Soc. 30 (2010) 1163-1170.

[6] Y. Tao, T. Xiong, C. Sun, L. Kong, X. Cui, T. Li, G.L. Song, Microstructure and corrosion performance of a cold sprayed aluminum coating on AZ91D magnesium alloy, Corros. Sci. 52 (2010) 3191-3197.

[7] S. Fleming, An Overview of magnesium based alloys for aerospace and automotive applications, [Dissertation] Rensselaer Polytechnic Institute, Hartford, CT August, 2012.

[8] A. Shahriari, H. Aghajani, Electrophoretic deposition of 3YSZ coating on AZ91D using an aluminum interlayer, Prot. Met. Phys. Chem. S. 53 (2017) 518-526.

[9] M. Ahmadi, H. Aghajani, Structural characterization of YSZ/Al_2O_3 nanostructured composite coating fabricated by electrophoretic deposition and reaction bonding, Ceram. Int. 44 (2018) 5988-5995.

[10] S. Candan, M. Unalb, E. Koc, Y. Turen, E. Candan. Effects of titanium addition on mechanical and corrosion behaviours of AZ91 magnesium alloy, J. Alloy. Compd. 509 (2011) 1958-1963.

[11] K. Yang, Z. Jiang, J. Chung, Electrophoretically Al-coated wire mesh and its application for catalytic oxidation of 1,2-dichlorobenzene, Surf. Coat. Tech. 168 (2003) 103-110.

[12] G. Lee, S. Pyun, C. Rhee, A study on electrophoretic deposition of Ni nanoparticles on pitted Ni alloy 600 with surface fractality, J. Colloid Interf. Sci. 308

(2007) 413-420.

[13] X. Guo, X. Li, H. Li, D. Zhang, C. Lai, W. Li, A Comprehensive investigation on the electrophoretic deposition (EPD) of Nano-Al/Ni energetic composite coatings for the combustion application, Surf. Coat. Tech. 265 (2015) 83-91.

[14] L. Besra, M. Liu, A review on fundamentals and applications of electrophoretic deposition (EPD), Prog. Mater. Sci. 52 (2007) 1-61.

[15] K. Liu, Q. Liu, Q. Han, G. Tu, Electrodeposition of Al on AZ31 magnesium alloy in TMPAC-AlCl$_3$ ionic liquids, T. Nonferr. Met. Soc. 21 (2011) 2104-2110.

[16] ASM International Handbook Committee, ASM Handbook of properties and selection: Nonferrous alloys and special-purpose materials, Vol. 2, ASM International, Ohio, 1990.

[17] J. Ma, W. Chen, Deposition and packing study of sub-micron PZT ceramics using electrophoretic deposition, Mater. Lett. 56 (2002) 721-727.

[18] N. Sato, M. Kawachi, K. Noto, N. Yoshimoto, M. Yoshizawa, Effect of particle size reduction on crack formation in electrophoretically deposited YBCO films, Physica C, 357-360 (2001) 1019-1022.

[19] Z. Wang, J. Shemilt, P. Xiao, Fabrication of ceramic composite coatings using electrophoretic deposition, reaction bonding and low temperature sintering, J. Eur. Ceram. Soc. 22 (2002) 183-189.

[20] L. Yang, X. Wu, D. Weng, Development of uniform and porous Al coatings on FeCrAl substrate by electrophoretic deposition, Colloid. Surface. A, 287 (2006) 16-23.

[21] S. Kim, S. Cho, J. Lee, S. Samal, H. Kim, Relationship between the process parameters and the saturation point in electrophoretic deposition, Ceram. Int. 38 (2012) 4617-4622.

[22] M.G. Fontana, Corrosion engineering, Third Edition, McGraw Hill, New York, 1987.

[23] A. Lasia, Electrochemical impedance spectroscopy and its applications, Springer-Verlag, New York, 2014.

# 18

# Modeling heat transfer of non-Newtonian nanofluids using hybrid ANN-Metaheuristic optimization algorithm

**Mohammad Hojjat**

*Department of Chemical Engineering, Faculty of Engineering, University of Isfahan, Isfahan, Iran*

## HIGHLIGHTS

- Designing an ANN and optimized it by using a PSO algorithm to predict the Nusselt number of non-Newtonian nanofluids

- Assessment of the ability of an artificial neural network to predict convective heat transfer experimental data

- Comparison of ANN-PSO with existing correlation

## GRAPHICAL ABSTRACT

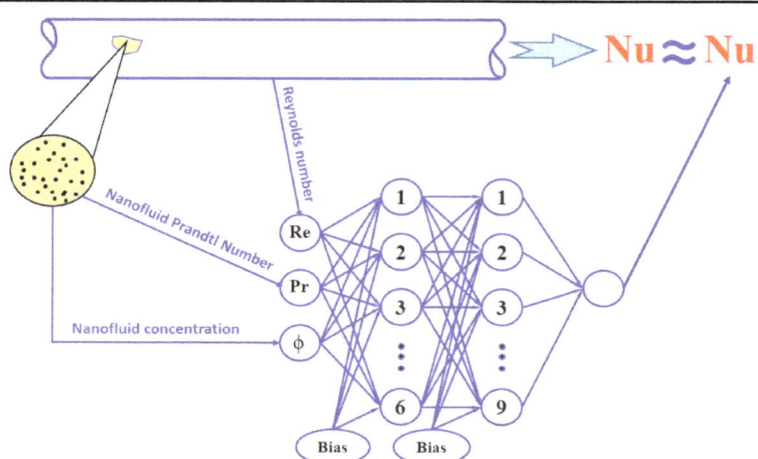

## ARTICLE INFO

*Keywords:*

Nanofluids
Non-Newtonian
Artificial neural network
Multi-layer perceptron
Particle swarm optimization

## ABSTRACT

An optimal artificial neural network (ANN) has been developed to predict the Nusselt number of non-Newtonian nanofluids. The resulting ANN is a multi-layer perceptron with two hidden layers consisting of six and nine neurons, respectively. The tangent sigmoid transfer function is the best for both hidden layers and the linear transfer function is the best transfer function for the output layer. The network was trained by a particle swarm optimization (PSO) algorithm. Nanofluid concentration, Reynolds number, and Prandtl number are input for the ANN and the nanofluid Nusselt number is its output. There exists an excellent agreement between the ANN predicted values and experimental data. The average and maximum differences between experimental data and those predicted by ANN are about 0.8 and 5.6 %, respectively. It was also found that ANN predicts the Nusselt number of nanofluids more accurately than the previously proposed correlation.

\* *Corresponding author:  ; E-mail address: m.hojjat@eng.ui.ac.ir*

## 1. Introduction

Heat transfer is the main challenge in modern systems with high heat fluxes such as nuclear power systems, high power lasers, space vehicles, and so on. Numerous heat transfer enhancement techniques have been proposed to overcome this challenge. Thermal characteristics of heat transfer fluids are major parameters affecting the performance of such equipment. Thus improvement of the thermal characteristics of heat transfer fluids can improve the performance of heat transfer systems. It is well known that dispersion of solid particles in fluids significantly enhances their thermal characteristics [1]. But problems concerning particles with large (milli- or micrometer) sizes, such as rapid settling, high erosion, clogging the channel with small dimensions and high pressure drop, have limited their use. Nanoparticles are uniformly suspended in base fluids to produce stable suspensions which do not have the mentioned problems. This new class of heat transfer fluids is called nanofluids [2]. Numerous studies have been carried out on thermophysical properties as well as the thermal and hydrodynamic behavior of nanofluids [3-10]. Results show that addition of nanoparticles enhances the thermal conductivity and convective heat transfer coefficient of the base fluid drastically. As new media for heat transfer, it is expected that nanofluids will create a revolution in heat transfer. A comprehensive investigation was conducted on thermal and hydrodynamic behavior of non-Newtonian nanofluids by Hojjat et al. [11-17]. Their results show that the thermal behavior of non-Newtonian nanofluids is superior to that of the base fluids. Before nanofluids can be used in practical applications we should increase our knowledge of principles governing the behavior of nanofluids. Results of most studies show that models and correlations of conventional heat transfer fluids do not predict thermophysical properties and thermal and hydrodynamic behavior of nanofluids well [3,9,18,19]. In other words, there exists no general model to predict the properties and behavior of nanofluids. Therefore, it is vital to find general models and correlations which can perfectly predict the properties and behavior of nanofluids.

Recently, data driven models based on experimental data, such as artificial neural network, fuzzy logic, and evolutionary optimization algorithm (genetic algorithm, particle swarm optimization, etc.), have been used to find general models for nanofluids behavior.

Rheological behavior of various nanofluids has been modelled by using an ANN [12,20-28]. Some investigators have used the ANN to predict the thermal conductivity of nanofluids [11,27-38]. Results reveal the high capability of ANN to predict nanofluids rheological behavior as well as thermal conductivity.

The effect of nanofluid on the cooling performance and pressure drop of a jacketed reactor has been experimentally investigated and modeled using an artificial neural network [39]. A multi-layer perceptron (MLP) neural network with one hidden layer containing ten neurons was used for convective heat transfer coefficient modeling and a MLP network with two hidden layers each contains six neurons was used for pressure drop. Both were trained by the Levenberg-Marquardt training algorithm. Reasonable agreement between experimental data and those predicted by ANNs is observed. Vaferi et al. [40] have proposed the best artificial neural network model for prediction of heat transfer coefficient of nanofluids in a circular tube subjected to various boundary conditions under different flow regimes. Results obtained from the ANN model have compared with some reliable correlations in the literature. They found that the performance of the proposed model was higher than other published works. Laminar convective heat transfer of $Al_2O_3$-water nanofluids flowing inside various flat tubes was investigated numerically using CFD methods by Safikhani et al. [41]. Simulation was carried out based on a two-phase model. They calculated heat transfer coefficient and pressure drop of nanofluids. Resulted data were modeled by a grouped method of data handling (GMDH) type ANN. Finally, the obtained GMDH model was used for Pareto based multi-objective optimization of nanofluid parameters in horizontal flat tubes by a non-dominated genetic algorithm. The resulting Pareto solution contains significant design information on nanofluids parameters in flat tubes. Kalani et al. [42] assessed the capability of two artificial neural networks of a radial basis function artificial neural network (RBF-ANN), MLP-ANN, and an adaptive fuzzy inference system (ANFIS) in modeling the complex non-linear relation between input and output parameters of a photovoltaic thermal nanofluid based collector system. Their results indicate that all three models have the ability to predict the performance of the mentioned system. However, the accuracy of the ANFIS and RBF-

ANN is higher in estimation of electrical efficiency and fluid outlet temperature, respectively.

Since the convective heat transfer coefficient of nanofluids is the main parameter influencing the performance of heat transfer equipment and fewer investigations have carried out on modeling this important parameter, in the present study turbulent flow forced convective heat transfer of non-Newtonian nanofluids flowing through a circular tube under constant wall temperature boundary condition was modeled by an optimized artificial neural network. It receives the nanofluid volume fraction, Prandtl number, and Reynolds number as input variables and gives the Nusselt number of nanofluids as output. A particle swarm optimization algorithm was used to determine the best values of the ANN parameters instead of conventional gradient based training algorithms.

## 2. Experiments

Experimental data are obtained from results of the author's previous work [16]. This experiment is briefly reviewed below. Turbulent forced convective heat transfer of non-Newtonian nanofluids flowing through a double-pipe heat exchanger was experimentally investigated. Nanofluids flow in the inner tube. Hot water circulated through the annular section at very high flow rates so it is reasonable to consider the boundary condition as constant temperature [16].

### 2.1. Nanofluids preparation

Nanofluids under consideration were suspensions of $\gamma$-Al$_2$O$_3$ (25 nm), TiO$_2$ (10 nm), CuO (30-50 nm) nanoparticles in 0.5 wt% CMC solution. Since ultrasonic vibrations altered the rheological behavior of a CMC solution, nanoparticles were first dispersed in deionized water and sonicated to obtain uniform suspensions. An appropriate amount of high concentration CMC solution was added to the suspensions and well mixed by a mechanical mixer to achieve the desired nanofluids [16].

A KD2 thermal property meter (Decagon Device Inc., USA) was used to measure the thermal conductivity of nanofluids. Rheological behavior of nanofluids was investigated using a rotational rheometer (HAAKE RV12). Results show that all nanofluids as well as the base fluid exhibit pseudoplastic behavior [12]. Other

physical properties were calculated at average bulk temperature according to the following equations [16]:

$$\rho_{nf} = \phi\rho_p + (1-\phi)\rho_{bf} \tag{1}$$

$$(\rho C_P)_{nf} = \phi(\rho C_P)_p + (1-\phi)(\rho C_P)_{nf} \tag{2}$$

Because of very low concentrations of the carboxy methyl cellulose, the physical properties of the base fluid are considered similar to those of pure water. Physical properties of nanoparticles are given in Table 1.

**Table 1.** Physical properties of the nanoparticles [43].

| Nanoparticle | Density (kg/m$^3$) | CP (J/kgK) |
|---|---|---|
| $\gamma$-Al$_2$O$_3$ | 3700 | 880 |
| TiO$_2$ | 3900 | 710 |
| CuO | 6350 | 535.6 |

### 2.2. Experimental procedure

A schematic diagram of the experimental set-up is shown in Figure 1 [16]. It is comprised of two loops. The first loop, including a container, a stainless steel gear pump, a bypass line, a flow meter, a cooler, several valves, and a test section, is related to nanofluids. The test section is a circular pipe with the length of 200 cm and inner diameter of 1 cm. The second cycle is related to hot water which is cooled in a double-pipe heat exchanger by the nanofluid. Six K-Type thermocouples mounted on the outer wall of the inner tube are used to measure the wall temperature. Two thermocouples are used to measure the inlet and outlet temperatures of the nanofluid. Details of experimental procedure and calculation of heat transfer coefficient and Nusselt number are explained in Ref. [16].

## 3. Particle swarm optimization

Particle swarm optimization is a population-based stochastic metaheuristic computational optimization algorithm inspired by bird flocking and fish schooling. PSO was first proposed by Kennedy and Eberhart in 1995 [44]. To implement PSO, first a population of particles is produced randomly on a D dimensional space of the problem. Every particle which can be a solution of the problem is characterized by its position, $X_i = (x_{i1}, x_{i2}, ..., x_{iD})$ and velocity, $V_i = (v_{i1}, v_{i2}, ..., v_{iD})$

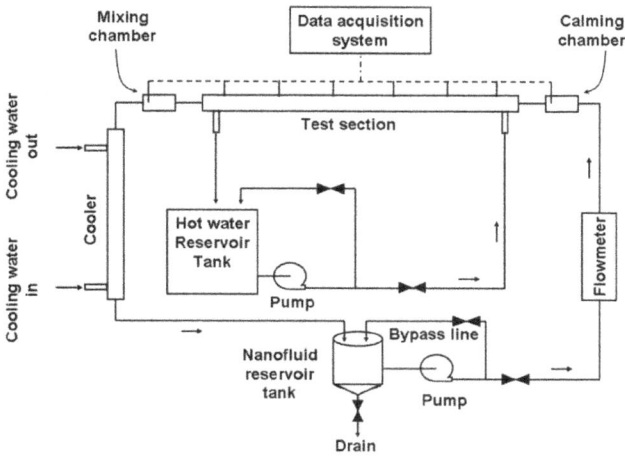

**Fig. 1.** Experimental setup of data results in [16].

vectors. For each particle the value of cost function is evaluated. The cost value of each particle is compared with its own best experience (local best) and the best experience of all other particles (global best). Each particle adjusts its velocity and position according to the following equations (Constriction coefficient):

$$v_{ij}(t+1)=\chi\left[v_{ij}(t)+\phi_1\left(x_{ij}^{lbest}(t)-x_{ij}(t)\right)+\phi_2\left(x_j^{gbest}(t)-x_{ij}(t)\right)\right] \quad (3)$$

$$x_{ij}(t+1) = x_{ij}(t) + v_{ij}(t+1) \quad (4)$$

where:

$$\chi=\frac{2\kappa}{\left|2-\phi-\sqrt{\phi(\phi-4)}\right|} \quad (5)$$

with:

$\phi = \phi_1 + \phi_2$

$\phi_1 = c_1 r_1$

$\phi_2 = c_2 r_2$

Equation (5) is used under the constraints that $\phi \geq 4$ and $\kappa \in [0,1]$.

The procedure is repeated until the stopping criterion is satisfied.

In the present study parameters were chosen as: $\phi = 4.1$, $\phi_1 = \phi_2 = 2.05$ and $\kappa = 1$.

## 4. Artificial Neural Network Design

ANN is a powerful tool inspired by the human nervous system and is capable of modelling complex functions. An ANN consists of an input, an output, and one or more hidden layers. Information of source is propagated into the neural network through the input layer. The output layer gives the results of information processing. The number of hidden layers depends on the complexity of the problem, but in most cases it was found that one or two hidden layers are sufficient [11,45-48]. Each layer consists of a number of neurons, which are the basic structural components of neural networks. The numbers of neurons in the input and output layers are equal to the numbers of input variables and targets, respectively. Several methods have been used by investigators to determine the number of neurons in the hidden layers, but they only produce general guidelines [49-52]. The optimum number of neurons in the hidden layers is often determined by trial and errors. The neural network is trained with different numbers of hidden neurons and the best number of hidden neurons is determined according to the values of one or more evaluating statistical criteria such as mean square of errors (MSE), mean absolute error (MAE), coefficient of determination ($R^2$), and so on. This kind of neural networks is often called multi-layer perceptron (MLP) neural network. The output of each neuron is sent to all neurons of the next layer through weighted connections. In each neuron input values are added with a bias, and then an activation function is applied on the resulting value to yield the output as:

$$y_i = f\left(\sum_{i=1}^{n} w_{ij}x_j + b_i\right) \quad (6)$$

where $y_i$ is the $i^{th}$ neuron's output, $x_j$ is the output of $j^{th}$ neuron in the previous layer, $w_{ij}$ is the weight, $b_i$ is the bias of the ith neuron, and f is the activation function. Weights are randomly selected at the beginning of the training process and then adjusted according to a training algorithm. Here the PSO algorithm is used as an alternative to traditional training algorithms such as Levenberg-Marquardt (LM), gradient descent (GD), and so on. The optimization algorithm minimizes the MSE as cost function.

Although not compulsory, input data are often normalized between 0 and 1 to avoid some numerical problems. This also causes the input data to be of the same order. The input dataset is divided into two parts: training data and test data. Training data are used to train the neural network according to the PSO algorithm. Test data are used to identify how well the ANN is trained.

The performance of ANNs may be assessed based on

some statistical criteria including mean squared error (MSE), maximum absolute relative deviation (Max ARD %), average absolute relative deviation (AARD %), and correlation coefficient (r), defined as:

$$MSE = \frac{1}{n}\sum_{i=1}^{n}\left(y_{iexp} - y_{ical}\right)^2 \tag{7}$$

$$Max\ ARD\ \% = Max_{i=1}^{n}\left(\left|\frac{y_{iexp} - y_{ical}}{y_{iexp}}\right| \times 100\right) \tag{8}$$

$$AARD\ \% = \frac{1}{n}\sum_{i=1}^{n}\left|\frac{y_{iexp} - y_{ical}}{y_{iexp}}\right| \times 100 \tag{9}$$

$$r = \frac{\sum_{i=1}^{n}\left[\left(y_{iexp} - \bar{y}_{exp}\right)\left(y_{ical} - \bar{y}_{cal}\right)\right]}{\sqrt{\sum_{i-1}^{n}\left(y_{iexp} - \bar{y}_{exp}\right)^2 \sum_{i=1}^{n}\left(y_{ical} - \bar{y}_{cal}\right)^2}} \tag{10}$$

where $y_{iexp}$, $y_{ical}$, $\bar{y}$, and n are experimental data, predicted data by ANN, mean value of data, and number of data points, respectively.

## 5. Artificial Neural Network Architecture

In the present study an optimal multi-layer perceptron neural network was designed to model the experimental data of turbulent flow of non-Newtonian nanofluids in a circular tube with constant wall temperature. Our previous experimental data are used to obtain the ANN model [16].

The optimum neural network architecture was determined by trial and error according to steps shown in Figure 2.

First experimental data were normalized between 0 and 1 according to equation (11):

$$X_{i,norm} = \frac{x_i - x_{i,min}}{x_{i,max} - x_{i,min}} \tag{11}$$

Then randomly divided into three parts: training data set (75%), validating data set (5%), and test data set (20%). The architecture of an ANN is normally determined by trial and error. First, the number of hidden layers and the number of neurons in each layer were set. Then the network was trained by a training algorithm. The activation functions of hidden and output layers were changed and the performance of ANN was assessed to specify the best activation functions. In order to determine the best number of hidden neurons, the network was trained with different numbers of hidden neurons and its performance was evaluated. After that, the number of hidden layers was changed to determine the best number of hidden layers. Next, the best activation function of added hidden layers was also determined by trial and error. To evaluate the network repeatability each network trained 15 times. Finally, instead of using conventional training algorithms for training the ANN, the PSO algorithm is used to determine the best values of ANN parameters (weights and biases).

The ANN architecture shown in Figure 3 consists of two hidden layers of 6 and 9 neurons, respectively. The activation functions of both hidden layers are hyperbolic tangent sigmoid transfer function (tansig), and that of the output layer is a linear transfer function (purelin). MSE was chosen as the network performance function. It evaluates the ANN performance according to the mean of squared errors. ANN receives nanofluid concentration, Reynolds number, and Prandtl number as input parameters and gives the resulting Nusselt number as output.

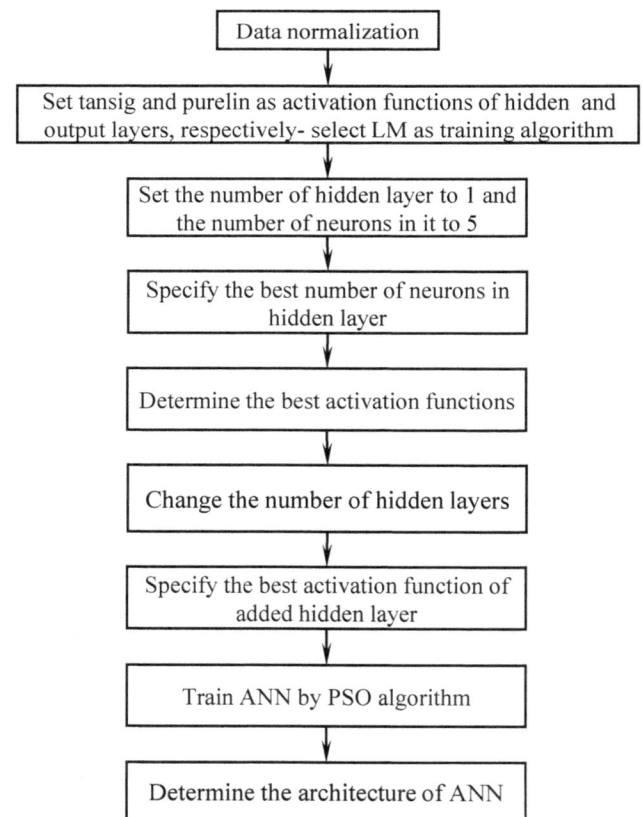

Fig. 2. Algorithm for optimization of ANN architecture.

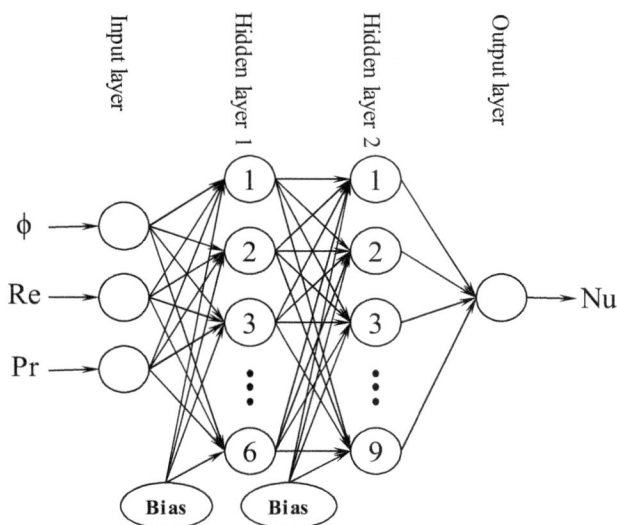

**Fig. 3.** Architecture of the ANN.

**Fig. 4.** ANN predicted Nu against experimental Nu in (a) training dataset and (b) test dataset.

## 5. Results

Figure 4 shows the ANN predicted values of Nu in comparison with the experimental data. As can be seen, there exists very good agreement between the experimental data and corresponding values predicted by the ANN model. The values of statistical criteria, given in Table 2, suggest the accuracy of the proposed ANN.

Values predicted by the ANN model are compared with experimental data in Figure 5. Excellent agreement between the ANN model and experimental data is obvious. Max ARD % and AARD % of ANN are about 5.6 and 0.8%, respectively.

The following correlation has been proposed for the Nusselt number of non-Newtonian nanofluids [16]:

$$\text{Nu} = 0.00115\, \text{Re}^{1.050}\, \text{Pr}^{0.693}\, (1 + \phi^{0.388}) \tag{12}$$
$$(2900 < \text{Re} < 8800 \text{ and } 39 < \text{Pr} < 71)$$

In Table 3 the artificial neural network is compared with the above mentioned correlation. It is clear that the ANN predicts the Nu of nanofluids better than the correlation. Max ARD % and AARD % of ANN are

almost half and one fifth of those of the correlation, respectively.

## 6. Conclusion

Nanofluids are a new class of heat transfer fluids that possess better thermal characteristics than the base fluids. Correlation of conventional heat transfer fluids cannot predict the behavior of nanofluids well. So finding new models for predicting the features of nanofluids is important. An artificial neural network has the ability of modeling nonlinear functions. In this

**Table 2.** Statistical criteria of the proposed ANN.

|  | Max ARD % | AARD % | r |
|---|---|---|---|
| Training data | 1.37729 | 0.316967 | 0.9999 |
| Test data | 5.638172 | 2.357044 | 0.9880 |

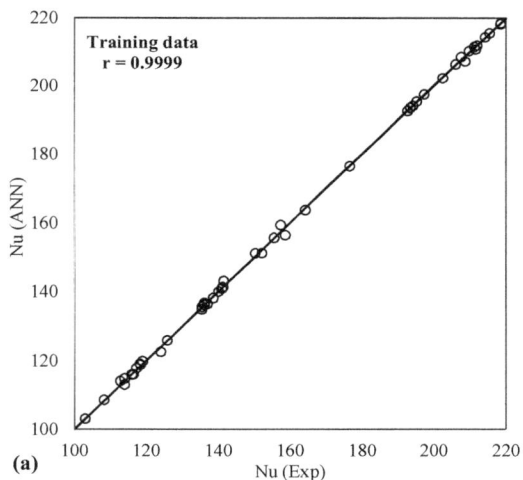

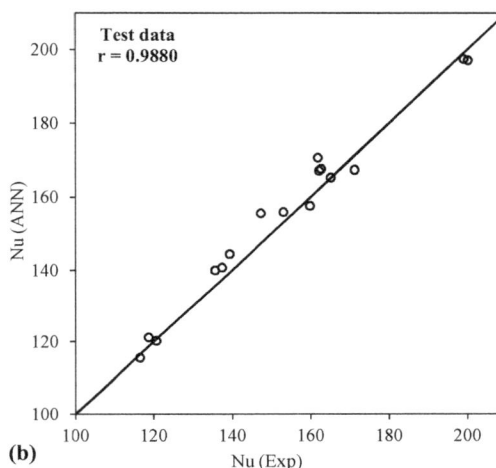

**Table 3.** Comparison of ANN and equation (12) for Nu of non-Newtonian nanofluid.

| Method | Max ARD % | AARD % |
|---|---|---|
| ANN | 5.64 | 0.83 |
| Proposed correlation [16] | 10.4 | 3.82 |

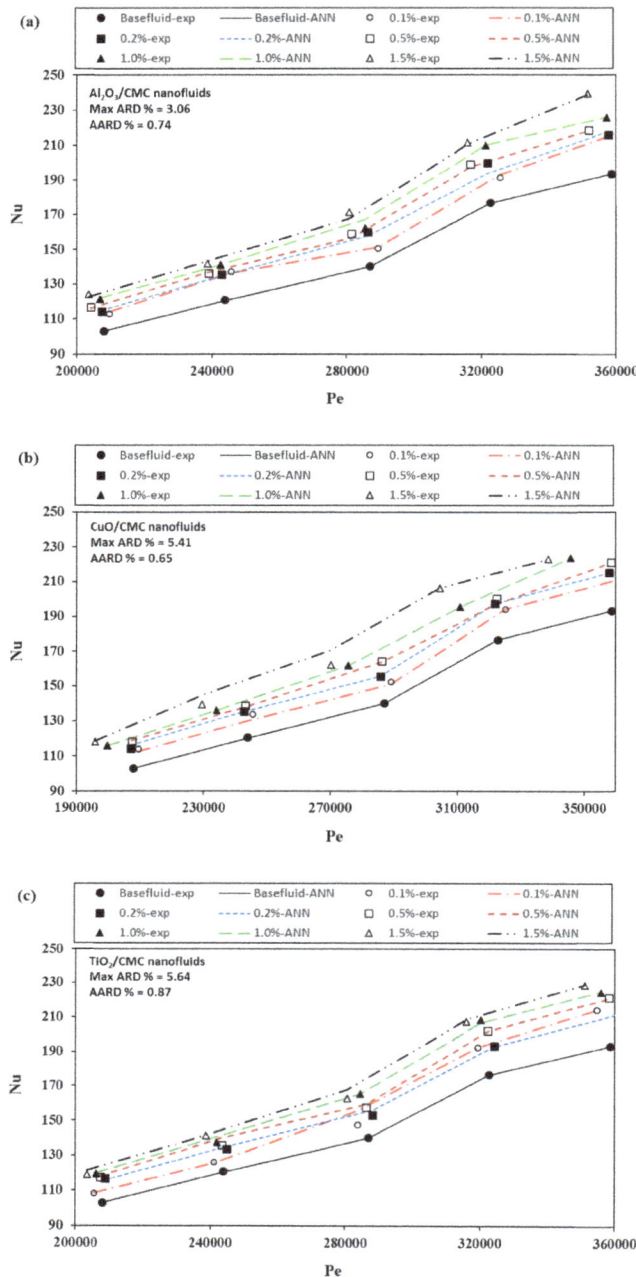

**Fig. 5.** Comparison between experimental data [16] and values predicted by the ANN (a)Al₂O₃/CMC, (b) CuO/CMC and (c)TiO₂/CMC nanofluids.

investigation an artificial neural network was designed and optimized by PSO algorithm to predict the Nusselt number of non-Newtonian nanofluids flowing through a circular tube subjected to constant wall temperature in a turbulent regime. The developed network consists of two hidden layers with six and nine neurons, respectively. Max ARD % and AARD % between the ANN predicted values and experimental data are 5.64 and 0.83, respectively; this indicates the excellent predictive ability of the ANN. Results also show that the

ANN is better than the previously proposed correlation. The maximum and average absolute relative deviations of the ANN are almost half and one fifth of the values predicted by the correlation in the literature.

## References

[1] J.C. Maxwell, Electricity and Magnetism, Clarendon Press, Oxford, UK, 1873.

[2] S.U.S. Choi, Enhancing thermal conductivity of fluids with nanoparticles, in: American Society of Mechanical Engineers, Fluids Engineering Division (Publication) FED, 1995, pp. 99-105.

[3] L. Yang, K. Du, A comprehensive review on heat transfer characteristics of $TiO_2$ nanofluids, Int. J. Heat Mass Tran. 108, Part A (2017) 11-31.

[4] K.S. Suganthi, K.S. Rajan, Metal oxide nanofluids: Review of formulation, thermo-physical properties, mechanisms, and heat transfer performance, Renew. Sust. Energ. Rev. 76 (2017) 226-255.

[5] R.M. Sarviya, V. Fuskele, Review on Thermal Conductivity of Nanofluids, Mater. Today-Proc. 4, Part A (2017) 4022-4031.

[6] J.A. Ranga Babu, K.K. Kumar, S. Srinivasa Rao, State-of-art review on hybrid nanofluids, Renew. Sust. Energ. Rev. 77 (2017) 551-565.

[7] K.A. Mohammed, A.R. Abu Talib, A.A. Nuraini and K.A. Ahmed, Review of forced convection nanofluids through corrugated facing step, Renew. Sust. Energ. Rev. 75 (2017) 234-241.

[8] K.Y. Leong, K.Z. Ku Ahmad, H.C. Ong, M.J. Ghazali, A. Baharum, Synthesis and thermal conductivity characteristic of hybrid nanofluids-A review, Renew. Sust. Energ. Rev. 75 (2017) 868-878.

[9] M. Gupta, V. Singh, R. Kumar and Z. Said, A review on thermophysical properties of nanofluids and heat transfer applications, Renew. Sust. Energ. Rev. 74 (2017) 638-670.

[10] R.B. Ganvir, P.V. Walke, V.M. Kriplani, Heat transfer characteristics in nanofluid- A review, Renew. Sust. Energ. Rev. 75 (2017) 451-460.

[11] M. Hojjat, S.G. Etemad, R. Bagheri, J. Thibault, Thermal conductivity of non-Newtonian nanofluids: Experimental data and modeling using neural network, Int. J. Heat Mass Trans. 54 (2011) 1017-1023.

[12] M. Hojjat, S.G. Etemad, R. Bagheri, J. Thibault, Rheological characteristics of non-Newtonian

nanofluids: Experimental investigation, Int. Commun. Heat Mass, 38 (2011) 144-148.

[13] M. Hojjat, S.G. Etemad, R. Bagheri, Laminar heat transfer of non-Newtonian nanofluids in a circular tube, Korean J. Chem. Eng., 27 (2010) 1391-1396.

[14] M. Hojjat, S.G. Etemad, R. Bagheri, J. Thibault, Laminar convective heat transfer of non-Newtonian nanofluids with constant wall temperature, Heat Mass Transfer, 47 (2011) 203-209.

[15] M. Hojjat, S.G. Etemad, R. Bagheri, J. Thibault, Convective heat transfer of non-Newtonian nanofluids through a uniformly heated circular tube, Int. J. Therm. Sci. 50 (2011) 525-531.

[16] M. Hojjat, S.G. Etemad, R. Bagheri, J. Thibault, Turbulent forced convection heat transfer of non-Newtonian nanofluids, Exp. Therm Fluid Sci. 35 (2011) 1351-1356.

[17] M. Hojjat, S.G. Etemad, R. Bagheri, J. Thibault, Pressure Drop of Non-Newtonian Nanofluids Flowing Through a Horizontal Circular Tube, J. Disper. Sci. Technol. 33 (2012) 1066-1070.

[18] R.V. Pinto, F.A.S. Fiorelli, Review of the mechanisms responsible for heat transfer enhancement using nanofluids, Appl. Therm. Eng. 108 (2016) 720-739.

[19] B. Farajollahi, S.G. Etemad, M. Hojjat, Heat transfer of nanofluids in a shell and tube heat exchanger, Int. J. Heat Mass Trans. 53 (2010) 12-17.

[20] M. Afrand, A. Ahmadi Nadooshan, M. Hassani, H. Yarmand, M. Dahari, Predicting the viscosity of multi-walled carbon nanotubes/water nanofluid by developing an optimal artificial neural network based on experimental data, Int. Commun. Heat Mass. 77 (2016) 49-53.

[21] E. Heidari, M.A. Sobati, S. Movahedirad, Accurate prediction of nanofluid viscosity using a multilayer perceptron artificial neural network (MLP-ANN), Chemometr. Intell. Lab. 155 (2016) 73-85.

[22] A. Alirezaie, S. Saedodin, M.H. Esfe, S.H. Rostamian, Investigation of rheological behavior of MWCNT (COOH-functionalized)/MgO-Engine oil hybrid nanofluids and modelling the results with artificial neural networks, J. Mol. Liq. 241 (2017) 173-181.

[23] A.S. Dalkilic, A. Çebi, A. Celen, O. Yıldız, O. Acikgoz, C. Jumpholkul, M. Bayrak, K. Surana, S. Wongwises, Prediction of graphite nanofluids' dynamic viscosity by means of artificial neural networks, Int. Commun. Heat Mass, 73 (2016) 33-42.

[24] M. Hemmat Esfe, M.R. Hassani Ahangar, M. Rejvani, D. Toghraie, M.H. Hajmohammad, Designing an artificial neural network to predict dynamic viscosity of aqueous nanofluid of $TiO_2$ using experimental data, Int. Commun. Heat Mass, 75 (2016) 192-196.

[25] G.A. Longo, C. Zilio, L. Ortombina, M. Zigliotto, Application of Artificial Neural Network (ANN) for modeling oxide-based nanofluids dynamic viscosity, Int. Commun. Heat Mass, 83 (2017) 8-14.

[26] M. Vakili, S. Khosrojerdi, P. Aghajannezhad, M. Yahyaei, A hybrid artificial neural network-genetic algorithm modeling approach for viscosity estimation of graphene nanoplatelets nanofluid using experimental data, Int. Commun. Heat Mass, 82 (2017) 40-48.

[27] M. Hemmat Esfe, S. Saedodin, N. Sina, M. Afrand, S. Rostami, Designing an artificial neural network to predict thermal conductivity and dynamic viscosity of ferromagnetic nanofluid, Int. Commun. Heat Mass, 68 (2015) 50-57.

[28] M. Hemmat Esfe, P. Razi, M.H. Hajmohammad, S.H. Rostamian, W.S. Sarsam, A.A. Abbasian Arani, M. Dahari, Optimization, modeling and accurate prediction of thermal conductivity and dynamic viscosity of stabilized ethylene glycol and water mixture Al2O3 nanofluids by NSGA-II using ANN, Int. Commun. Heat Mass, 82 (2017) 154-160.

[29] M. Afrand, D. Toghraie, N. Sina, Experimental study on thermal conductivity of water-based Fe3O4 nanofluid: Development of a new correlation and modeled by artificial neural network, Int. Commun. Heat Mass, 75 (2016) 262-269.

[30] E. Ahmadloo, S. Azizi, Prediction of thermal conductivity of various nanofluids using artificial neural network, Int. Commun. Heat Mass, 74 (2016) 69-75.

[31] A. Aminian, Predicting the effective thermal conductivity of nanofluids for intensification of heat transfer using artificial neural network, Powder Technol. 301 (2016) 288-309.

[32] M.A. Ariana, B. Vaferi, G. Karimi, Prediction of thermal conductivity of alumina water-based nanofluids by artificial neural networks, Powder Technol. 278 (2015) 1-10.

[33] M. Hemmat Esfe, M. Afrand, W-M. Yan and M. Akbari, Applicability of artificial neural network and

nonlinear regression to predict thermal conductivity modeling of $Al_2O_3$-water nanofluids using experimental data, Int. Commun. Heat Mass, 66 (2015) 246-249.

[34] M. Hemmat Esfe, S. Wongwises, A. Naderi, A. Asadi, M.R. Safaei, H. Rostamian, M. Dahari, A. Karimipour, Thermal conductivity of $Cu/TiO_2$-water/EG hybrid nanofluid: Experimental data and modeling using artificial neural network and correlation, Int. Commun. Heat Mass, 66 (2015) 100-104.

[35] G.A. Longo, C. Zilio, E. Ceseracciu, M. Reggiani, Application of artificial neural network (ANN) for the prediction of thermal conductivity of oxide-water nanofluids, Nano Energy, 1 (2012) 290-296.

[36] M. Vafaei, M. Afrand, N. Sina, R. Kalbasi, F. Sourani, H. Teimouri, Evaluation of thermal conductivity of MgO-MWCNTs/EG hybrid nanofluids based on experimental data by selecting optimal artificial neural networks, Physica E, 85 (2017) 90-96.

[37] S.H. Rostamian, M. Biglari, S. Saedodin, M. Hemmat Esfe, An inspection of thermal conductivity of CuO-SWCNTs hybrid nanofluid versus temperature and concentration using experimental data, ANN modeling and new correlation, J. Mol. Liq. 231 (2017) 364-369.

[38] M. Tahani, M. Vakili, S. Khosrojerdi, Experimental evaluation and ANN modeling of thermal conductivity of graphene oxide nanoplatelets/deionized water nanofluid, Int. Commun. Heat Mass, 76 (2016) 358-365.

[39] A.M. Ghahdarijani, F. Hormozi, A.H. Asl, Convective heat transfer and pressure drop study on nanofluids in double-walled reactor by developing an optimal multilayer perceptron artificial neural network, Int. Commun. Heat Mass, 84 (2017) 11-19.

[40] B. Vaferi, F. Samimi, E. Pakgohar, D. Mowla, Artificial neural network approach for prediction of thermal behavior of nanofluids flowing through circular tubes, Powder Technol. 267 (2014) 1-10.

[41] H. Safikhani, A. Abbassi, A. Khalkhali, M. Kalteh, Multi-objective optimization of nanofluid flow in flat tubes using CFD, Artificial Neural Networks and genetic algorithms, Adv. Powder Technol. 25 (2014) 1608-1617.

[42] H. Kalani, M. Sardarabadi, M. Passandideh-Fard, Using artificial neural network models and particle swarm optimization for manner prediction of a photovoltaic thermal nanofluid based collector, Appl. Therm. Eng. 113 (2017) 1170-1177.

[43] M. Hojjat, Experimental investigation on convective heat transfer of non-Newtonian nanofluids in circular tube with different thermal boundary conditions, PhD Thesis, Isfahan University of Technology, Iran, 2010.

[44] J. Kennedy, R. Eberhart, Particle swarm optimization, in: Neural Networks, 1995. Proceedings., IEEE International Conference on Neural Networks, IEEE, 1995, pp. 1942-1948.

[45] P. Orbanić, M. Fajdiga, A neural network approach to describing the fretting fatigue in aluminium-steel couplings, Int. J. Fatigue, 25 (2003) 201-207.

[46] R. Eslamloueyan, M.H. Khademi, A neural network-based method for estimation of binary gas diffusivity, Chemometr. Intell. Lab. 104 (2010) 195-204.

[47] C.S. Lee, W. Hwang, H.C. Park, K.S. Han, Failure of carbon/epoxy composite tubes under combined axial and torsional loading 1. Experimental results and prediction of biaxial strength by the use of neural networks, Compos. Sci. Technol. 59 (1999) 1779-1788.

[48] H.S. Rao, A. Mukherjee, Artificial neural networks for predicting the macromechanical behaviour of ceramic-matrix composites, Comp. Mater. Sci. 5 (1996) 307-322.

[49] I. Kaastra, M. Boyd, Designing a neural network for forecasting financial and economic time series, Neurocomputing, 10 (1996) 215-236.

[50] D.R. Hush, Classification with neural networks: a performance analysis, in: IEEE 1989 International Conference on Systems Engineering, 1989, pp. 277-280.

[51] I. Kanellopoulos, G.G. Wilkinson, Strategies and best practice for neural network image classification, Int. J. Remote Sens. 18 (1997) 711-725.

[52] M.E. Haque, K.V. Sudhakar, ANN back-propagation prediction model for fracture toughness in microalloy steel, Int. J. Fatigue, 24 (2002) 1003-1010.

# 19

# Effect of process parameters on quality properties and drying time of hawthorn in a vibro-fluidized bed dryer

**Elham Farzan, Mahmood Reza Rahimi, Vahid Madadi Avargani***

*[1] Department of Chemical Engineering, Yasouj University, Yasouj, Iran*

## HIGHLIGHTS

- The kinetic drying of hawthorn in a fluidized bed dryer with and without vibration was studied.

- Effect of vibration intensity, drying air temperature and flow rate on drying rate and shrinkage of hawthorn was investigated.

- All the drying process occurred in the falling rate period and no constant rate period was observed in the drying of hawthorn.

- The vibration intensity, drying air temperatures and flow rate had no significant effect on the shrinkage of hawthorn but has a significant effect on its drying rate.

- The shrinkage varies linearly with respect to moisture content, and the reduction in radial dimension of hawthorn samples was around 40% at the end of the drying process.

## ARTICLE INFO

*Keywords:*
Vibro-fluidized bed dryer
Hawthorn drying
Shrinkage
Drying rate
Mathematical modelling

## GRAPHICAL ABSTRACT

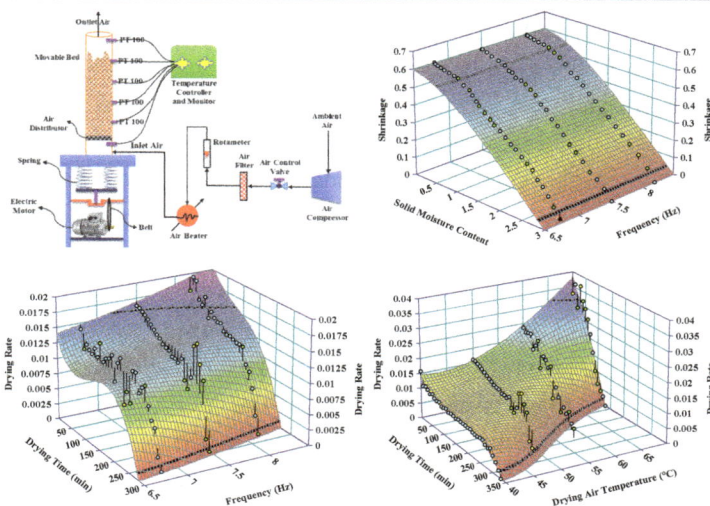

## ABSTRACT

The drying kinetics of hawthorn in a pilot-scale experimental fluidized bed dryer with and without vibration was investigated. The effect of operating parameters, such as vibration intensity, drying air flow rate, and temperature, on the drying rate and shrinkage of hawthorn was studied. The hawthorn fruit was dried at various drying air temperatures ranging from 40-70°C and drying air volume flow rates ranging from 22-30 m$^3$/h with the vibration intensity ranging from 6.8 to 8.2 Hz. The entire drying process occurred in the falling rate period and no constant rate period was observed in the drying of hawthorn. Four mathematical drying models investigating the drying behavior of hawthorn were evaluated and then the experimental moisture data were fitted in these models. The quality of the models fitting was assessed using the coefficient of determination, chi-square and root mean square error. The logarithmic and Page models for drying rate and the Ratti and Vazquez models for shrinkage were found to be the most suitable for describing the drying and shrinkage curves of hawthorn. The results showed that the vibration intensity, drying air temperature and flow rate has no significant effect on the shrinkage of hawthorn. All mentioned parameters had a significant effect on the drying rate of hawthorn, but the effect of drying air temperature was considerably more compared to the other parameters. It was observed that shrinkage varies linearly with respect to moisture content, and the reduction in radial dimension of hawthorn samples was around 40% at the end of the drying process.

*\* Corresponding author. ; E-mail address: v.madadi@yu.ac.ir*

## 1. Introduction

Agricultural products are often initially relatively high in moisture content and are highly vulnerable to pollution and corruption. Spoilage is variable depending on the type of material, and products can be rendered completely unusable in a short period of time. Therefore, the drying of agricultural product is very important and a variety of dryers can be used. In the drying process the method of drying greatly effects the product quality by reducing the moisture content to an acceptable level and extending the shelf life of products [1,2].

There are many different drying techniques and procedures used to reduce water from foodstuffs and increase the shelf life of these food materials. Convective drying is the most generally employed method due to its simplicity and cost effectiveness. Fluidized bed dryers are widely used in drying processes, especially in the food industry, because of their drying uniformity, relative low cost of operation and high rate of drying [3].

Nowadays, Vibro-Fluidized Bed Dryers (V-FBD) are highly regarded in food drying processes since they have greater efficiency than conventional types of fluidized bed dryers. V-FBDs can be used as a strategy to improve the fluidization quality of materials [4], increase the bed uniformity as a result of using vibration in the fluidized beds, and to avoid some problems like channeling and de-fluidization, all which improve drying of materials in fluidized bed dryers. The vibrating fluidized bed dryer is a modification of a common fluidized bed dryer, in which the vibration energy is used to transfer the material from a packed to fluidized state. Vibration can be produced by mechanical shaking of the entire apparatus [5,6].

One of the most important physical changes commonly seen in the early stages of material drying is a reduction in the size of the material's outer shell also called shrinkage [7,8]. Fruits are porous in nature and can experience shrinkage during the drying processes, which can affect the drying time. This shrinkage can be controlled in V-FBDs by various parameters such as intake air temperature and velocity and frequency of the engine device [9]. Fruits have high initial moisture content, and due to evaporation of the moisture contained in the fruit the shrinkage phenomenon is an undesirable change that can occur during drying. The shrinkage of material can affect the mass transfer diffusion coefficient

in the drying process and consequently the drying time and rate [7,10-12]. Also, from the food industry standpoint, shrinkage can affect the quality of food drying. The amount of shrinkage is different for each material and depends on the initial moisture content and the structure of the material. The material shrinkage can be controlled by several operational and structural parameters in drying processes such as inlet air temperature and velocity and frequency of the bed in a V-FBD [8,13,14].

Marring et al. studied the effect of vibration on a fluidized bed dryer for potato starch drying and observed that potato starch, which did not fluidize with aeration alone, could become well fluidized when vibration was applied [15]. Jinescu et al. reported that the vibrational state of the bed allows an increase in the drying rate due to the consequent increase in the specific area of gas–solid contact [16]. Karbassi et al. reported that bed uniformity, as a result of using vibration in fluid beds, can be achieved by preventing bubbles in the bed [17]. Also, the vibration helps to overcome interparticle forces, consequently improving the fluidization of material [18]. Mori et al. studied the drying behavior of sawdust particles in a vibro-fluidized bed dryer and found that using a vibro-fluidized dryer significantly reduced the required air velocity for drying sawdust [19].

Sadeghi and Khoshtaghaza studied the drying behavior of tea particles in vibro-fluidized bed dryers and found that the vibration system decreases the minimum fluidization velocity of tea particles in the bed, and this velocity was reduced by increasing the vibration intensity [20]. Lima and Ferreira examined fluidized and vibro-fluidized shallow beds for drying fresh leaves and concluded that the MCB dryer with vibration to produce a convenient configuration of drying can be used [21]. Silva Costa et al. analyzed the use of a hybrid CSTR/neural network model to describe the highly coupled heat and mass transfer during paste drying with inert particles in vibro-fluidized beds. The dynamics behavior of the outlet air temperature and relative humidity are described well by the CSTR lumped model for most evaluated conditions [22].

Nunes et al. investigated the fluid dynamics and polymeric coating of sodium bicarbonate in a vibro fluidized bed. In all experimental conditions there was an improvement in sodium bicarbonate fluidization, and a reduction in clusters and preferential channels compared to CFB [23]. Fyhr and Kemp presented a multiparticle model for fluidized bed dryers to study two different materials (zeolite and wheat), and the model predictions showed a good agreement with experimental data. Moreover, the calculations showed that the model can be used for transient conditions as well [24].

Zanoelo considered a theoretical and experimental study of simultaneous heat and mass transport resistances in a shallow fluidized bed dryer of mate leaves. An analysis of the joint confidence regions of both external heat and effective mass transfer coefficients was carried out and a significant effect of the equivalent particle diameter and temperature on these model parameters was observed [25].

The aim of this present work is to investigate the effect of some parameters such as vibration intensity, drying air temperature and flow rate on the shrinkage, drying time and drying kinetics of hawthorn in a fluidized bed dryer with and without vibration. In addition, the experimental moisture data is fitted to various mathematical models available from the literature.

## 2. Materials and methods

### 2.1. Hawthorn samples preparation

In this work, which is designed to dry hawthorn in a fluidized bed dryer, the fresh hawthorn samples were picked from a local tree at Yasouj University, situated in south-west Iran. The hawthorn samples were stored in a refrigerator for 24 h at $4\pm1°C$. The weight of the samples was determined by means of an electronic balance (AND model GF600, Japan) with an accuracy of 0.001 g. A photo of the hawthorn samples used in the experiment is shown in Figure 1. Based on device capacity and fundamentals of the drying process, the operating range of the study parameters are found in Table 1. According to the design of experiments (DOEs) in Design Expert Software, the number of optimized experiments and the magnitude of study parameters were obtained and are given in Table 2. The weight of three hawthorn fruit samples and the averaged dimensions of the dried samples in three directions were measured by means of a sliding gauge at different time intervals during each run. Experiments to determine the influence of process variables on shrinkage and drying time of hawthorn in a fluidized bed dryer with and without vibration were performed. The average diameter and

**Fig. 1.** Photo of hawthorn samples used in the experiments

**Table 1.** The operating range of considered parameters in the fluidized bed dryer

| Variable | Symbol | Operating Range |
|---|---|---|
| Inlet air temperature (°C) | T | 40-70 |
| Ambient air temperature (°C) | $T_a$ | $28 \pm 2$ |
| Ambient air relative humidity (%) | RH | $27 \pm 3$ |
| Air flow rate (m³/h) | Q | 22-30 |
| Frequency (Hz) | f | 6.8-8.2 |

the shrinkage of samples in the experiments are given by Eqs. (1) and (2), respectively, as below [26-28]:

$$D = \sqrt[3]{ABC} \tag{1}$$

$$Sh = (1 - \frac{D_t}{D_0}) \times 100 \tag{2}$$

where, the parameters $A$, $B$ and $C$ in Eq. (1) are the dimensions of the dried samples in three directions, and $D_t$ and $D_0$ in Eq. (2) are the instant and initial diameter of the samples, respectively.

### 2.2. Drying apparatus

A pilot-scale vibro-fluidized bed dryer was set up for performing the drying experiments. The schematic diagram of the experimental apparatus is shown in Figure 2. The dryer is made of a plexiglass column with a 0.1 m inside diameter, a height of 0.8 m, and a fitted perforated metal plate at the bottom as an air distributor. Drying air is supplied from a high-pressure air compressor and its pressure is adjusted by a regulator. The air is then passed through a rotameter before entering an electrical heater to raise the temperature of the drying air. Six PT100 thermocouples are used to measuring the air temperature along the bed length and

**Table 2.** Experimental design of hawthorn drying process in the fluidized bed dryer Experiment

| # | Repetitions | Inlet air temperature (°C) | Air flow rate (m³/h) | Frequency (Hz) | Amount of sample (gr) |
|---|---|---|---|---|---|
| 1 | 1 | 40 | 24 | 7.15 | 250 |
| 2 | 1 | 50 | 22 | 7.50 | 250 |
| 3 | 3 | 50 | 26 | 7.50 | 250 |
| 4 | 1 | 60 | 28 | 7.85 | 250 |
| 5 | 1 | 60 | 24 | 7.15 | 250 |
| 6 | 1 | 30 | 26 | 7.50 | 250 |
| 7 | 1 | 60 | 26 | 7.15 | 250 |
| 8 | 1 | 50 | 26 | 8.20 | 250 |
| 9 | 1 | 40 | 28 | 7.15 | 250 |
| 10 | 1 | 50 | 26 | 6.80 | 250 |
| 11 | 1 | 70 | 26 | 7.50 | 250 |
| 12 | 1 | 60 | 24 | 7.85 | 250 |
| 13 | 1 | 50 | 30 | 7.50 | 250 |
| 14 | 1 | 40 | 24 | 7.85 | 250 |
| 15 | 1 | 40 | 28 | 7.85 | 250 |

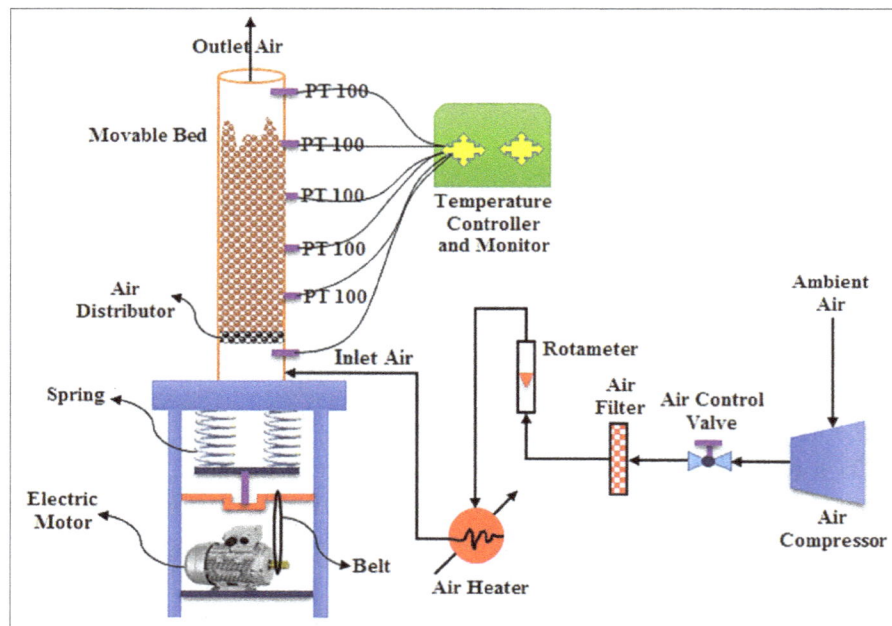

**Fig. 2.** Schematic diagram of the pilot plant vibro-fluidized bed dryer

a temperature controller with a temperature indicator is used for regulating and showing the temperatures of the drying air medium entering the bed within ±2°C. The inlet and outlet air temperature, velocity and humidity are measured by means of an electronic temperature/humidity meter (Model Testo 480, Germany, with a vane measurement probe of Ø 3.9" and a humidity and temperature probe of Ø 0.5", see Figure 3). The sample moisture (on a dry basis) was determined by measuring the wet and dry weight of the sample and using Eq. (3) as below:

$$\frac{W_w \quad W_d}{} \tag{3}$$

where, $M_d$ is the solid moisture content on a dry basis, and $W_w$ and $W_d$ are the wet and dry weight of solid samples, respectively.

The bed was vibrated through movement of the entire bed. To operate in the vibro-fluidized mode, the displacement was controlled by an electric motor with variable rotation, an eccentric mechanism was used to adjust the amplitude of vibration, and a mechanical

**Fig. 3.** The device which is used to measure air humidity, temperature and velocity

controller using a belt and two springs located at the axle of the electric motor allowed for adjustment of the vibration frequency as shown in Figure 2.

## 3. Mathematical modelling

Four simplified drying models, given in Table 3, have been used to describe the drying kinetics of hawthorn. Four simplified shrinkage models, given in Table 4, have been used to describe the shrinkage of hawthorn in the present work.

In these models, $MR$ represents the dimensionless moisture ratio and can be given by the following equation:

$$MR = \frac{M_t}{M_0} \tag{4}$$

where, $M_t$ is the moisture content at time t, and $M_0$ is the initial moisture content. A non-linear regression analysis was performed using TableCurve 3D software (Version 4.0) to fit the selected mathematical models with the experimental data. The coefficient of determination, $R^2$, is one of the primary criteria in order to evaluate the fit quality of these models. In addition, the reduced chi-square ($\chi^2$) and root mean square error ($RMSE$) were used to determine suitability of the fit. The coefficient of determination and reduced chi-square and RMSE can be calculated as follows:

$$R^2 = 1 - \frac{\sum_{i=1}^{N}\left(P_{pred,i} - P_{exp,i}\right)^2}{\sum_{i=1}^{N}\left(\overline{P}_{pred.i} - P_{exp,i}\right)^2} \tag{5}$$

$$\chi^2 = \frac{\sum_{i=1}^{N}\left(P_{exp.i} - P_{pred,i}\right)^2}{N-Z} \tag{6}$$

**Table 3.** Mathematical models applied to drying curves

| No. | Model name | Model |
|-----|-----------|-------|
| 1 | Newton [1] | $MR = \exp\left(-Kt\right)$ |
| 2 | Page [2] | $MR = \exp\left(-Kt\right)^n$ |
| 3 | Henderson and Pabis [3] | $MR = a\exp\left(-Kt\right)$ |
| 4 | Logarithmic [4] | $MR = a\exp\left(-Kt\right)+C$ |

**Table 4.** Mathematical models applied to shrinkage curves

| No. | Model name | Model |
|-----|-----------|-------|
| 1 | Mayor and Sereno [5] | $Sh = k_1 + k_2\left(MR\right) + k_3\left(MR\right)^2$ |
| 2 | Ratti [6] | $Sh = k_4 + k_5 M_t + k_6 M_t^2 + k_7 M_t^3$ |
| 3 | Mulet et al. [5] | $Sh = k_8 + k_9\left(\frac{M_t}{1+M_t}\right) + \exp\left(k_{10}\frac{M_t}{1+M_t}\right)$ |
| 4 | Vazquez et al. [7] | $Sh = k_{11} + k_{12}M_t + k_{13}M_t^{3/2} + k_{14}\exp\left(k_{15}M_t\right)$ |

$$RMSE = \sqrt{\frac{\left(\sum_{i=1}^{N}\left(P_{exp,i} - P_{pred,i}\right)^2\right)}{N}} \tag{7}$$

where, $P_{exp,i}$ is the $i^{th}$ experimental data such as the moisture ratio ($MR_{exp,i}$) and shrinkage ($Sh_{exp,i}$), $P_{pred,i}$ is the $i^{th}$ predicted value, $N$ is the number of observations and Z is the number of constants in a model. The higher the $R^2$ values and lower the ($\chi^2$) and $RMSE$ values, the better the goodness of fit [29-32].

## 4. Results and discussions

### 4.1. Analysis of drying curves

Several drying experiments were carried out under different operating conditions to investigate the effect of operating parameters, such as drying air flow rate and temperature, on shrinkage and drying time of hawthorn with and without vibration in a fluidized bed. The changes in diameter of the drying sample and the weight of test samples were determined, and consequently the shrinkage and moisture content of the samples were estimated. The drying curves [$MR=MR(t)$] of the hawthorn dried in a fluidized bed dryer with and without vibration for different values of drying air temperature and flow rate are as shown in Figures 4-6. As it is seen from these figures, the moisture content decreases continuously with drying time. The vibration intensity, the drying air temperature and flow rate have a significant effect on the drying kinetics of hawthorn. Increasing vibration intensity, drying air temperature and flow rate reduced the drying time of the hawthorn, and the drying air temperature had the most significant effect on this reduction. The shortest drying time in the experiments was observed for the V-FBD with the greatest vibration intensity (f=8.2 Hz) and a drying air temperature equal to 70°C.

Figure 4 shows the effect of drying air flow rate on the drying time of hawthorn in a fluidized bed dryer with and without vibration. As can be seen, drying time decreases as drying air flow rate increases due to increases in the convective heat and mass transfer coefficients. The variation of moisture ratio versus drying time for different drying air temperatures is shown in Figure 5. The drying air temperature had a more significant effect on drying time compared to the drying air flow rate and vibration intensity. The driving force for heat transfer from the gas phase to solid particle is the difference between the gas and solid temperatures, which causes moisture to be transferred from a solid particle to the gas phase. By increasing drying air temperature, the heat and mass transfer increase; hence, the drying time is significantly decreased. When the drying air temperature increased from 40 to 70°C, the drying time decreased about 60%. Figure 6 shows the effect of vibration intensity on drying time of hawthorn in the bed. The results show that an increase of about 20% in the vibration intensity decreased the drying time about 20%. This can be explained by an increase in the gas-solid contact area due to the prevention of particles sticking to each other, and as a result increases the heat and mass transfer between the gas phase and solid particles. Similar results have been reported by previous researchers [16,33].

### 4.2. Analysis of shrinkage curves

Variation of hawthorn shrinkage during drying time with and without vibration at different drying air temperature is shown in Figure 7. The results show that the vibration intensity has no significant effect on the shrinkage of hawthorn. Figure 7 also shows that the effect of drying air temperature on drying time is more significant compared to vibration intensity. When the drying air temperature increased, the drying time decreased significantly and at the end of the drying

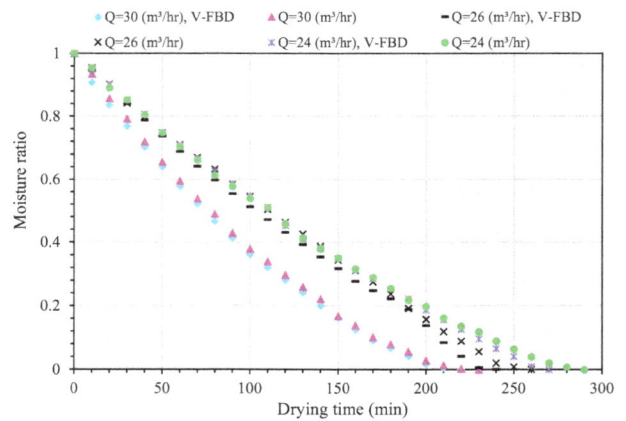

**Fig. 4.** Moisture ratio versus drying time of hawthorn at different drying air flow rates (drying air temperature=50°C and vibration intensity=7.5 Hz in vibro mode)

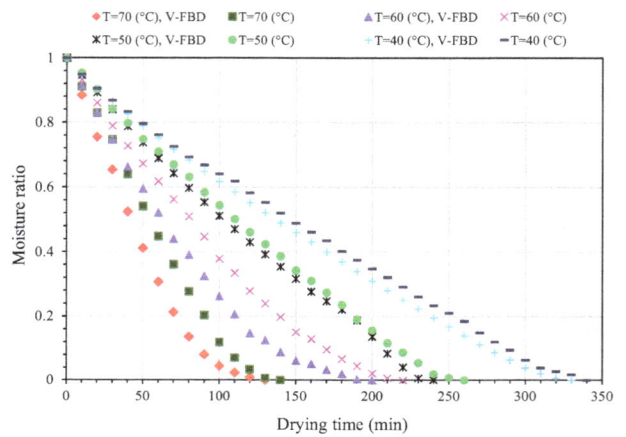

**Fig. 5.** Moisture ratio versus drying time of hawthorn at different drying air temperatures (drying air flow rate=26 m³/h and vibration intensity=7.5 Hz in vibro mode)

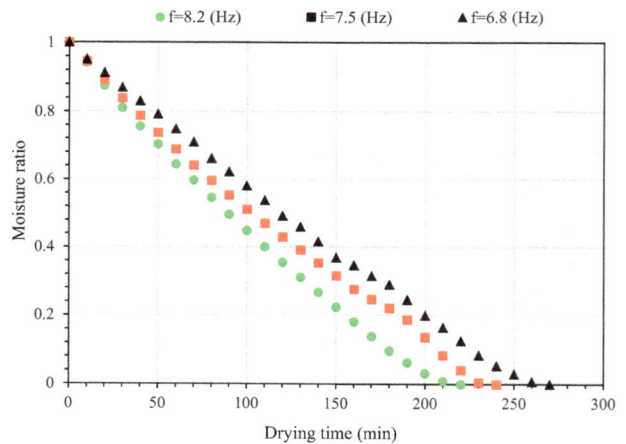

**Fig. 6.** Moisture ratio versus drying time of hawthorn at different vibration intensity (drying air flow rate=26 m³/h and drying air temperature=50°C)

process the amount of hawthorn shrinkage is the same for various drying air temperatures with and without

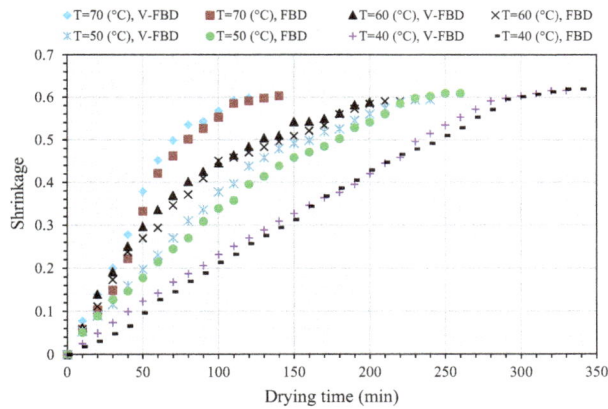

Fig. 7. Shrinkage variation versus drying time of hawthorn at drying air temperatures with and without vibration (drying airflow rate= 26 m³/h)

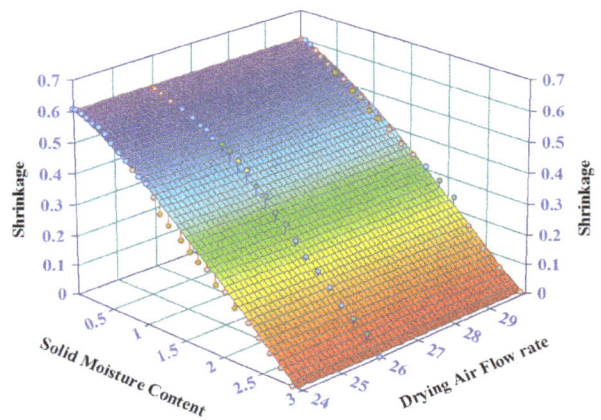

Fig. 8. Variation of hawthorn shrinkage versus solid moisture content at different vibration intensity (drying air flow rate=26 m³/h and drying air temperature=50°C)

For more clarity, other types of 3D graphs which investigate the effect of the studied parameters on shrinkage and drying rate are reported. The effect of solid moisture content, drying air flow rate and temperature on shrinkage of hawthorn dried in the fluidized bed dryer with and without vibration are shown in Figures 8-10. According to the results shown in these figures, the vibration intensity, drying air temperature and flow rate have no considerable effect on shrinkage of hawthorn in the fluidized bed dryer. Figures 8-10 show that at the end of the process when the solid moisture content is at its minimum value, the amount of shrinkage is at its maximum value. Although the solid moisture content and consequently the drying rate vary for different drying air conditions and various vibration intensities, the amount of shrinkage for various vibration intensities, drying air temperatures and flow rates remains the same for a constant solid moisture content. While the rate of shrinkage for higher drying air temperatures is greater than lower temperatures, at the end of the process the amount of the shrinkage is the same for all drying temperatures. According to Figure 8, when the solid moisture has a specified value the amount of shrinkage is the same, even for various bed vibration intensities.

The shrinkage profile of hawthorn with respect to solid moisture content for various studying parameters are shown in Figures 8-10. They clearly depict that shrinkage of the sample occurs almost linearly with solid moisture content, or in the other words, with drying time. Sometimes due to severe drying conditions, rapid shrinkage of sample surface occurs which leads to cracking of the surface. Surface cracking due to non-uniform shrinkage leads to the formation of unbalanced

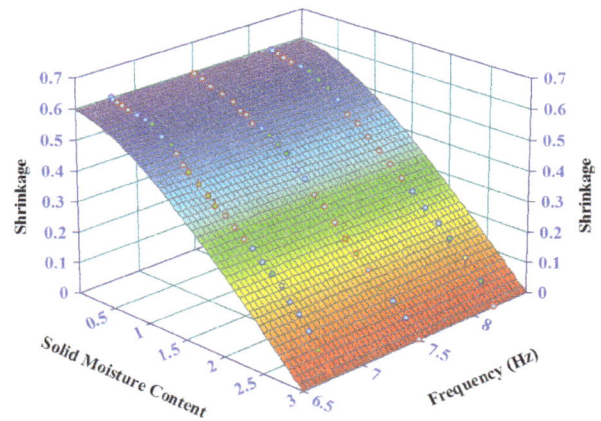

Fig. 9. Variation of hawthorn shrinkage versus solid moisture content at different drying air flow rates (vibration intensity=7.5 Hz and drying air temperature=50°C)

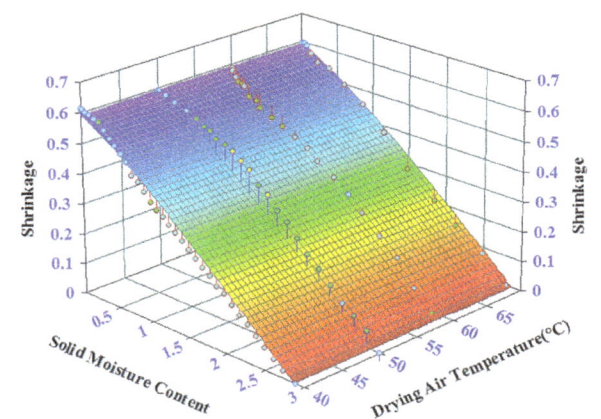

Fig. 10. Variation of hawthorn shrinkage versus solid moisture content at different drying air temperatures (drying air flow rate=26 m³/h and vibration intensity=7.5 Hz)

stresses and failure of the material. In this study, results show that the shrinkage of hawthorn in the fluidized bed dryer with and without vibration is approximately uniform, and as a result the surface cracking phenomenon does not occur.

In addition, Figures 8-10 show that up to 60% shrinkage was observed in the hawthorn sample during the entire process. Conversely, the diameter of the hawthorn sample only reduced to 40% of its original diameter at the end of the drying process.

### 4.3. Analysis of drying rate curves

The drying rate ($DR$) is defined as the amount of the evaporated moisture from the solid body over drying time. The drying rates of hawthorn were calculated by the following equation [2]:

$$DR = \frac{M_{t+dt} - M_t}{t} \qquad (8)$$

where, $M_{t+dt}$ and $M_t$ are the solid moisture content at time $t$ and $t+dt$, respectively, based on a dry solid basis (kg water/ kg dry solid) and $t$ is the drying time (min).

Figures 11-15 show the variation of drying rate for hawthorn in a fluidized bed dryer with and without vibration versus drying time at different vibration intensities, drying air temperatures and flow rates. The results in these figures show that the drying rate decreases continuously with drying time due to a reduction in solid moisture content. The drying rate curves show that there was no constant drying period during the drying of hawthorn. All of the drying rate curves show that the drying of hawthorn occurs in the falling rate period; solid internal molecular diffusion is the dominant mechanism in this period of drying. Some similar results were reported in previous works for other agricultural products such as olive pomace [2, 34-36], organic tomatoes [37], green peppers [38], and figs [39].

The results in Figures 11-13 indicate that the drying rate in a fluidized bed with vibration is greater than the drying rate without vibration. As mentioned before, when vibration is applied to the dryer the efficiency of gas-solid contact increases because the particles are prevented from sticking to each other and the minimum fluidization velocity and fluidization pressure drops decrease and homogeneity and stability of the fluidized bed layers increase. Finally, as a result, the heat and mass transfer between the gas phase and solid particles, and consequently the drying rate increases in the vibro-fluidized bed mode. The results in Figure 13 indicate that increasing the vibration intensity from 6.8 to 8.2 Hz for

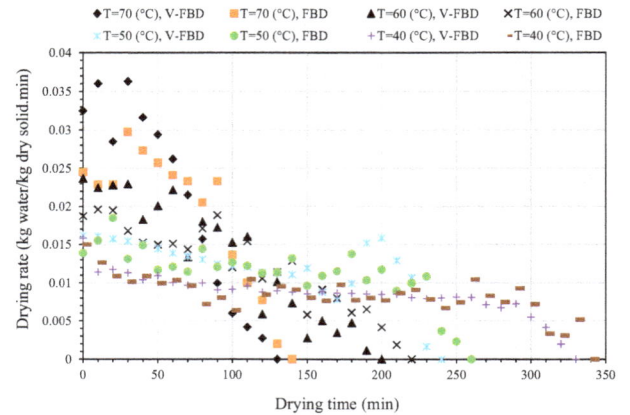

**Fig. 11.** Drying rate versus drying time at different drying air temperatures with and without vibration (drying air flow rate=26 m³/h, vibration intensity=7.5 Hz)

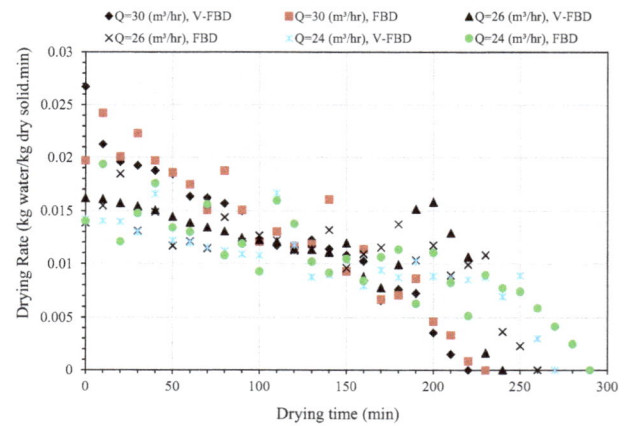

**Fig. 12.** Drying rate versus moisture content at different drying air flow rates with and without vibration (drying air temperature=50°C, vibration intensity=7.5 Hz)

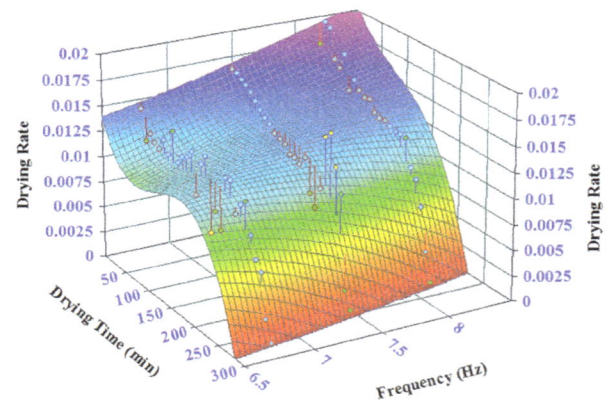

**Fig. 13.** Drying rate versus drying time of hawthorn at different vibration intensity (drying air flow rate=26 m³/h and drying air temperature=50°C)

a drying air temperature and flow rate equal to 50°C and 26 m³/h, respectively, reduced the drying time by about 20%. In addition, at the beginning of the process, when the solid moisture content is high, the drying rate for vibration intensity equal to 8.2 Hz is about 33% greater than when the vibration intensity is equal to 6.8 Hz.

The effect of drying air flow rate on the drying rate of hawthorn in a fluidized bed dryer with and without vibration is illustrated in Figures 12 and 14. The results show that by increasing the drying air flow rate the drying rate increases, and consequently the drying time decreases. This can be explained as when the drying air flow rate increases the convective heat and mass transfer coefficient increase and the amount of heat transfer from the gas phase to the solid phase and the moisture transfer from the solid phase to the gas phase increase. The results in Figure 14 show that when drying air temperature and vibration intensity are equal to 50°C and 7.5 Hz, respectively, increasing the drying air flow rate from 26 to 30 m³/h reduces the drying time of hawthorn about 20%.

One of the most important operating parameters in the drying process is the drying air temperature. In the drying of materials, the drying air temperature is selected based on many essential factors such as the thermal properties of the materials and the desired quality of the dried materials. In food industries, this is more important than other operating parameters. In this study, a range of 40 to 70°C was selected for the experiments. In drying processes with warm air, the moisture is transfer from the solid body to the solid surface and then to the gas phase by heat transfer from the gas phase to the solid surface. Therefore, in a drying process with warm air, the convective heat transfer is a function of the convective heat transfer coefficient and the difference between the solid-gas interface and gas bulk temperatures.

The effect of drying air temperature on the drying rate and consequently drying time of hawthorn in the fluidized bed dryer with and without vibration is shown in Figures 11 and 15. It can be seen from these figures that the drying air temperature had a more significant effect on the drying time of hawthorn in the fluidized bed compared to the drying air flow rate and vibration intensity. According to the results in Figure 15, increasing the drying air temperature from 40 to 70°C decreases the drying time about 60%.

### 4.4. Models of drying and shrinkage curves

The amount of hawthorn shrinkage and moisture

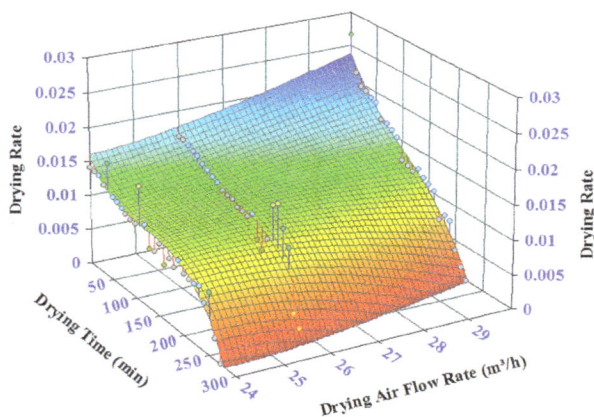

**Fig. 14.** Drying rate versus drying time of hawthorn at different drying air flow rates (drying air temperature=50°C and vibration intensity=7.5 Hz)

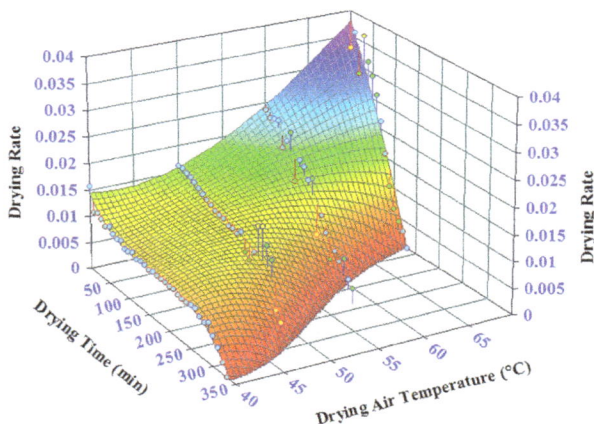

**Fig. 15.** Drying rate versus drying time of hawthorn at different drying air temperatures (drying air flow rate=26 m³/h and vibration intensity=7.5 Hz)

ratio at different times were measured by the methods mentioned above. The obtained experimental data from the drying of hawthorn in the fluidized bed dryer for various conditions were fitted to four mathematical models listed in Tables 1 and 2 for moisture ratio and hawthorn shrinkage, respectively. Values of the determination coefficient ($R^2$), the reduced chi-square ($\chi^2$) and the root mean square error ($RMSE$) for different drying air temperatures determined by non-linear regression analysis are necessary to suggest a proper equation to predict the suitable behavior of hawthorn drying in a fluidized bed dryer, see Tables 5 and 6. In accordance with the results shown in these tables, the $R^2$, $\chi^2$ and $RMSE$ values for the moisture ratio ranges from 0.950462 to 0.999370, 0.000072 to 0.004009 and 0.007881 to 0.079668, respectively. The $R^2$, $\chi^2$ and $RMSE$ values for shrinkage ranges from 0.824923 to 0.999634, 0.000080 to 0.006971 and 0.008507 to

**Table 5.** Statistical results of mathematical models fitting for moisture ratio at different drying air temperatures

| Model | T (°C) | $R^2$ | $\chi^2$ | RMSE |
|---|---|---|---|---|
| | 70 | 0.956640 | 0.003824 | 0.060345 |
| Newton | 60 | 0.962428 | 0.004009 | 0.061791 |
| | 50 | 0.966199 | 0.003643 | 0.058906 |
| | 40 | 0.950462 | 0.002845 | 0.052054 |
| | 70 | 0.997413 | 0.000310 | 0.016759 |
| Page | 60 | 0.997086 | 0.000487 | 0.021001 |
| | 50 | 0.990251 | 0.001029 | 0.030511 |
| | 40 | 0.981352 | 0.002807 | 0.050394 |
| | 70 | 0.965764 | 0.003138 | 0.053280 |
| Henderson and | 60 | 0.972556 | 0.003087 | 0.052850 |
| Pabis | 50 | 0.973009 | 0.003066 | 0.052669 |
| | 40 | 0.956158 | 0.007015 | 0.079668 |
| | 70 | 0.995653 | 0.000724 | 0.024916 |
| Logarithmic | 60 | 0.995563 | 0.000526 | 0.021233 |
| | 50 | 0.999370 | 0.000075 | 0.008038 |
| | 40 | 0.999339 | 0.000072 | 0.007881 |

**Table 6.** Statistical results of mathematical models fitting for hawthorn shrinkage at different drying air temperatures

| Model | T (°C) | $R^2$ | $\chi^2$ | RMSE |
|---|---|---|---|---|
| | 70 | 0.996735 | 0.000134 | 0.011308 |
| Mayor and | 60 | 0.994478 | 0.000170 | 0.012741 |
| Sereno | 50 | 0.995845 | 0.000157 | 0.012234 |
| | 40 | 0.982852 | 0.002917 | 0.051239 |
| | 70 | 0.999434 | 0.000083 | 0.008694 |
| Ratti | 60 | 0.999634 | 0.000080 | 0.008507 |
| | 50 | 0.996822 | 0.000127 | 0.010700 |
| | 40 | 0.996951 | 0.000111 | 0.010025 |
| | 70 | 0.907306 | 0.003994 | 0.060113 |
| Mulet *et al.* | 60 | 0.841852 | 0.005139 | 0.068188 |
| | 50 | 0.824923 | 0.006971 | 0.079417 |
| | 40 | 0.851827 | 0.005224 | 0.068748 |
| | 70 | 0.997652 | 0.000107 | 0.009566 |
| Vazquez *et al.* | 60 | 0.992577 | 0.000255 | 0.014773 |
| | 50 | 0.996324 | 0.000158 | 0.053245 |
| | 40 | 0.998003 | 0.000090 | 0.008804 |

0.079417, respectively. The highest values for the determination coefficient and the lowest values for the reduced chi-square and root mean square error indicate a good fit in all cases.

From the drying curves in the models considered the Page and logarithmic models gave values of $R^2$ higher than 0.99 for almost all drying air temperatures, except the value of $R^2$ was 0.981352 in the Page model at a drying air temperature of 40°C. Furthermore, the values of *RMSE* and chi-square for the drying curves in the Page and logarithmic models were less than the two others. The drying curve analysis shows that at higher drying air temperatures the Page model predicted better than the logarithmic and the logarithmic model predicted better at lower temperatures. The logarithmic and Page models appear to be the most appropriate in describing the drying rate of hawthorn under the experimental conditions studied.

Among the models considered for hawthorn shrinkage, Ratti and Vazquez models gave values of $R^2$ above 0.99 for all drying air temperatures in the range of (40-70°C). The Ratti and Vazquez models gave comparatively higher values for $R^2$ in all drying air temperatures, whereas $\chi^2$ and *RMSE* values were lower. Thus, these models appear to be the mosappropriate in describing the shrinkage of hawthorn under the experimental conditions studied. Among the others investigated mathematical models, the Newton, Henderson and Pabis models for drying rate and the Mulete model for shrinkage appear to be the least suitable models for the drying and shrinkage behavior of hawthorn under the experimental conditions studied.

## 5. Conclusions

In this study, the effect of drying air temperature and flow rate ranging from 40 to 70°C and 22 to 30 m³/h, respectively, on the drying rate and shrinkage of hawthorn in a fluidized bed dryer with and without vibration was investigated. The drying of hawthorn took place only in the falling rate period. Behavior of shrinkage changes along the the drying process was demonstrated and the effect of the mentioned parameters on hawthorn shrinkage was studied. A uniform radial shrinkage, up to 60% of the original radius at the end of drying process, was observed to happen linearly with respect to the solid moisture content. It is observed that the vibration intensity, drying air temperature and flow rate have no significant effect on the shrinkage of hawthorn. The effect of the mentioned parameters on the drying rate of hawthorn was investigated and results show that the drying air temperature has a more significant effect on the drying rate of hawthorn than the other parameters. For drying air flow rate and vibration intensity equal to 26 m³/h and 7.5 Hz, respectively, increasing the drying air temperature from 40 to 70°C decreased the drying time by about 60%.

The drying and shrinkage curves were fitted to four different mathematical drying models, and the logarithmic and Page models for drying rate and the Ratti and Vazquez models for shrinkage were found to be the most suitable for describing the drying and shrinkage curves of hawthorn under the experimental conditions studied

## Nomenclature

| | |
|---|---|
| $Sh$ | Shrinkage, (dimensionless) |
| $D_0$ | Initial diameter of product, (mm) |
| $D_t$ | Diameter of the dried product, (mm) |
| $W_w$ | Initial weight of undried product, (gr) |
| $W_d$ | Weight of dry matter in product, (gr) |
| $M_d$ | Moisture (on dry basis), (kg water/kg dry solid) |
| $D$ | Average diameter, (mm) |
| $A, B, C$ | Three orthogonal diameters, (mm) |
| $MR$ | Moisture ratio, (dimensionless) |
| $M_t$ | Moisture content at time t, (kg water/kg dry matter) |
| $M_0$ | Initial moisture content, (kg water/kg dry matter) |
| $R^2$ | Coefficient of determination |
| $Z$ | Number of constants in a model |
| $P_{pred,i}$ | i[th] predicted value |
| $P_{exp,i}$ | i[th] experimental value |
| $\chi^2$ | chi-square, (dimensionless) |
| $N$ | Number of experiments |
| $RMSE$ | Root mean square error, (dimensionless) |
| $k_1$-$k_{15}$ | Constants in models, (dimensionless) |
| $a, n, c$ | Coefficients in models, (dimensionless) |
| $t$ | Drying time, (min) |
| $T$ | Drying air temperature, (°C) |

## References

[1] J. Aprajeeta, R. Gopirajah, C. Anandharamakrishnan, Shrinkage and porosity effects on heat and mass transfer during potato drying, J. Food Eng. 144 (2015) 119-128.

[2] S. Meziane, Drying kinetics of olive pomace in a fluidized bed dryer, Energ. Convers. Manage. 52 (2011) 1644-1649.

[3] I. Białobrzewski, M. Zielińska, A.S. Mujumdar, M. Markowski, Heat and mass transfer during drying of a bed of shrinking particles–Simulation for carrot cubes dried in a spout-fluidized-bed drier, Int. J. Heat Mass Tran. 51 (2008) 4704-4716.

[4] E. Jaraiz, S. Kimura, O. Levenspiel, Vibrating beds of fine particles: estimation of interparticle forces from expansion and pressure drop experiments, Powder Technol. 72 (1992) 23-30.

[5] R. Moreno, R. Rios, H. Calbucura, Batch vibrating fluid bed dryer for sawdust particles: experimental results, Dry. Technol. 18 (2000) 1481-1493.

[6] M. Stakić, T. Urošević, Experimental study and simulation of vibrated fluidized bed drying, Chem. Eng. Process. 50 (2011) 428-437.

[7] M. Prado, Drying of dates (Phoenix Dactyulifera L.) to obtain dried date (passa), Campinas, UNICAMP, 1998.

[8] L. Mayor, A. Sereno, Modelling shrinkage during convective drying of food materials: a review, J. Food Eng. 61(3) (2004) 373-386.

[9] I. Sjöholm, V. Gekas, Apple shrinkage upon drying, J. Food Eng. 25 (1995) 123-130.

[10] N. Wang, J. Brennan, Changes in structure, density and porosity of potato during dehydration, J. Food Eng. 24 (1995) 61-76.

[11] W. Senadeera, B.R. Bhandari, G. Young, B. Wijesinghe, Influence of shapes of selected vegetable materials on drying kinetics during fluidized bed drying, J. Food Eng. 58 (2003) 277-283.

[12] B.A. Souraki, D. Mowla, Axial and radial moisture diffusivity in cylindrical fresh green beans in a fluidized bed dryer with energy carrier: Modeling with and without shrinkage, J. Food Eng. 88 (2008) 9-19.

[13] G. Hashemi, D. Mowla, M. Kazemeini, Moisture diffusivity and shrinkage of broad beans during bulk drying in an inert medium fluidized bed dryer assisted by dielectric heating, J. Food Eng. 92 (2009) 331-338.

[14] E. Marring, A. Hoffmann, L. Janssen, The effect of vibration on the fluidization behaviour of some cohesive powders, Powder Technol. 79 (1994) 1-10.

[15] G. Jinescu, C. Tebrencu, E. Ionescu, M. Petrescu, C. Jinescu, Hydrodynamic aspects at vibrated-fluidized drying of polydisperse powdery materials, in International Drying Symposium IDS2000, 2000.

[16] A. Karbassi, Z. Mehdizadeh, Drying rough rice in a fluidized bed dryer, J. Agr. Sci. Tech.-Iran. 10 (2010) 233-241.

[17] D. Kunii, O. Levenspiel, Fluidization Engineering. Butterworth-Heinemann, Boston, 1991.

[18] S. Mori, Vibro-fluidization of group-c particles and its industrial application, in AIChE Symp. Ser. 1990.

[19] M. Sadeghi, M. Khoshtaghaza, Vibration effect on particle bed aerodynamic behavior and thermal performance of black tea in fluidized bed dryers, J. Agr. Sci. Tech.-Iran. 14 (2012) 781-788.

[20] R.d.A.B. Lima, M. do Carmo Ferreira, Fluidized and vibrofluidized shallow beds of fresh leaves, Particuology 9 (2011) 139-147.

[21] A.S. Costa, F.B. Freire, J. Freire, M. Ferreira, Modelling drying pastes in vibrofluidized bed with inert particles, Chem. Eng. Process. 103 (2016) 1-11.

[22] J.F. Nunes, F.C.A. de Alcântara, V.A. da Silva Moris, S.C. dos Santos Rocha, Fluid dynamics and coating of sodium bicarbonate in a vibrofluidized bed, Chem. Eng. Process. 52 (2012) 34-40.

[23] C. Fyhr, I.C. Kemp, Mathematical modelling of batch and continuous well-mixed fluidised bed dryers, Chem. Eng. Process. 38 (1999) 11-18.

[24] E.F. Zanoelo, A theoretical and experimental study of simultaneous heat and mass transport resistances in a shallow fluidized bed dryer of mate leaves, Chem. Eng. Process. 46 (2007) 1365-1375.

[25] R. Amiri Chayjan, M. Kaveh, Physical parameters and kinetic modeling of fix and fluid bed drying of terebinth seeds, J. Food Process. Pres. 38 (2014) 1307-1320.

[26] S. Aral, A.V. Beşe, Convective drying of hawthorn fruit (Crataegus spp.): Effect of experimental parameters on drying kinetics, color, shrinkage, and rehydration capacity, Food Chem. 210 (2016) 577-584.

[27] S. Mercier, S. Villeneuve, M. Mondor, L.-P. Des Marchais, Evolution of porosity, shrinkage and density of pasta fortified with pea protein concentrate during drying, LWT-Food Sci. Technol. 44 (2011) 883-890.

[28] İ. Doymaz, Convective drying kinetics of strawberry, Chem. Eng. Process. 47 (2008) 914-919.

[29] C. Hii, C. Law, M. Cloke, Modeling using a new thin layer drying model and product quality of cocoa, J. Food Eng. 90 (2009) 191-198.

[30] C. Ertekin, O. Yaldiz, Drying of eggplant and selection of a suitable thin layer drying model, J. Food Eng. 63 (2004) 349-359.

[31] H.O. Menges, C. Ertekin, Thin layer drying model for treated and untreated Stanley plums, Energ. Convers. Manage. 47 (2006) 2337-2348.

[32] O. Molerus, K.-E. Wirth, Heat transfer in fluidized beds, Vol. 11, Springer Science & Business Media, 2012.

[33] F. Göğüş, M. Maskan, Air drying characteristics of solid waste (pomace) of olive oil processing, J. Food Eng. 72 (2006) 378-382.

[34] I. Doymaz, O. Gorel, N.A. Akgun, Drying Characteristics of the Solid By-product of Olive Oil Extraction, Biosyst. Eng. 88 (2004) 213-219.

[35] N.A. Akgun, I. Doymaz, Modelling of olive cake thin-layer drying process, J. Food Eng. 68 (2005) 455-461.

[36] K. Sacilik, R. Keskin, A.K. Elicin, Mathematical modelling of solar tunnel drying of thin layer organic tomato, J. Food Eng. 73 (2006) 231-238.

[37] E.K. Akpinar, Y. Bicer, Mathematical modelling of thin layer drying process of long green pepper in solar dryer and under open sun, Energ. Convers. Manage. 49 (2008) 1367-1375.

[38] S.J. Babalis, V.G. Belessiotis, Influence of the drying conditions on the drying constants and moisture diffusivity during the thin-layer drying of figs, J. Food Eng. 65 (2004) 449-458.

# 20

# Modeling of splat particle splashing data during thermal spraying with the Burr distribution

**Hanieh Panahi[1,*], Saeid Asadi[2]**

[1] *Department of Mathematics and Statistics, Lahijan Branch, Islamic Azad University, Lahijan, Iran*

[2] *Department of Mechanical Engineering, Payame-Noor University (PNU), Tehran, Iran*

## HIGHLIGHTS

- Inference about the splat particle splashing data which sprayed with a normal angle.

- Perform the number of model selection tests to determine the appropriate probability model under complete and progressive censored sample.

- Study of different methods for predicting the missing splat particle splashing data.

## GRAPHICAL ABSTRACT

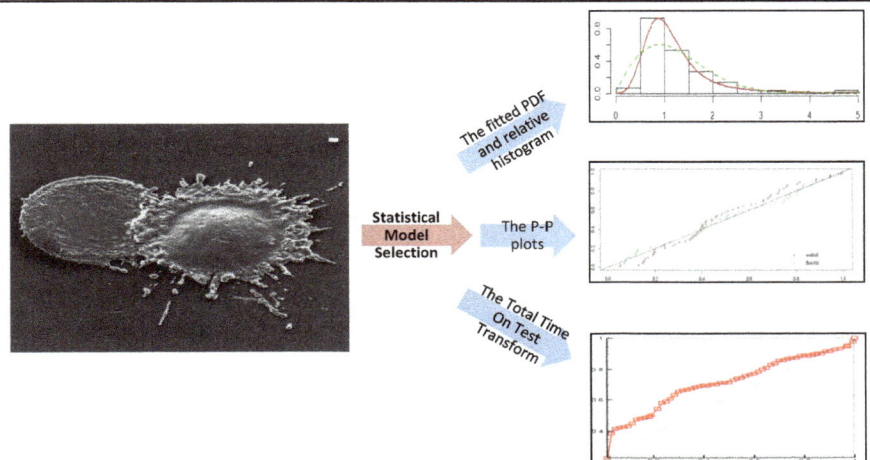

## ARTICLE INFO

*Keywords:*
Burr Type XII
Censored data
Splat particle
Splashing
Progressively censoring
Thermal spray

## ABSTRACT

Splashing of splat particles is one of the most important phenomena in industrial processes such as thermal spray coating. The data relative to the degree of splashing of splats sprayed with a normal angle are commonly characterized by the Weibull distribution function. In this present study, an effort has been made to show that the Burr distribution is better than the Weibull distribution for presenting the distribution of the degree of splashing. For this purpose, the Burr Type XII distribution and Weibull distribution are compared using different criteria. Furthermore, because of the great importance of statistical prediction of censored data in reducing costs and improving quality of the coating process, we consider different predictors of this data based on a progressively censored sample. For computing the prediction values we obtain the maximum likelihood estimates using the Expectation-Maximization (EM) algorithm. An important implication of the present study is that the Burr Type XII distribution more appropriately described the degree of splashing data. Therefore, the Burr Type XII can be used as an alternative distribution that adequately describes the splashing data and thereby predicts the censored data.

* Corresponding author:  ; E-mail address: panahi@liau.ac.ir

## 1. Introduction

Thermal sprayed coatings are widely used to protect components exposed to corrosion, wear or heat. The mechanical properties of coatings are known to depend strongly on the shape of splat particles formed by individual particles as they impact and freeze. As a good coated surface is extremely important in industry, one of the most important phenomena, due to its impact on the deterioration of the coated surface, is particle splashing. Splashing occurs when a single particle breaks up on impact producing secondary, or satellite, particles. Figure 1 illustrates splashing via a sequence of photographs of the impact of molten tin particles onto a hot surface [1]. Splashing degrade coating quality since they leave voids in the deposit, increasing its porosity and reducing its strength. The physical mechanisms of splashing are still not completely understood and splash study is an extremely interesting and attractive phenomenon. Moreover, prediction of particle splashing can potentially reduce the cost of the development of new coating considerably.

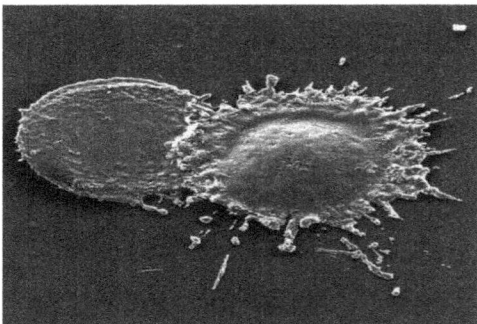

**Fig. 1.** Splashing of molten tin droplet on a stainless steel surface [1].

The first study of particle fingering and splashing is the remarkable work of Worthington [2,3] published more than a century ago. He drew interesting pictures from direct visual observations of the impact of a mercury drop on a glass plate. More than a half century after Worthington, Engel [4] studied the impact of water droplets onto various surfaces, with application to the erosion of aircraft components due to rain drop impact. A small solid surface roughness has been found to have an important influence on the limiting conditions of the onset of splashing [5]. Montavon et al. [6] studied the effects of spray angle on the morphology of thermally sprayed particles impinging on polished substrates. They evaluated the degree of splashing of splats as a function of their equivalent diameter for 90° and 30°

spray angles. Thoroddsen and Sakakibara [7] considered the evolution of the fingering pattern at the edge of drop during the impact of a water drop on a glass plate. They observed that systematic changes in frontal shapes take place during the expansion. Hardalupas et al. [8] examined the impact of a stream of particles onto stainless steel, to examine the influence of surface curvature. Aziz and Chandra [1] studied the impact and solidification of molten tin particles on a stainless steel surface. They photographed particle impact and measured splat diameter and liquid-solid contact angle from these photographs and used a simple energy conservation model to predict the maximum spread of particles during impact.

Asadi et al. [9] extended the numerical and analytical model of the inclined impact of a plasma particle on a solid surface in a thermal spray coating process. The effects of particle velocity, impact angle, and ambient gas pressure (or density) on the threshold of splashing and the motion of the ambient gas surrounding the particle were examined by Liu et al. [10]. Asadi [11] applied a modified computational fluid dynamics and molecular kinetic theory (CFD-MK) method to model the spread and splash of nanoparticle impact on a flat surface. Li et al. [12] estimated impact energy stored in the splash structures via a theoretical model and several morphological parameters. They found that the particle size and the impact velocity displayed similar proportional trends with respect to the splashing height, but did not accompany the secondary particle separation; also the increase of pool temperature dramatically intensified the splashing effect, with the fusiform secondary particle detached from a central jet. Liang et al. [13] examined spreading and splashing processes during a single liquid particle impact on an inclined wetted surface by using a high-speed digital camera. They observed that both surface tension and viscosity can greatly affect the spreading and splashing behaviors. Liang et al. [14] studied rebound and spreading behaviors during a single particle impact on wetted cylindrical surfaces and discussed deformation factor with the critical Weber numbers. While much research has been done on the study of particles splashing, less attention has been paid to the distribution modeling and statistical prediction of this phenomenon.

In the present work, we consider the model selection and prediction of splat particle splashing data obtained by Montavon et al. [6] . We observed that the Weibull distribution has been used as the statistical distribution for modeling the engineering data. We want to answer this question, "Is there a more appropriate statistical distribution?". Thus, we use different methods, such as Kolmogorov-Smirnov (K-S) distance, Akaike information criterion [15], Baysian information criterion [16], and the total time on test (TTT) transform and maximum likelihood criterion, to show the appropriateness of the Burr distribution in the particle splashing data. Since the experimenter may not always obtain complete information on the data in many experimental studies, data obtained from such experiments are called censored data. Type I and Type II censoring schemes are the two most common and popular censoring schemes, but these censoring schemes do not have the flexibility of allowing the removal of units at points other than the terminal point of the experiment. For this reason, in the last few years the progressive censoring scheme has received considerable attention in applied science. This scheme allows one to remove experimental units at points other than the terminal point of the experiment. Several authors have considered different aspects of this censoring scheme; see for example [17-20]. Prediction of censored observation based on the current available data is one of the important problems in engineering experiments. We know that in experiments some of the splashing data are missing. Thus, the second purpose of this paper is to predict future splashing data under the progressive censoring scheme. We obtain the conditional median predictor and prediction interval based on the pivotal method. For obtaining the prediction method, we substitute the unknown parameters with their maximum likelihood estimates under the progressive censoring scheme. It is observed the maximum likelihood estimators (MLE's) cannot be obtained in closed form. So, we propose to use the EM algorithm to compute the maximum likelihood estimators. The EM algorithm is a very powerful and useful tool for analyzing the censored data.

## 2. Two rival models

In this section we briefly describe Burr Type XII and Weibull distributions as the rival models.

### 2.1. Burr Type XII distribution

Burr [21] introduced twelve cumulative distribution functions with the primary purpose of fitting distributions to real data. One of the most important of them is the Burr Type XII distribution. The cumulative distribution function and probability density function of the Burr Type XII are given by, respectively

$$F^{(\alpha,\beta)}(x) = \begin{cases} 1-(1+x^\beta)^{-\alpha}; & x \ge 0, \alpha > 0, \beta > 0 \\ 0 & \text{otherwise} \end{cases} \quad (1)$$

$$f^{(\alpha,\beta)}(x) = \begin{cases} \alpha\beta x^{\beta-1}(1+x^\beta)^{-(\alpha+1)}; & x \ge 0, \alpha > 0, \beta > 0 \\ 0 & \text{otherwise} \end{cases}$$

Here $\alpha$ and $\beta$ are the two shape parameters. The shape of the hazard rate function of the Burr Type XII distribution depends only on parameter $\beta$. For $\beta>0$, the hazard rate is eventually decreasing. For $\beta>0$, the hazard rate is a unimodal function whereas for $\beta\le0$, it is decreasing. Thus the shape $\beta$ parameter plays an important role in the distribution. Its capacity to assume various shapes often permits a good fit when used to describe biological, clinical, engineering or other experimental data. It also approximates the distributional form of normal, lognormal, gamma, logistic, and several Pearson-Type distributions. For instance, the normal density function may be approximated as a Burr Type XII distribution with $\beta=4.8544$ and $\alpha=6.2266$ and the gamma distribution with shape parameter 16 can be approximated as a Burr Type XII distribution with $\beta=3$ and $\alpha=6$, and the log-logistic distribution is a special case of the Burr Type XII distribution. Extensive work has been done on the Burr Type XII distribution, see for example [22-25].

### 2.2. Weibull distribution

The Weibull distribution is one of the most popular distributions in analyzing lifetime data. The two parameter Weibull distribution (W) with the shape parameter $\alpha>0$ and scale parameter $\beta>0$ has the probability density function as;

$$f^{(\alpha,\beta)}(x) = \begin{cases} \alpha\beta x^{\beta-1}\exp(-\beta x^\alpha); & x \ge 0, \alpha > 0, \beta > 0 \\ 0 & \text{otherwise.} \end{cases} \quad (2)$$

where, $\alpha$ and $\beta$ are the shape and scale parameters, respectively. If $x\sim Weibull(\alpha,\beta)$, then the cumulative distribution function, reliability function and hazard function are

$$F^{(\alpha,\beta)}(x) = \begin{cases} 1-\exp(-\beta x^\alpha); & x \ge 0, \alpha > 0, \beta > 0 \\ 0 & \text{otherwise.} \end{cases}$$

$$R(x) = \exp(-\beta x^\alpha), \quad x \ge 0, \alpha > 0, \beta > 0$$

and

$$h(x) = \alpha\beta x^{\beta-1}, \quad x \geq 0, \alpha > 0, \beta > 0$$

respectively. A detailed discussion of the W distribution has been provided by [26].

## 3. Different Criteria for model selection

In this section we describe different available criteria for choosing the best fitted model to a given dataset. Suppose there are two families, say, $F = \{f^\theta(.), \theta \in R^p\} = (f)$ and $G = \{g^{\theta'}(.), \theta' \in R^q\} = (g)$, the problem is to choose the correct family for a given dataset $\{x_1,...,x_n\}$. The following criteria can be used for model selection.

### 3.1. Kolmogorov-Smirnov (K-S) distance

The K-S distance is one of the important distances between two distribution functions, say $F$ and $G$, and it can be described as follows;

$$D(F,G) = \sup_{-\infty < x < \infty} |F(x) - G(x)| \qquad (3)$$

To implement this procedure, a candidate from each parametric family that has the smallest K-S distance should be found and then the different best fitted distributions should be compared.

### 3.2. Akaike's information criterion

Consider a sample of independently identically distributed (i.i.d.) random variables, $X_1,...,X_n$ having the probability density function $h(.)=h$. The Kullback-Leibler (KL) information in favor of $h$ against $f^\theta$ is defined as

$$KL(h, f^\theta) = E_h\left(\log\frac{h(X)}{f^\theta(X)}\right) = \int_{-\infty}^{\infty} h(x)\log\frac{h(x)}{f^\theta(x)}\, dx$$

We have $KL(h,f^\theta) \geq 0$ and $KL(h,f^\theta)=0$, which implies that $h=f^\theta$. The KL divergence is often intuitively interpreted as a distance between the two probability measures, but this is not mathematically a distance; in particular, the KL divergence is not symmetric. Akaike [15] introduced the Akaike information criterion (AIC) to select the best model under parsimony. The goal of AIC is to minimize the KL divergence of the selected model from the true model. Notice that the relevant part

of the KL divergence is $E_h(\log f^\theta(X))$ which has an estimator as

$$\frac{1}{n}\sum_{i=1}^{n} \log f^{\hat{\theta}_n}(x_i) \qquad (4)$$

where, $\hat{\theta}_n = (\hat{\alpha}_n, \hat{\beta}_n)$ is the maximum likelihood estimator (MLE) of $\theta = (\alpha,\beta)$. It can be considered as an estimator of the divergence between the true density and the model. Akaike introduced his criterion to model selection as

$$AIC^f(\hat{\theta}_n) = -2\sum_{i=1}^{n} \log f^{\hat{\theta}_n}(x_i) + 2p \qquad (5)$$

where, $p$ is the number of parameters in the model. Now choose the family $F$ if $AIC^f < AIC^g$ otherwise choose family $G$. For computing the maximum likelihood estimators of unknown parameters of the mentioned distributions (Burr Distribution and Weibull distribution), one can use the inbuilt packages like *nlm()* and *optim()* of the R-software [27].

### 3.3. Bayesian information criterion

The Bayesian information criterion (BIC) is one of the important criteria for determining the best model for a given data. One major difference of this criterion is the different penalty term that it uses. Thus BIC [16] is defined as

$$BIC^f(\hat{\theta}_n) = -2\sum_{i=1}^{n} \log f^{\hat{\theta}_n}(x_i) + p\log n \qquad (6)$$

where, $p$ and $n$ are the number of parameters and sample size, respectively. The BIC is based on Bayesian probability and can be applied to models estimated by the maximum likelihood method. We choose family $F$ if $BIC^f < BIC^g$; otherwise we choose family $G$.

### 3.4. The Total Time On Test (TTT) transform

The total time on test (TTT) transform is a convenient tool for examining the nature of the hazard rate and accordingly checking for the adequacy of a model to represent the failure behavior of the data. The TTT transform of a probability distribution with absolutely continuous distribution function $F(.)$ is given by

$$\varphi_F(x) = H_F^{-1}(x) / H_F^{-1}(1)$$

where, $H_F^{-1}(x) = \int_0^{F^{-1}(x)} [1 - F(u)]du; \quad 0 \leq u \leq 1$ . The corresponding empirical version of the scaled TTT transform is defined as

$$\varphi(i/n) = \frac{H_n^{-1}(i/n)}{H_n^{-1}(1)} = \frac{\sum_{j=1}^{i}(n-j+1)(x_{j:n}-x_{j-1:n})}{\sum_{j=1}^{n}(n-j+1)(x_{j:n}-x_{j-1:n})}; \quad i=1,...,n , \quad x_{0:n}=0 \tag{7}$$

It has been shown by Aarset [28] that the TTT transform is convex (concave) if the hazard rate is decreasing (increasing). In addition, for a distribution with a bathtub (unimodal) failure rate the scaled TTT transform is first convex (concave) and then concave (convex). In this example, the scaled TTT transform of the data shows that the empirical hazard function is unimodal.

### 3.5. Maximum Likelihood Criterion (MLC)

Suppose, $\hat{\theta}_n$ and $\hat{\theta}'_n$ are the MLEs of $\theta$ and $\theta'$, respectively. The maximum likelihood criterion is defined as

$$T(\hat{\theta}_n,\hat{\theta}'_n) = \sum_{i=1}^{n}\ln\left(\frac{f^{\hat{\theta}_n}(x_i)}{g^{\hat{\theta}'_n}(x_i)}\right) \tag{8}$$

Then we choose $(f)$ or $(g)$ as the preferred model if $T(\hat{\theta}_n,\hat{\theta}'_n)$ is greater than zero or less than zero, respectively.

### 4. Results and discussion

In this section, we consider the degree of splat particle splashing. A particle splashes when it hits a solid body with adequate rate. Immediately after a molten particle impacts on a surface a skinny liquid film jets out radially from under it. The degree of splat splashing is defined as

$$\text{Degree of splat spalshing} = \frac{S}{12.56637}.R^2 \tag{9}$$

where, S is the area of the selected feature and R is the perimeter to area ratio. The degrees of particle splashing data are reported in different spray angles. We use the data of particle normal impact on a solid surface. The mean, standard deviation and the coefficient of skewness are calculated as 1.2052, 0.7104 and 2.3326, respectively. The measure of skewness indicates that the data are positively skewed. For comparison purposes, we have fitted Burr XII and Weibull distributions to the complete observation. The plot of the empirical and the fitted cumulative distribution functions for these

distributions and the fitted probability distribution functions (PDFs) and the relative histogram for the degree of splashing are presented in Figures 2 and 3, respectively. Theses plots indicate that the fitted Burr XII distribution is better than the fitted Weibull distribution. The estimated parameter values, AIC values, BIC values Kolmogorov-Smirnov (K-S) distances and the corresponding p-value are presented in Table 1. From the K-S distances, AIC, BIC values, MLC and p-values of Table 1, it is quite clear that the Burr XII model with estimated parameters   provides a much better fit than the Weibull distributions. We also present the percentile-percentile (P-P) plots of the Burr XII and Weibull distributions for the degree of splashing data in Figure 4. This plot shows a strong relationship supporting the appropriateness of the Burr XII distribution. Also, we consider a graphical method based on total time on test (TTT) transform. The plot of the scaled TTT transform of this data set, Figure 5, indicates that the empirical hazard function is unimodal; therefore, it is reasonable to use a BXII distribution to analyze the data.

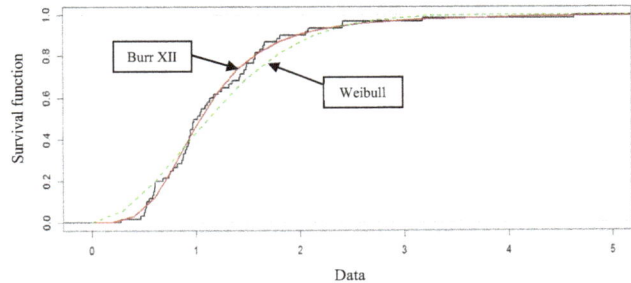

**Fig. 2.** Empirical function and the fitted functions for degree of splashing.

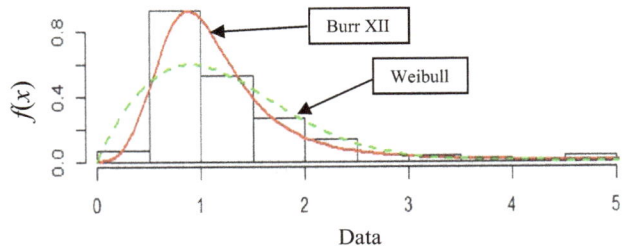

**Fig. 3.** The fitted PDF and relative histogram for the degree of splashing.

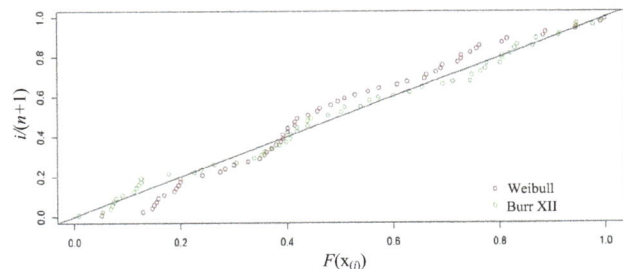

**Fig. 4.** The P-P plots for degree of splashing.

**Table 1.** Estimated parameters, K-S distances and AIC values for different distribution functions of the degree of splashing data.

| Distribution | Estimated parameters | K-S (p-value) | AIC | BIC | MLC |
|---|---|---|---|---|---|
| Weibull | $\alpha$=0.56168 , $\beta$=1.8533 | 0.1134 (0.4237) | 113.6794 | 117.86810 | 7.50859 |
| Burr XII | $\alpha$=0.89478 , $\beta$=3.66013 | 0.0744 (0.8944) | 98.4622 | 102.65090 | |

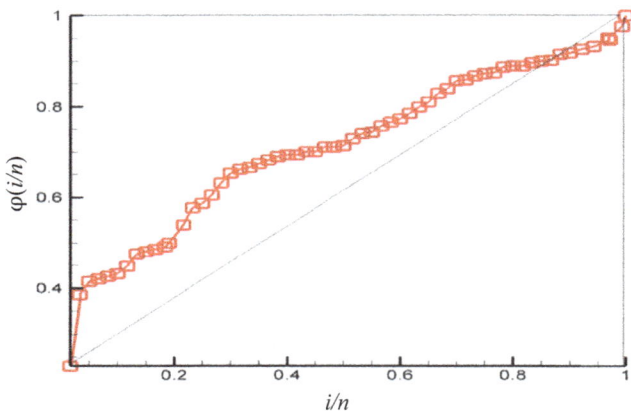

**Fig. 5.** The scaled TTT transform of degree of splashing.

## 5. Different Prediction Methods

Often do not have all the data in engineering sciences. In other words, some of the data are missing (censoring). Prediction of censored data based on observed data is a very interesting topic in applied science. We know that the splashing data follows a progressive censoring scheme. This censoring can be described as follows. Consider an experiment in which $n$ units are placed on an experimental test and only $m(<n)$ are completely observed until failure. The censoring occurs progressively in $m$ stages. These $m$ stages offer failure times of the $m$ observed units. At the time of the first failure (the first stage) $X_{1:m:n}$, $r_1$ of the $n-1$ units are randomly removed (censored) from the experiment. Similarly, at the time of the second failure (the second stage) $X_{2:m:n}$, $r_2$ of the $n-2-r_1$ units are randomly removed (censored) from the experiment. Finally, at the time of the $m^{th}$ failure (the mth stage) $X_{m:m:n}$, all the remaining $r_m=n-m-(r_1+r_2+...+r_{m-1})$ units are removed from the experiment. We will refer to this as the progressive censoring scheme $(r_1, r_2, ..., r_m)$. Now, let $X_{1:m:n}, X_{2:m:n} ..., X_{m:m:n}$ denote a progressively censored sample from the Burr XII model, with $(r_1, r_2, ..., r_m)$ being the progressive censoring scheme. For simplicity, we replace $X_{i:m:n}$ by $X_i$ throughout the paper. Our interest is to predict $Z=X_{i,(s)}$; $s=1, 2, ..., r_i$; $i=1, 2, ..., m$ in all $m$ stages of censoring based on the observed progressively censored sample $x=(x_1, ..., x_m)$. For this purpose, first we estimate the

unknown parameters of Burr XII distribution using the maximum likelihood method. The likelihood function based on a progressive censored sample from $BXII(\alpha,\beta)$ is given by

$$l(\alpha,\beta|data) \propto \alpha^m \beta^m \prod_{i=1}^{m} x_i^{\beta-1}(1+x_i^\beta)^{-(\alpha(r_i+1)+1)}$$

and the corresponding log likelihood function is

$$L(\alpha,\beta|data) = \log l(\alpha,\beta|data) \propto m\log\alpha + m\log\beta \qquad (10)$$

$$+ (\beta-1)\sum_{i=1}^{m}\log x_i - \sum_{i=1}^{m}\{\alpha(r_i+1)+1\}\log(1+x_i^\beta)$$

Taking derivatives with respect to $\alpha$ and $\beta$ of (10) and putting then equal to zero we obtain

$$\frac{\partial L}{\partial\alpha} = \frac{m}{\alpha} - \sum_{i=1}^{m}(r_i+1)Ln(1+x_i^\beta) = 0 \qquad (11)$$

$$\frac{\partial L}{\partial\beta} = \frac{m}{\beta} + \sum_{i=1}^{m}\log(x_i) - \sum_{i=1}^{m}\frac{\{\alpha(r_i+1)+1\}x_i^\beta\log(x_i)}{1+x_i^\beta} = 0 \quad (12)$$

Maximum likelihood estimates of $\alpha$ and $\beta$, say $\hat{\alpha}$ and $\hat{\beta}$ respectively, can be obtained by solving these two likelihood equations. But the explicit solutions of (11) and (12) cannot be obtained. We propose to use the EM algorithm to compute the MLEs of the unknown parameters which involves solving two one dimensional optimization problems rather than one two dimensional problem (see Appendix). Now, we want to evaluate the different methods of the prediction of $Z=X_{i,(s)}$; $s=1, 2, ..., r_i$; $i=1, 2, ..., m$ [29]. We know that the conditional distribution of $Z$ given $\underline{X}$ is just the distribution of $Z$ given $X_i=x_i$ due to the Markovian property of progressively censored order statistics. This indicates that the density of $Z$ given $\underline{X}=\underline{x}$ is the same as the density of the $s^{th}$ order statistic out of $r_i$ units from the left truncated distribution with the density function

$$\hbar*(z) = \frac{f(z)}{1-F(x_i)}; \qquad z > x_i \qquad (13)$$

where

$$f(z) = \alpha\beta z^{\beta-1}(1+z^\beta)^{-\alpha-1}$$

and

$$F(x_i) = 1 - (1 + x_i^{\beta})^{-\alpha}$$

Thus, the conditional density of $Z = X_{i,(s)}$ given $X_i = x_i$ for the Burr distribution is given by

$$f(z|data, \alpha, \beta) = \frac{r_i!}{(s-1)!(r_i - s - d)!} \alpha \beta z^{\beta-1}(1+z^{\beta})^{-(\alpha+1)}$$

$$\left[(1+x_i^{\beta})^{-\alpha} - (1+z^{\beta})^{-\alpha}\right]^{s-1} \times (1+z^{\beta})^{-\alpha(r_i-s)}(1+x_i^{\beta})^{\alpha r_i} \quad (14)$$

### 5.1. Conditional Median Predictor

The median of the distribution of $Z = X_{i,(s)}$ given $X_i = x_i$ whose density is given in (14), is called the conditional median predictor (CMP) [30]. On the other hand, a statistic $\hat{Z}$ is called a conditional median predictor, if

$$P(Z \le \hat{Z} | \underline{X} = \underline{x}) = P(Z \ge \hat{Z} | \underline{X} = \underline{x})$$

So

$$P_{\alpha,\beta}\left(Z \le \hat{Z} | \underline{X} = \underline{x}\right) = P_{\alpha,\beta}\left(1 - (\frac{1+Z^{\beta}}{1+x_i^{\beta}})^{-\alpha} \ge 1 - (\frac{1+\hat{Z}^{\beta}}{1+x_i^{\beta}})^{-\alpha} | \underline{X} = \underline{x}\right)$$

It is clear that the distribution of $1 - \left(\frac{1+Z^{\beta}}{1+x_i^{\beta}}\right)^{-\alpha}$ given

$X_i = x_i$ is a $Beta(s, r_i - s + 1)$ distribution with a pdf of

$$f(w) = \frac{w^{d-1}(1-w)^{n-s-d}}{Beta(d, n-s-d+1)}; \quad 0 < w < 1$$

So, the conditional median predictor can be written as

$$\hat{Z}_{CMP} = \left[\left((1 - Med(B))^{-1/\alpha}(1+x_i^{\beta})\right) - 1\right]^{\frac{1}{\beta}} \quad (15)$$

Where, $B$ has a Beta distribution with shape parameters $s$ and $r_i - s - 1$, respectively.

### 5.2. Pivotal Quantity

In this case, our interest is to predict $Z = X_{i,(s)}$ using the pivotal method. So, we choose $W = 1 - \left(\frac{1+Z^{\beta}}{1+x_i^{\beta}}\right)^{-\alpha}$ as a

pivotal quantity for obtaining the prediction interval for $Z$. Therefore, the $100(1 - \wp)\%$ prediction interval for the order statistic $Z$ is given by eq. (16), where, $B_{\wp/2}$ is the percentile of the Beta distribution with parameters $s$ and $r_i - s + 1$, respectively. Prediction interval can be obtained by substituting the unknown parameters with

$$L_{Pivot} = \left[\left(\left(1 - B_{\wp/2}\right)^{1/} (1+x_i)\right) - 1\right]^{-}$$

$$(16)$$

$$U_{Pivot} = \left[\left(\left(1 - B_{1-\wp/2}\right)^{1/} (1+x_i)\right) - 1\right]^{-}$$

their MLEs. Now, we consider the different prediction methods for predicting the censored splashing data. We propose that $m = 46$ and consider the following two censoring schemes:

**Censoring Scheme 1**: (19*0, 3, 10, 10*0, 1, 14*0). We obtain the following progressively censored sample: 0.2783, 0.4688, 0.5054, 0.5128, 0.5201, 0.5274, 0.5494, 0.5860, 0.5934, 0.6007, 0.6080, 0.6080, 0.6813, 0.7399, 0.7545, 0.7838, 0.8278, 0.8644, 0.8791, 0.8864, 0.9010, C, 0.9304, 0.9377, C, 0.9523, C, 0.9743, 0.9743, 0.9816, 1.0183, 1.0476, 1.0549, 1.0915, 1.1135, 1.1355, 1.1721, 1.2161, 1.2527, C, C, C, 1.4285, C, 1.4871, 1.4945, C, C, C, C, 1.6483, 1.6630, C, 1.8095, 2.0586, C, 2.4102, C, 3.1794, 4.6300.

**Censoring Scheme 2**: (45*0, 14). We obtain the following progressively censored Sample: 0.2783, 0.4688, 0.5054, 0.5128, 0.5201, 0.5274, 0.5494, 0.5860, 0.5934, 0.6007, 0.6080, 0.6080, 0.6813, 0.7399, 0.7545, 0.7838, 0.8278, 0.8644, 0.8791, 0.8864, 0.9010, 0.9157, 0.9304, 0.9377, 0.9377, 0.9523, 0.9523, 0.9743, 0.9743, 0.9816, 1.0183, 1.0476, 1.0549, 1.0915, 1.1135, 1.1355, 1.1721, 1.2161, 1.2527, 1.3186, 1.3553, 1.4212, 1.4285, 1.4652, 1.4871, 1.4945, C, C, C, C, C, C, C, C, C, C, C, C, C, C.

Here, C denotes the censored data. In both the schemes we have estimated the unknown parameters using the MLEs. For computing the MLEs we have used the EM algorithm. For schemes 1 and 2, the MLEs of $(\alpha, \beta)$ for the Burr XII distribution are (1.09106, 3.37267), (1.52112, 4.33760), respectively. The results for different prediction methods and different censoring schemes are presented in Tables 2 and 3. From these tables, it is observed that the prediction methods work well.

## 6. Conclusion

In this paper, we compare Weibull distribution and Burr XII distribution for the degree of particle splashing

**Table 2.** Values of interval prediction method for $X_{i,(s)}$; $s=1, ..., r_i$, (for scheme 1, $i=20, 21, 32$ and for scheme 2, $i=46$) and their real values.

| Scheme 1 | | | Scheme 2 | | |
|---|---|---|---|---|---|
| $X_{i,(s)}$ | Real values | PI (Pivot) | $X_{i,(s)}$ | Real values | PI (Pivot) |
| $X_{20,(1)}$ | 0.9157 | (0.8832, 0.9517) | $X_{46,(1)}$ | 1.5604 | (1.5287, 1,5935) |
| $X_{20,(2)}$ | 0.9377 | (0.9245, 0.9734) | $X_{46,(2)}$ | 1.5750 | (1.5335, 1.5865) |
| $X_{20,(3)}$ | 1.3186 | (1.2789, 1.3395) | $X_{46,(3)}$ | 1.5750 | (1.5335, 1.5865) |
| $X_{21,(1)}$ | 1.3553 | (1.3228,1.3497) | $X_{46,(4)}$ | 1.6190 | (1.5758, 1,7169) |
| $X_{21,(2)}$ | 0.9523 | (0.9410, 0.9844) | $X_{46,(5)}$ | 1.6483 | (1.6355, 1.6697) |
| $X_{21,(3)}$ | 1.4212 | (1.4052, 1.4637) | $X_{46,(6)}$ | 1.6630 | (1.6067, 0.6954) |
| $X_{21,(4)}$ | 1.4652 | (1.4180, 1.4909) | $X_{46,(7)}$ | 1.7802 | (1.7525, 1,8240) |
| $X_{21,(5)}$ | 1.5604 | (1.5356, 1.6011) | $X_{46,(8)}$ | 1.8095 | (1.7843, 1,8278) |
| $X_{21,(6)}$ | 1.7802 | (1.7525, 1.8240) | $X_{46,(9)}$ | 2.0586 | (2.0168, 2.1137) |
| $X_{21,(7)}$ | 1.5750 | (1.5199, 1.5747) | $X_{46,(10)}$ | 2.0805 | (1.9956, 2.1754) |
| $X_{21,(8)}$ | 2.0805 | (1.9956, 2.1754) | $X_{46,(11)}$ | 2.4102 | (2.3768, 2.4096) |
| $X_{21,(9)}$ | 1.5750 | (1.5199,1.5747) | $X_{46,(12)}$ | 2.4175 | (2.3766, 2.4562) |
| $X_{21,(10)}$ | 2.4175 | (2.3766, 2.4562) | $X_{46,(13)}$ | 3.1794 | (3.1547, 3.2128) |
| $X_{32,(1)}$ | 1.6190 | (1.5699, 1.7068) | $X_{46,(14)}$ | 4.6300 | (4.6036, 4.4637) |

**Table 3.** Values of point prediction method for $X_{i,(s)}$; $s=1, ..., r_i$, (for scheme 1, $i=20, 21, 32$ and for scheme 2, $i=46$) and their real values.

| Scheme 1 | | | Scheme 2 | | |
|---|---|---|---|---|---|
| $X_{i,(s)}$ | Real values | PI (CMP) | $X_{i,(s)}$ | Real values | PI (CMP) |
| $X_{20,(1)}$ | 0.9157 | 0.9076 | $X_{46,(1)}$ | 1.5604 | 1.5546 |
| $X_{20,(2)}$ | 0.9377 | 0.9265 | $X_{46,(2)}$ | 1.5750 | 1.5637 |
| $X_{20,(3)}$ | 1.3186 | 1.3245 | $X_{46,(3)}$ | 1.5750 | 1.5637 |
| $X_{21,(1)}$ | 1.3553 | 1.3463 | $X_{46,(4)}$ | 1.6190 | 1.6045 |
| $X_{21,(2)}$ | 0.9523 | 0.9511 | $X_{46,(5)}$ | 1.6483 | 1.6247 |
| $X_{21,(3)}$ | 1.4212 | 1.4286 | $X_{46,(6)}$ | 1.6630 | 1.6739 |
| $X_{21,(4)}$ | 1.4652 | 1.4625 | $X_{46,(7)}$ | 1.7802 | 1.7748 |
| $X_{21,(5)}$ | 1.5604 | 1.5587 | $X_{46,(8)}$ | 1.8095 | 1.8137 |
| $X_{21,(6)}$ | 1.7802 | 1.7819 | $X_{46,(9)}$ | 2.0586 | 2.0455 |
| $X_{21,(7)}$ | 1.5750 | 1.5688 | $X_{46,(10)}$ | 2.0805 | 2.0732 |
| $X_{21,(8)}$ | 2.0805 | 2.0654 | $X_{46,(11)}$ | 2.4102 | 2.4269 |
| $X_{21,(9)}$ | 1.5750 | 1.5567 | $X_{46,(12)}$ | 2.4175 | 2.4020 |
| $X_{21,(10)}$ | 2.4175 | 2.4030 | $X_{46,(13)}$ | 3.1794 | 3.1663 |
| $X_{32,(1)}$ | 1.6190 | 1.5941 | $X_{46,(14)}$ | 4.6300 | 4.6451 |

in thermal spray. Different plots and statistical criteria were used to identify the best fitted distribution for this data. Using several statistical criteria, like minimum K-S distance, minimum AIC value and minimum BIC value, the Burr XII distribution function appears to be a more appropriate statistical distribution function for this data. One important problem in engineering sciences is the prediction of future observations. So, we reported the different prediction values of future observations and observed that these methods of prediction work well. Finally, we should mention that our results can be extended for the degree of particle splashing observed at other spray angles.

## Appendix

The EM algorithm is an efficient iterative procedure to compute the maximum likelihood estimate in the presence of missing data and consists of an expectation step (E-step) and a maximization step (M-step). First, let us denote the observed and censored data by $X = (X_1, ..., X_m)$ and $Z = (Z_1, ..., Z_m)$, respectively, where each $Z_j$ is $1 \times r_j$ vector $Z_j = (Z_{j1}, ..., Z_{jr_j})$ for $j=1,..., m$ and they are not observable. The censored data vector $Z$ can be thought of as missing data $W = (X,Z)$ and represents the complete data set. Therefore, the log-likelihood function $L_c = (W; \alpha, \beta)$ of the complete data after ignoring the constants can be written as:

$$L_c(W;\alpha,\beta) = n\log\alpha + n\log\beta + (\beta-1)\sum_{i=1}^{m}\log x_i - (\alpha+1)\sum_{i=1}^{m}\log(1+x_i^\beta)$$

$$+(\beta-1)\sum_{i=1}^{m}\sum_{k=1}^{r_i}\log z_{ik} - (\alpha+1)\sum_{i=1}^{m}\sum_{k=1}^{r_i}\log(1+z_{ik}^\beta) \qquad (17)$$

**E-step**:

This step involves the computation of the conditional expectation of the log-likelihood with respect to the incomplete data given the observed data. For this purpose, we compute the pseudo log-likelihood function as:

$$L_s(\alpha,\beta) = E\left(L_c(W;\alpha,\beta)\middle|X\right) = n\log\alpha + n\log\beta + (\beta-1)\sum_{i=1}^{m}\log x_i$$

$$-(\alpha+1)\sum_{i=1}^{m}\log(1+x_i^\beta) + (\beta-1)\sum_{i=1}^{m}\sum_{k=1}^{r_i}E\left[\log z_{ik}\middle|z_{ik} > x_i\right]$$

$$-(\alpha+1)\sum_{i=1}^{m}\sum_{k=1}^{r_i}E\left[\log(1+z_{ik}^\beta)\middle|z_{ik} > x_i\right] \qquad (18)$$

where

$$E\left(\log Z_{ik}\middle|Z_{ik} > c\right) = \frac{\alpha\beta}{1-F_X(c;\alpha,\beta)}\int_c^\infty x^{\beta-1}\left(1+x^\beta\right)^{-(1+\alpha)}(\log x)dx$$

and

$$E\left(\log(1+Z_{ik}^\beta)\middle|Z_{ik} > c\right) = \frac{\alpha\beta}{1-F_X(c;\alpha,\beta)}\int_c^\infty x^{\beta-1}\left(1+x^\beta\right)^{-(\alpha+1)}$$

$$\log(1+x^\beta)dx = \alpha^{-1} + \log\left(1+c^\beta\right)$$

We denoted $E\left(\log Z_{ik}\middle|Z_{ik} > c\right)$ and $E\left(\log(1+Z_{ik}^\beta)\middle|Z_{ik} > c\right)$ by A(c, $\alpha$, $\beta$) and B(c, $\alpha$, $\beta$), respectively.

**M-step:**

This step includes the maximization of the pseudo log-likelihood function (18). Therefore, if at the $k^{th}$ stage the estimate of $(\alpha, \beta)$ is $(\alpha^{(k)}, \beta^{(k)})$, then $(\alpha^{(k+1)}, \beta^{(k+1)})$ can be obtained by maximizing

$$L_c^*(W;\alpha,\beta) = n\log\alpha + n\log\beta + (\beta-1)\left[\sum_{i=1}^{m}\log x_i\right] - (\alpha+1)\left[\sum_{i=1}^{m}\log(1+x_i^\beta)\right]$$

$$+(\beta-1)\sum_{i=1}^{m}r_i A(x_i,\alpha^{(k)},\beta^{(k)}) - (\alpha+1)\sum_{i=1}^{m}r_i B(x_i,\alpha^{(k)},\beta^{(k)}) \qquad (19)$$

with respect to $\alpha$ and $\beta$. Notice that the maximization of (19) can be obtained by different methods, such as, Kundu and Pradhan [31].

## Acknowledgements

This research was supported in part by the Research Deputy of Payame-Noor University project. This support is gratefully acknowledged.

## References

[1] S.D. Aziz, S. Chandra, Impact, recoil and splashing of molten metal droplets, Int. J. Heat Mass Tran. 43 (2000) 2841-2857.

[2] A.M. Worthington, The splash of a drop, Society for Promoting Christian Knowledge, London, 1895.

[3] A. Worthington, On the forms assumed by drops of liquids falling vertically on a horizontal plate, Proc. R. Soc. Lond., 25 (1876) 261-272.

[4] O.G. Engel, Waterdrop collisions with solid surfaces, J. Res. NBS, 5 (1955) 281-298.

[5] C. Stow, M. Hadfield, An experimental investigation of fluid flow resulting from the impact of a water drop with an unyielding dry surface, Proc. R. Soc. Lond. A 373 (1981) 419-441.

[6] G. Montavon, S. Sampath, C. Berndt, H. Herman, C. Coddet, Effects of the spray angle on splat morphology during thermal spraying, Surf. Coat. Technol. 91 (1997) 107-115.

[7] S. Thoroddsen, J. Sakakibara, Evolution of the fingering pattern of an impacting drop, Phys. Fluids 10 (1998) 1359.

[8] Y. Hardalupas, A. Taylor, J. Wilkins, Experimental investigation of sub-millimetre droplet impingement on to spherical surfaces, Int. J. Heat Fluid Fl. 20 (1999) 477-485.

[9] S. Asadi, M. Passandideh-Fard, M. Moghiman, Numerical and analytical model of the inclined impact of a droplet on a solid surface in a thermal spray coating process, Iran. J. Surf. Eng. 4 (2008) 1-14.

[10] J. Liu, H. Vu, S.S. Yoon, R.A. Jepsen, G. Aguilar, Splashing phenomena during liquid droplet impact, Atomization Spray. 20 (2010) 297-310.

[11] S. Asadi, Simulation of nanodroplet impact on a solid surface, Inter. J. Nano Dim. 3 (2012) 19-26.

[12] H. Li, S. Mei, L. Wang, Y. Gao, J. Liu, Splashing phenomena of room temperature liquid metal droplet striking on the pool of the same liquid under ambient air environment, Int. J. Heat Fluid Fl. 47 (2014) 1-8.

[13] G. Liang, Y. Guo, Y. Yang, N. Zhen, S. Shen, Spreading and splashing during a single drop impact on an inclined wetted surface, Acta Mech. 224 (2013) 2993-3004.

[14] G. Liang, Y. Yang, Y. Guo, N. Zhen, S. Shen, Rebound and spreading during a drop impact on wetted cylinders, Exp. Therm. Fluid Sci. 52 (2014) 97-103.

[15] H. Akaike, Information theory and an extension of the maximum likelihood principle, Proceeding of the Second International Symposium on Information Theory, Akademinai Kiado, Budapest, 1973, pp. 267-281.

[16] G. Schwarz, Estimating the dimension of a model, Ann. Stat. 6 (1978) 461-464.

[17] B. Pradhan, D. Kundu, On progressively censored generalized exponential distribution, Test 18 (2009) 497-515.

[18] A. Asgharzadeh, R. Valiollahi, Point Prediction for the Proportional Hazards Family under Progressive Type-II Censoring, J. Iran. Stat. Soc. 9 (2010) 127-148.

[19] R.R.A. Awwad, M.Z. Raqab, I.M. Al-Mudahakha, Statistical inference based on progressively type II censored data from Weibull model, Commun. Stat. Simulat. 44 (2015) 2654-2670.

[20] H. Ng, Parameter estimation for a modified Weibull distribution, for progressively type-II censored samples, IEEE T. Reliab. 54 (2005) 374-380.

[21] I.W. Burr, Cumulative frequency functions, Ann. Math. Stat. 13 (1942) 215-232.

[22] M.K. Rastogi, Y.M. Tripathi, Estimating the parameters of a Burr distribution under progressive type II censoring, Stat. Methodol. 9 (2012) 381-391.

[23] H. Panahi, A. Sayyareh, Parameter estimation and prediction of order statistics for the Burr Type XII distribution with Type II censoring, J. Appl. Stat. 41 (2014) 215-232.

[24] A. Abd-Elfattah, A.S. Hassan, S. Nassr, Estimation in step-stress partially accelerated life tests for the Burr type XII distribution using type I censoring, Stat. Methodol. 5 (2008) 502-514.

[25] B. Abbasi, S.Z. Hosseinifard, D.W. Coit, A neural network applied to estimate Burr XII distribution parameters, Reliab. Eng. Syst. Safe. 95 (2010) 647-654.

[26] N.L. Johnson, S. Kotz, N. Balakrishnan, Continuous Univariate Distributions, Wiley Ser. Prob. Stat., vol. 1, 1994.

[27] R Development Core Team, R: A Language and Environment for Statistical Computing. Vienna, Austria : The R Foundation for Statistical Computing, (2011). http://www.R-project.org/.

[28] M.V. Aarset, How to identify a bathtub hazard rate, IEEE T. Reliab. 36 (1987) 106-108.

[29] A. Asgharzadeh, R. Valiollahi, Point Prediction for the Proportional Hazards Family under Progressive Type-II Censoring, J. Iran. Stat. Soc. 9 (2010) 127-148.

[30] M.Z. Raqab, H.N. Nagaraja, On some predictors of future order statistics, Metron, 53 (1995) 185-204.

[31] D. Kundu, B. Pradhan, Estimating the parameters of the generalized exponential distribution in presence of hybrid censoring, Commun. Stat.-Theor. M. 38 (2009) 2030-2041.

# 21

# The influence of cellulose pulp and cellulose microfibers on the flexural performance of green-engineered cementitious composites

**Fatemeh Masoudzadeh, Mohammad Fasihi, Masoud Jamshidi***

*School of Chemical, Petroleum and Gas Engineering, Iran University of Science and Technology, Tehran, Iran*

## HIGHLIGHTS

- Mechanical and chemical treatment of Kraf paper to produce cellulose pulp (CP) and cellulose microfibers (CMF)

- Investigation of flexural properties of cementitious composites reinforced by CP and CMF

- Comparison of flexural behavior cement composites reinforced by CP, CMF and PVA fiber

- Preparation and study of hybrid engineered cementitious composites (ECC) by using a mixture of CMF and PVA fiber

## GRAPHICAL ABSTRACT

## ARTICLE INFO

*Keywords:*
Cement
Engineered cementitious composite
Cellulose
Green composite
Flexural behavior
Microstructure

## ABSTRACT

The aim of this study was to investigate the flexural behavior of engineered cementitious composites (ECCs) reinforced by cellulose pulp (CP) and cellulose microfibers (CMF). The reinforcements were obtained from chemical-mechanical treatments of Kraft paper and used in ECC mix design. Results showed that cement reinforced by CP exhibited a strain-hardening behavior in the three-point bending test, while CMF led to a brittle behavior in cement composites. Moreover, different hybrid combinations of polyvinyl alcohol (PVA) and CMF achieved quite a high strength while maintaining a high level of flexural toughness. A combination of 0.5 vol% CMF and 1.5 vol% PVA resulted in a significant increase in flexural toughness and a slight improvement in flexural strength. The properties of this hybrid composite were comparable with one containing 2 vol% of PVA fiber.

\* *Corresponding author: ; E-mail address: mjamshidi@iust.ac.ir*

# 1. Introduction

In the past few decades, many studies have been conducted to modify the brittle behavior of cement materials to a more ductile one, mainly by incorporating various fibers. For this purpose, different types of man-made fibers and natural fibers were used in cement composites [1]. Environmental issues have created a great incentive to do extensive research on eco-friendly materials such as fibers obtained from natural resources. Natural fibers have many advantages over synthetic fibers as they are accessible, sustainable, non-hazardous and have low cost. Therefore, natural fibers were investigated as a potential alternative to synthetic fibers in many researches [2].

Natural fibers are obtained from plants, animals, and minerals. Plant fibers are individually comprised of four main chemical components namely cellulose, hemicelluloses, lignin and pectin. The mechanical properties of plant fibers strongly depend on their cellulose content [3,4]. In recent decades, some research has been performed on using natural fibers in cement composites. In a comparative study of plant fibers and PVA in cement-based composite, Juarez et al. [5] concluded that natural fibers can be an alternative candidate for PVA fibers to reduce the crack formation caused by plastic shrinkage. Khorami et al. [6] examined the effect of three types of agricultural wastes including bagasse, wheat and eucalyptus fibers on the properties of cement composites. Their results showed that natural fibers improved the energy absorption and flexural strength of the cement composites. They also added Kraft cardboard wastes into the cement matrix and observed strain-softening behavior in the prepared cementitious composites [7]. In another study sisal fibers were incorporated into cement mortar and led to strain-hardening behavior [8]. This illustrated the capability of these fibers in structural applications. In addition, a cement composite with a strain-hardening behavior was obtained by adding bleached pine fiber to the cement [9]. Singh et al. [10] used a combination of polypropylene fiber and steel fiber to improve the toughness of cementitious composites.

Today, efforts to achieve sustainable consumption patterns, especially in construction and building materials, are growing. Engineered cementitious composites (ECCs), normally fine-grained concrete reinforced with PVA fibers with a strain-hardening

behavior, are not an exception. Over the past few decades, many studies have been performed to develop a green engineered cementitious composite for infrastructure systems by replacing its ingredients (i.e. cement and fibers) with industrial wastes and/or natural fibers [11]. The first study on green ECC was performed by Lee [12,13]. Some studies focused on replacement of cement with industrial wastes or mineral fillers to produce green ECC [12,14]. Moreover, the performance of various synthetic fibers such as polyvinyl alcohol (PVA) [15], polyethylene (PE) [16], polypropylene (PP) [15,17,18], acrylic and nylon fibers [18-20] used in ECCs have been studied. However, there are limited reports on the application of natural fibers as reinforcement in cement composites.

In this study, the flexural behavior of ECCs reinforced by cellulose pulp and cellulose microfibers were studied and compared with that prepared by PVA fibers. In addition, the properties of hybrid ECC reinforced by a mixture of cellulose microfiber and PVA fiber was investigated.

# 2. Experimental

## 2.1. Materials

Ordinary Portland cement type I was supplied by Tehran Cement Co., Iran. The chemical composition of the cement is seen in Table 1. Brown Kraft paper was provided from Roxcel GmbH, India. Fly ash was supplied by Dirk G., India. Sand was provided from local companies. Polycarboxylate ether Paya 210 from Payazhic Co. (Iran) was used as super plasticizer. The particle size distribution of the sand is shown in Figure 1. PVA with the characteristics reported in Table 2 was used as reinforcement.

**Table 1.** Chemical composition of the cement (% wt)

| $SiO_2$ | $Al_2O_3$ | $Fe_2O_3$ | CaO | $Na_2O$ | MgO | $SO_2$ | Other |
|---------|-----------|-----------|------|---------|-----|--------|-------|
| 21.4 | 5.3 | 3.2 | 63.4 | 0.3 | 3 | 1 | 1.4 |

## 2.2. Preparation of cellulose pulp and cellulose microfibers

In order to produce cellulose pulp, the Kraft paper was cut into small pieces and soaked in water for 24 h. Then, it was stirred for 1.5 hr at 500 rpm by a mechanical mixer to obtain a homogeneous pulp suspension. This

**Table 2.** Physical and mechanical properties of PVA fiber

| Fiber type | Fiber diameter (μm) | Length (mm) | Tensile strength (MPa) | Elongation (%) | Modulus of elasticity (CN/dtex) |
|---|---|---|---|---|---|
| PVA | 20 | 6 | 1600 | 6.92 | 300 |

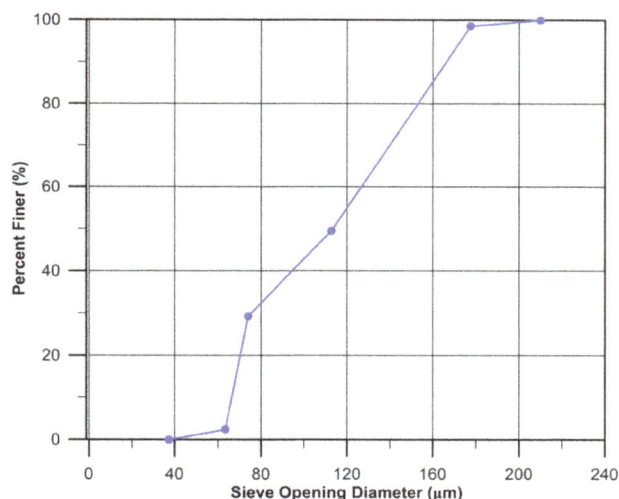

**Fig.1.** Cumulative particle size distribution of the sand

paste was used as cellulose pulp in the ECC mix design seen in Table 3. Cellulose microfiber was prepared in a mechanical-chemical treatment using the acid hydrolysis method [21,22]. To do this, first the cellulose pulp was immersed in sodium hydroxide solution (10 wt%) for 2 hr at 80°C under stirring. Then, the fibers were neutralized several times by washing with distilled water. The fibers were subjected to, acidic hydrolysis by sulfuric acid solution (30 wt%) for 3 hr at 80°C. The fibers were neutralization again after the hydrolysis process. Finally, the fiber suspension was sonicated for 15 min with a sonicator (UHP-400 from Ultrasonic Technology Development Co., Iran). Figure 2 presents the SEM images of the fibers before and after mechanical-chemical treatments.

## 2.3. Cementitious composite preparation

A standard mix design, ECC-M45, was used to prepare the ECC samples [23]. The composition of ECC is presented in Table 3. The fibers used in this mix design were either PVA, CP or CMF.

A control sample without fiber was also prepared for comparison. Moreover, some hybrid fiber reinforced ECCs were prepared using different combinations of PVA and CMF. The total fiber content in the hybrid composites was 2 vol%. For each formulation three plate specimens with dimensions of 9×100×210 mm were cast using a suitable mold. Figure 3 presents an image of the applied mold. The samples were removed

**Table 3.** ECC mix design proportions

| Cement (g) | Flay ash (g) | Sand (g) | Water (g) | Fiber (% vol) | Super plasticizer (g) |
|---|---|---|---|---|---|
| 1 | 1.2 | 0.8 | 0.57 | 0-2 | 0.01-0.02 |

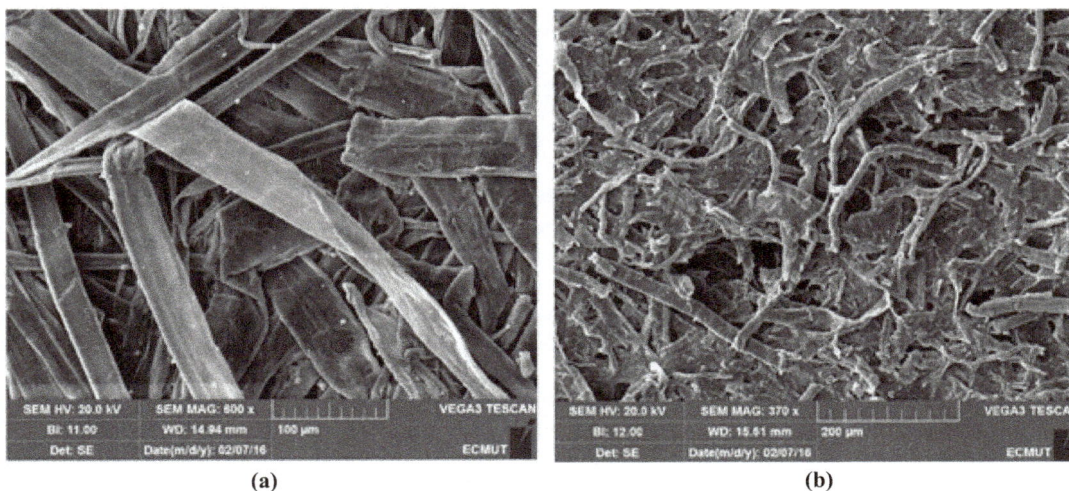

**Fig.2.** SEM images of (a) Kraft and (b) cellulose microfibers

from the mold after two days rest and kept at ambient temperature and 95±5% relative humidity for four weeks before any tests.

## 2.4. Testing method

A three-point bending test was conducted using a SANTAM-STM150 universal machine according to EN-12467 standard test method. The flexural stress ($\sigma$) and strain ($\varepsilon$) were obtained by the following equations:

$$\sigma = \frac{3FL}{2bh^2} \tag{1}$$

$$\varepsilon = \frac{6Dh}{L^2} \tag{2}$$

where, $F$ is the bending force, $L$ is the bearing distance, $b$ and $h$ are the width and thickness of the ECC sheet, respectively, and $D$ is the displacement of the loading bar. Toughness energy was calculated from the area under the stress-strain curve. The result of mechanical properties was obtained from averaging over five measurements.

The microstructure of samples were observed by using scanning electron microscopy (SEM, Vega 3 Tescan, Czech).

## 3. Results and Discussions

### 3.1. Cement composites reinforced by CP and CMF

Figure 4 presents the bending test results of composites with different volume fractions of CP. As can be clearly seen, the ductility of the composites containing cellulose pulp increased compared to the control sample. The increment of ductility of the composite was proportional to the fiber content.

Claramuntet et al. [24] stated that the toughness and ductility of cement composites increased by adding

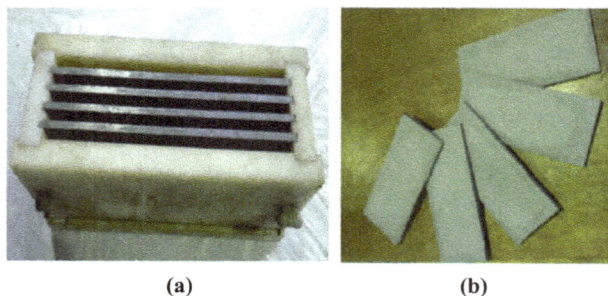

**(a)**                                          **(b)**

Fig. 3.The mold used for preparing composite samples

softwood pulp. In this study, at 0.5% CP content, the composites displayed a reduction in flexural strength while their ductility increased. At higher concentrations of CP, the strength remained almost constant, but ductility obviously increased.

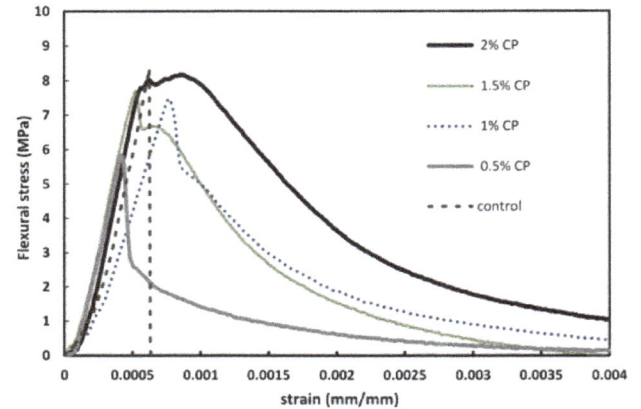

Fig. 4. Stress-strain curve of cement composites containing cellulose pulp

The flexural stress-strain curves of CMF reinforced composites are shown in Figure 5. The addition of microfiber to the cement not only decreased the flexural strength, but also lowered its strain to break and ductility. CP, which included long fibers, enhanced the ductility of the composites, while CMF failed to reinforce the cement because of its short fiber length. The aspect ratio and length of fibers in CMF decreased due to the fracturing of fiber during the mechanical-chemical treatment.

Figure 6 presents SEM images of the cross-section of specimens containing CP and CMF after the bending test. The image presents the pull-out mechanism for the composite containing CP, indicating non-strong

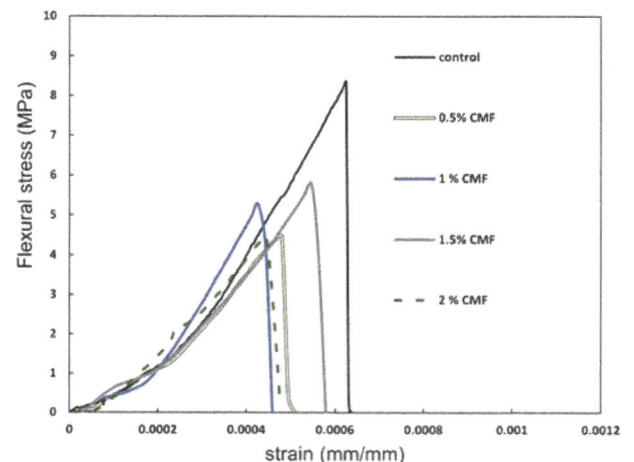

Fig. 5. Stress-strain curve of cementitious composites containing cellulose microfiber

adhesion between the untreated cellulose and cement matrix. The long fibers in CP can bridge a crack owing to their length and so improve the ductility of the composite. On the other hand, in the CMF reinforced cements, no pull-out of fiber from the cement was observed, indicating good adhesion between the treated fibers and cement matrix. However, CMF could not resist against crack growth by crack bridging due to their short length.

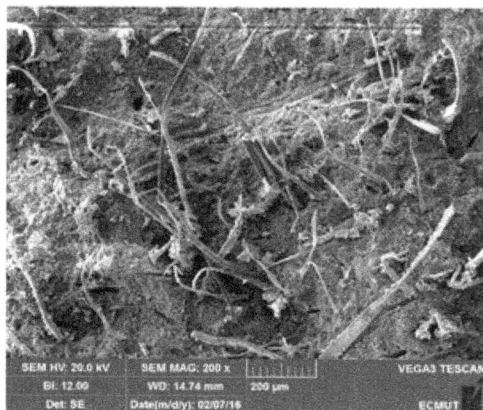

(a)

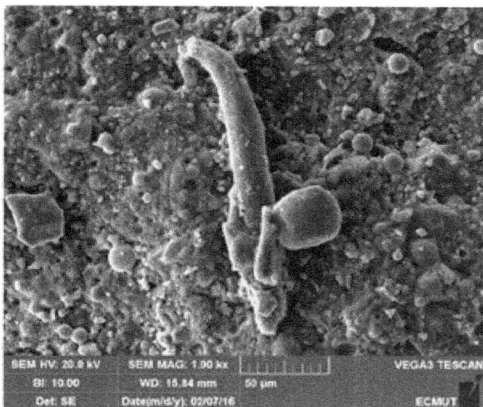

(b)

**Fig. 6** SEM image of cement composites containing (a) CP and (b) CMF.

### 3.2. Cellulose vs. PVA fibers in the cement composites

PVA is the main polymer often used to modify cement. In order to study the performance of cellulose reinforced composites, cement composites including PVA fiber were also prepared and their properties were then compared with the cellulose-reinforced composites. Figure 7 presents the results of the flexural behavior of ECC prepared with up to 2% PVA fiber. As observed in this figure, strain-softening behavior was observed in the composite containing 0.5% PVA. By increasing the amount of fiber up to 2% both strength and toughness

were increased and a strain-hardening behavior was observed.

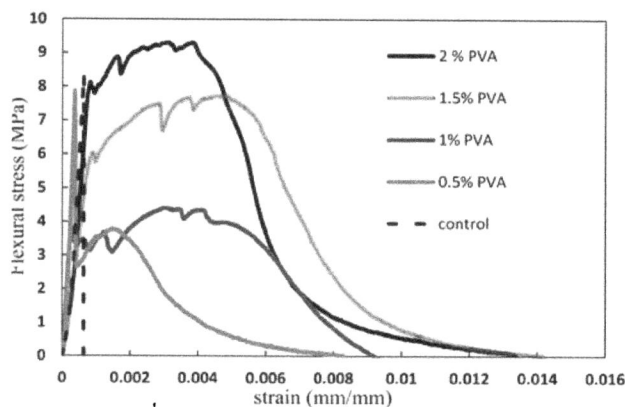

**Fig. 7.** Stress-strain curve of ECC containing PVA fiber

A comparison between the strength and toughness of the ECC containing PVA fiber and the composites reinforced by CP are depicted in Fig. 8. First-crack strength refers to the point at which the stress-strain curve becomes non-linear. Post-crack (second peak) strength is the maximum strength after the first-crack peak. The specimens produced by 0.5% and 1% CP only showed the first-crack peak as a result of semi strain-softening behavior. By increasing the fiber content to 1.5% and 2%, strain-hardening behavior and a post-crack peak appeared due to the higher surface contact between fibers and matrix. Also, in the ECC including PVA, increasing fiber volume fraction led to a reduction of the first-crack and an increase in the post-crack strength.

Figure 9 represents a comparison between the properties of CP, CMF, and PVA fiber composites at 2% of fiber volume fraction. The flexural strength of the composite containing CP was comparable with that of the PVA reinforced composite. While, the flexural strength of the CMF composite was about 40% lower. However, the toughness energy of the CP and CMF composites were much smaller than the PVA composite.

### 3.3. Composites prepared by hybrid fibers

Although CP made more improvement than CMF in the mechanical properties of cement, in hybrid form it showed a weak processability due to large bundles and aggregates, which did not facilitate its uniform dispersion in the cement mixture. On this basis, CP was not use in hybrid composites. Composites containing hybrid fibers were prepared by using CMF along with

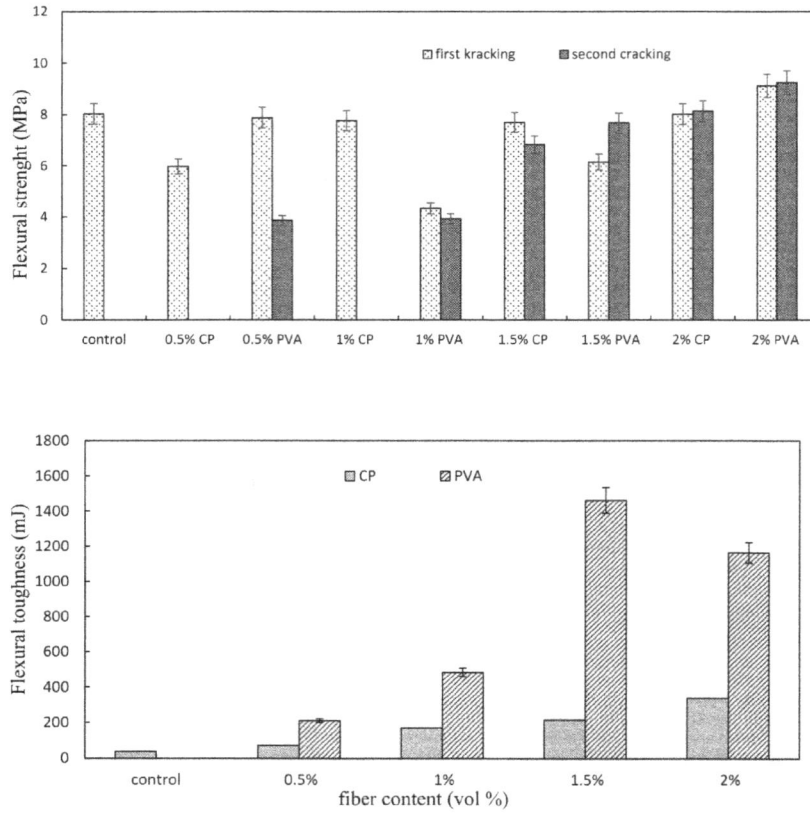

**Fig. 8.** Flexural properties of composites containing different content of PVA and CP

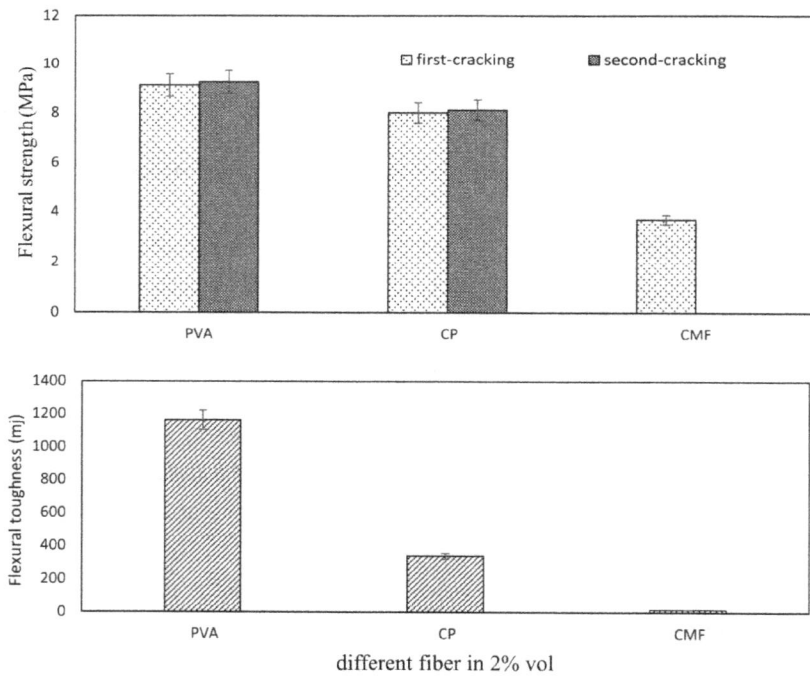

**Fig. 9.** Flexural properties of different cement composites containing 2 vol% fiber.

a total fiber concentration of 2 vol%. Figure10, illustrates the flexural properties of the hybrid composites. The flexural strength of the hybrid composites decreased as the CMF content increased. The results of flexural

first- and post-crack strength and flexural toughness of the hybrid composites are displayed in Figure 11. As shown, the samples containing 1.5/0.5 and 1/1 volume percent of PVA/CMF represented a strain-hardening behavior similar to PVA-ECC. The hybrid composite including 0.5% CMF showed just a little lower flexural strength compared with 2% PVA ECC. In addition, the toughness energy of this hybrid fiber's ECC was on the level of pure PVA ECC. So, PVA fiber could be partially replaced by CMF in ECC mix design.

## 4. Conclusions

In this study, green cement composites were prepared by CP, CMF, and a hybrid of PVA fiber and CMF. By adding CP to the cement, the flexural strength did not change considerably, but toughness energies were significantly increased. The composite containing 2% CP presented up to nine times more toughness energy compared with the control sample. Although, CMF did not improve the mechanical properties of cement, the mixture of PVA and CMF in an ECC mix design created comparable properties with PVA-ECC at 2 vol% fiber. This indicated that in ECCs, PVA fiber could be replaced by CMF to some extent.

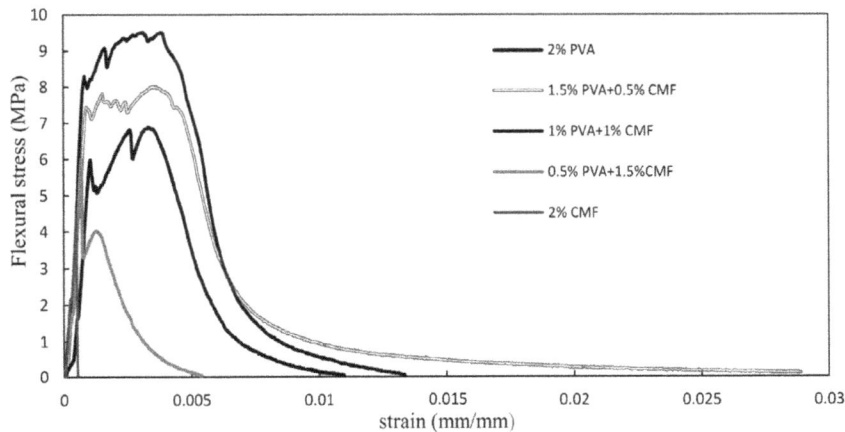

Fig. 10. Stress-strain curve of hybrid ECCs.

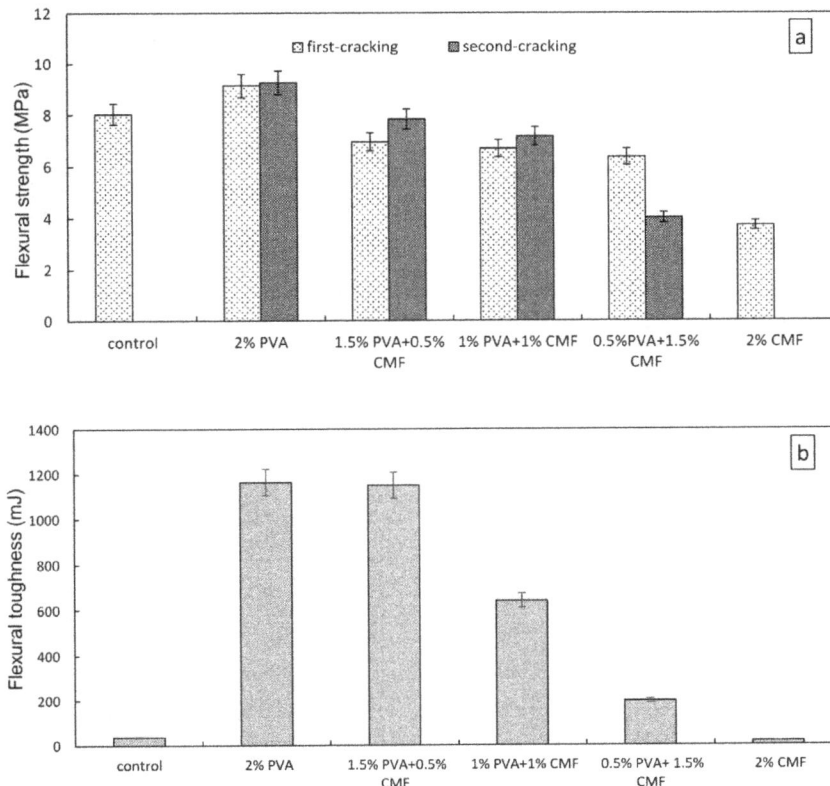

Fig. 11. The flexural properties of hybrid composites.

## References

[1] M.G.S. Beltran, E. Schlangen, Fibre-matrix interface properties in a wood fibre reinforced cement matrix, Proceedings of FraMCoS-7, Korea Concrete Institute, (2010) 1425-1430.

[2] N. Ardanuy, J. Claramunt, R.D.T. Filho, Cellulosic fiber reinforced cement-based composites: A review of recent research, Constr. Build. Mater. 79 (2015) 115-128.

[3] F. Pacheco-Torgal, S. Jalali, Cementitious building materials reinforced with vegetable fibres: A review, Constr. Build. Mater. 25 (2011) 1575-1581.

[4] H.M. Akil, M.F. Omar, A.A.M. Mazuki, S. Safiee, Z.A.M Ishak, A.A. Bakar, Kenaf fiber reinforced composites: A review, Mater. Design. 32 (2011) 4107-4121.

[5] C.A. Juareza, G. Fajardo, S. Monroy, A. Duran-Herrera, P. Valdeza, C. Magniont, Comparative study between natural and PVA fibers to reduce plastic shrinkage cracking in cement-based composite, Constr. Build. Mater. 91 (2015) 164-170.

[6] M. Khorami, J. Sobhani, An experimental study on the flexural performance of agro-waste cement composite boards, Int. J. Civ. Eng. 11 (2013) 207-216.

[7] M. Khorami, E. Ganjian, The effect of limestone powder, silica fume and fibre content on flexural behaviour of cement composite reinforced by waste Kraft pulp, Constr. Build. Mater. 46 (2013) 142-149.

[8] F.A. Silva, B. Mobasher, R.D.T. Filho, Advances in natural fiber cement composites: A material for the sustainable construction industry, 4th Colloquium on Textile Reinforced Structures (CTRS4) , Technische Universität Dresden. (2009) 377-388.

[9] G.H.D. Tonoli, M.N. Belgacem, J. Bras, M.A. Pereira-da-Silva, F.A. Rocco Lahr, H. Savastano, Impact of bleaching pine fibre on the fibre/cement interface, J. Mater. Sci. 47 (2012) 4167-4177.

[10] S.P. Singh, A.P. Singh, V. Bajaj, Strength and flexural toughness of concrete reinforced with steel-polypropylene hybrid fibres, Asian J. Civ. Eng. 11 (2010) 495-507.

[11] M. Lepech, G.A. Keoleian, Development of green engineered cementitious composites for sustainable infrastructure systems, Conference Proceeding International Workshop on Sustainable Development and Concrete Technology, Tsinghua University. (2004) 181-192.

[12] V.C. Li, S. Wang, C. Wu, Tensile strain-hardening behavior of polyvinyl alcohol-engineered cementitious composite (PVA-ECC), ACI Mater. J. 98 (2001) 483-492.

[13] V.C. Li, On engineered cementitious composites (ECC). A review of the material and its applications, J. Adv. Concr. Technol. 1 (2003) 215-230.

[14] X. Huang, R. Ranade, W. Ni, V.C. Li, Development of green engineered cementitious composites using iron ore tailings as aggregates, Constr. Build. Mater. 44 (2013) 757-764.

[15] H.R. Pakravan, M. Latifi, M. Jamshidi, Ductility improvement of cementitious composites reinforced with polyvinyl alcohol-polypropylene hybrid fibers, J. Ind. Tex. 45 (2016) 637-651.

[16] V.C. Li, Engineered cementitious composites-tailored composites through micromechanical modeling, Canadian Society of Civil Engineers. (1998) 64-97.

[17] H.R. Pakravan, M. Jamshidi, M. Latifi, M. Neshastehriz, Application of polypropylene nonwoven fabrics for cement composites reinforcement, Asian J. Civ. Eng. 12 (2011) 551-562.

[18] M.R. Abdi, H. Mirzaeifar, Effects of Discrete Short Polypropylene Fibers on Behavior of Artificially Cemented Kaolinite, Int. J. Civ. Eng. 14 (2016) 253-262.

[19] M. Halvaei, M. Jamshidi, M. Latifi, Effect of Polymeric Fibers on Mechanical Properties of Engineered Cementitious Composite (ECC), Proceeding of the 12th Asian Textile conference. (2015) 1769-1775.

[20] M. Halvaei, M. Jamshidi, M. Latifi, Application of low modulus polymeric fibers in engineered cementitious composites, J. Ind. Text. 43 (2012) 511-524.

[21] G.H.D. Tonoli, EM Teixeira, A.C. Corrêa, J.M. Marconcini, L.A. Caixeta, M.A. Pereira-da-Silva, Cellulose micro/nanofibres from eucalyptus Kraft pulp: Preparation and properties, Carbohyd. Polym. 89 (2012) 80-88.

[22] Y. Li, G. Li, Y. Zou, Q. Zhou, X. Lian, Preparation and characterization of cellulose nanofibers from partly mercerized cotton by mixed acid hydrolysis, Cellulose. 21 (2014) 301–309.

[23] E.G. Nawy, Concrete Construction Engineering Handbook, CRC Press, New York, 2008.

[24] J. Claramunt, M. Ardanuy, F. Parés, and H. Ventura, Mechanical performance of cement mortar composites reinforced with cellulose fibres, 9th International conference on Composite Science and Technology, University of Bath, (2013) 477-484.

# 22

# Dynamic modelling of hardness changes of aluminium nanostructure during mechanical ball milling process

**Vahid Bidarian[1,*], Esmaeil Koohestanian[2], Maryam Omidvar[3]**

[1] Young Researchers and Elite Club, Quchan Branch, Islamic Azad University, Quchan, Iran

[2] Department of Chemical Engineering, University of Sistan and Baluchestan, Zahedan, Iran

[3] Department of Chemical Engineering, Quchan Branch, Islamic Azad University, Quchan, Iran

## HIGHLIGHTS

- Elastic and plastic deformation effects during collision were investigated.

- The combination of NFD and powder hardness model shows better results.

- This model is compared to Maurice et.al. The results show a 35% increase in model accuracy.

## GRAPHICAL ABSTRACT

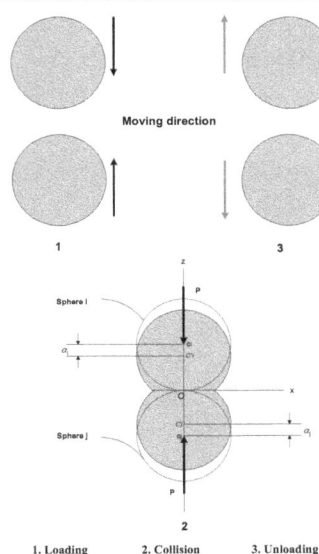

1. Loading    2. Collision    3. Unloading

## ARTICLE INFO

*Keywords:*
Ball milling
Modelling
Nanostructure
NFD model
Aluminium powder

## ABSTRACT

In this research, the feasibility of using mathematical modelling in the ball milling process has been evaluated to verify the hardness changes of an aluminium nanostructure. Considering the model of normal force displacement (NFD), the radius of elastic-plastic and normal displacement of two balls were computed by applying analytical modelling and coding in MATLAB. Properties of balls and aluminium powder were entered into the software as input data. The impact radius and then the hardness of powder were calculated accordingly. The changes of aluminium powder hardness resulting from the collision of two spherical balls during the synthesis of an aluminium nanostructure were analytically derived and compared with experimental data obtained from the literature. Calculation of results accuracy shows the model has a better agreement with the experimental data at the beginning than the results from Maurice et al. ($R^2= 0.68$ in this model).This research innovation is to combine the NFD model with hardness formulation to calculate final hardness.

* *Corresponding author:  ; E-mail address: bidarian.vahid@yahoo.com*

## 1. Introduction

Ball milling (BM) is a technique used to synthesize non-crystallized and non-equilibrium metals. In recent decades, a number of nanostructures have been produced by this method [1-4]. The main mechanism involved in BM is the repetition of the cold welding process and particle refraction, leading to the production of nanostructures over time [5]. As crystal defects (such as dislocations and atomic vacuums) are greatly increased during the sharp collisions of balls and powders, the original powders are crushed several times, leading to nano-scale particles [6]. The BM method enables the accruing of nanoparticles at ambient temperature by accelerating the kinetics of many chemical reactions and changing metallurgical modes [7]. Therefore, many materials and structures in solid state can be produced by this method [8]. The simple single-stage ball mills and relatively low cost process will allow the production of more materials and alloys. In recent years, BM has been widely used to synthesize new magnetic metals such as permanent magnets [9,10], iron oxide and micro powders [11]. In spite of studies on many systems, only some parts of the phenomena involved in the BM process have been understood. Considering most of the studies still take place in the laboratory setting, BM modeling aims to identify the affecting factors in this process and optimize the process of particles production by reducing the number of experiments [12]. These models may be used as instruments of process control.

Many models have been developed by various researchers at different levels of precision for the milling process. Some of these models are based on deformation of solids during particle/particle and particle/ball collisions [14]. For example, Hertz offered some solutions for modeling the elastic contact between loaded spheres [15], and Mindlin et al. subsequently developed this model for elastic-plastic contact [16]. Vu-Quoc demonstrated the deficiency in models based on the theory of elastic collisions by showing the main effect of plastic deformation [17]. He proved that application of these models to stimulate dry again flows in most contacts, included plastic deformation, can lead to inaccurate results. The model of normal force displacement (NFD), as suggested by Walton [18], is based on finite element analysis (FEA) and plastic deformation calculation in which a constant elastic deformation rate is applied during the spheres collision.

However, experimental results have shown that this parameter depends on the balls' speed before collision. Therefore, the given model is inconsistent with experimental results [19]. Thornton suggested a NFD model with the calculation of elastic and plastic deformation, producing different rebound coefficients for spheres of different speeds [20]. Maurice et al. presented a comprehensive model for BM [21-23]. This model consists of calculating the forces, tension, hardness, collision time, and other required parameters for the process modeling.

Mio et al. studied the effect of rotating direction and rate of rotating speed of chamber to disk in a planetary ball mill [24]. They demonstrated that collision energy may be increased by increasing the rate of chamber speed to disk speed during the initial stages and then it will be reduced to about the critical speed ratio. They also reported that an effective BM process must be performed at critical speed [24].

In this research, a non-linear NFD model has been used to model the collision of two spherical balls. The balls obtain force from the disk and chamber that are used. The ultimate force on the two balls is determined and the limiting factor of the collision is then calculated. The contact radius and indentation rate of the two balls are calculated by changing the normal force from zero to a calculated value. The hardness changes of the particles and the collision time during the milling process are calculated using these parameters. The limiting factor in each collision is the collision force, and the limiting factor of the process is the final hardness of particles.

## 2. The model algorithm

The following algorithm is applied to model the process.

1. Enter values of the rotating velocity of disk, chamber and ball radius, radius of the chamber and other constants.

2. Divide each collision by 3000 to calculate the radius and normal displacement, and then divide the loading and unloading forces by 1500.

3. Calculate the elastic radius, plastic radius, elastic-plastic radius and normal displacement using the algorithm given in Figure 1.

4. Calculate powder hardness.

5. Compare the calculated powder hardness from the model to final hardness.

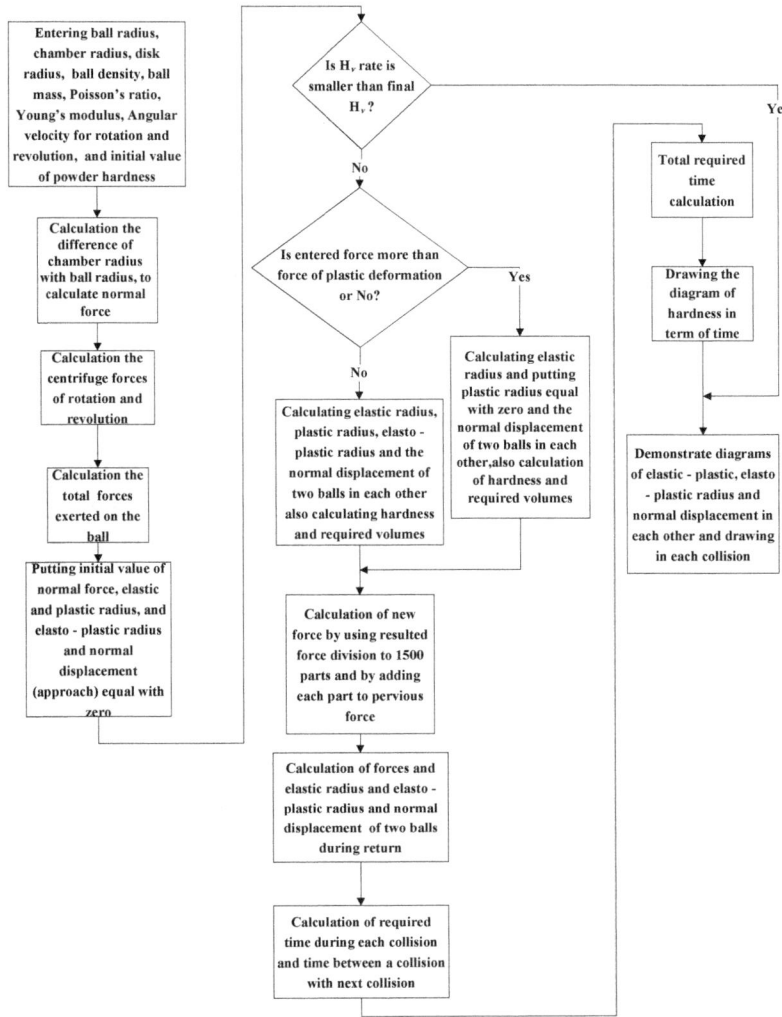

**Fig. 1.** Box diagram of calculating the powder hardness during BM process.

The mentioned algorithm continues until powder hardness equals the final hardness of the sample. The complete algorithm is shown in Figure 1.

### 2.1. Calculation of collision force

The total force exerted on the ball is obtained by calculating each force exerted on the ball [24]:

$$F_p = ml_c \omega^2 \tag{1}$$

$$F_r = m(R - l_c)\Omega^2 \tag{2}$$

In the above relations $F_p$, $F_r$, and $l_c$ are the centrifugal force of rotation and revolution and $l_c$ is the radius difference of chamber and ball, while $\omega$ and $\Omega$ are the angular velocity for rotation and revolution.

### 2.2. Model NFD

According to Figure 2, the kinetic energy of the balls will be converted to deformation energy, while approaching their centers. The energy of deformation is considered as a reduction of distance center to their original center. In order to understand the phenomena, in Figure 2 the real radius of contact has been exaggerated compared to the balls radius.

Corresponding to this energy conversion, tension is considered as the resistance against plastic and elastic deformation. In this model, balls and powder will experience plastic and elastic deformation during collision. In the early stages of contact, powder and balls are deformed elastically, but ultimately deform plastically.

In most collisions, the duration of elastic deformation of powders is shorter than plastic deformation. Thus,

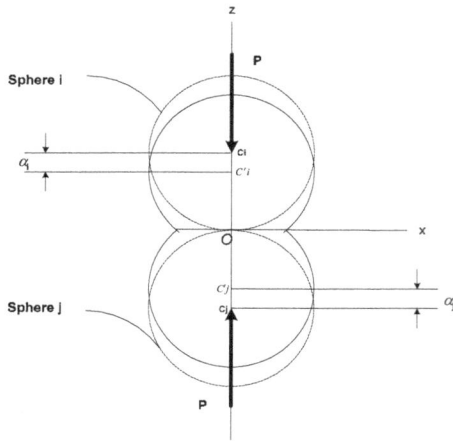

**Fig. 2.** Two balls contacting with each other and convert kinetic energy to deformation energy during collision of two balls [14].

, the longest-duration collision is considered as plastic deformation. Normal displacement (Figure 2) and deformation can be stated as a function of the radius as follows [22]:

$$a(r) = Rv\left(\frac{\rho_b}{H_v}\right)^{1/2} - \frac{r^2}{R} \tag{3}$$

where $r$ is the distance from center of contact, $R$ is the ball radius, $v$ is the relative velocity of balls during collision, $\rho_b$ is the ball density, and $H_v$ is the powder hardness. The rate of ball normal displacement has been calculated from equation (4):

$$\alpha = \frac{\left(a_{ep}\right)^2}{2R_p^*} \tag{4}$$

In this equation, $a_{ep}$ is the elasto-plastic radius. The equivalent radius of contact area curve ($R_p^*$) is defined by the following equation [14]:

$$\frac{1}{R_p^*} = \frac{1}{R_{pi}} + \frac{1}{R_{pj}} \tag{5}$$

With regard to the plastic deformation as an irreversible process, ($a_{res}$) is formed after collision. In order to calculate the normal displacement during the separation of two balls, the following equation is used [14]:

$$a - a_{res} = \frac{\left(a_e\right)^2}{\left(C_R\right)_{p=p_{max}} R} \tag{6}$$

where $a_{res}$ is calculated from Eq. (7) [14]:

$$a_{res} = a_{max} - \frac{\left(a^e\right)_{max}^2}{\left(C_R\right)_{(p=p_{max})} R} \tag{7}$$

In this equation, $a^e_{max}$ and $a_{max}$ are the normal displacement of maximum force and elastic radius rate, respectively. $C_R$ is the coefficient of radius adjustment to calculate plastic deformation [14]:

$$C_R(p) = \begin{cases} 1 & \text{for } P \le P_y \\ 1 + K_c\left(P - P_y\right) & \text{for } P > P_y \end{cases} \tag{8}$$

In the equation above, $p_y$ is the threshold force, in which elastic radius is converted to plastic radius. In the relationship $K_c$ is constant, and is determined according the ball characteristics [14]. Regardless of the type of milling, the BM process is described by collision between powder and milling components; hence, the model of a planetary ball mill has been used. There are several geometric possibilities to treat these collisions. For example, the powder may be trapped among the balls or may end up between ball and milling chamber.

Since the possibility of crushing the powder between the balls is more than that of between the balls and milling chamber [21], this type of collision has been examined in this study. It should be noted that for a different collision (for example a collision between the chamber wall and ball) only the geometric factor is different.

In this model, only crushing and deformation of powder by direct collision has been considered, and other collisions have been neglected. This analysis is used when a little powder covers the balls.

Process parameters and material property data are provided in Tables 1 and 2, respectively.

**Table 1.** The used parameters in model [23]

| Process Parameter | Al (MAP1) |
|---|---|
| Ball diameter (mm) | 6.35 |
| Ball density (g/cm³) | 7.8 |
| Charge ratio | 6.3 |
| $h_0$ (μm) | 100 |
| Impact angle (deg) | 0 |
| Impact velocity (m/s) | 4.0 |
| Impact frequency (s⁻¹) | 7.28 |

Figures 3-5 show the NFD curves and related coefficients of restitution generated by our elasto-plastic NFD model. Figure 5 shows the elastic-plastic radius of collision in terms of the normal displacement of two balls into each other during a collision and the return of

**Table 2.** The physical characteristics of aluminum used in the model [23]

| Property | |
|---|---|
| Modulus (GPa) | 72 |
| Tensile strength (MPa) | 90 |
| Strain to fracture — | 0.34 |
| Fracture toughness (MPa√m ) | 20 |
| Density (g/cm) | 2.7 |
| Starting hardness (kg/mm$^2$) | 40 |
| Dispersoid diameter (μm) | 0.1 |
| Starting particle size (μm) | 75 |
| Starting shape factor | 0.95 |
| K$^*$ (MPa) | 104 |
| n$^*$ | 0.37 |
| Weight fraction | - |
| Mass ratio of dispersoid | 0, 0.01 |
| Dispersoid hardness (kg/mm$^2$) | 550 |

$$\sigma = \sigma_0 + K_\varepsilon^n, \quad H_v = 3\sigma$$

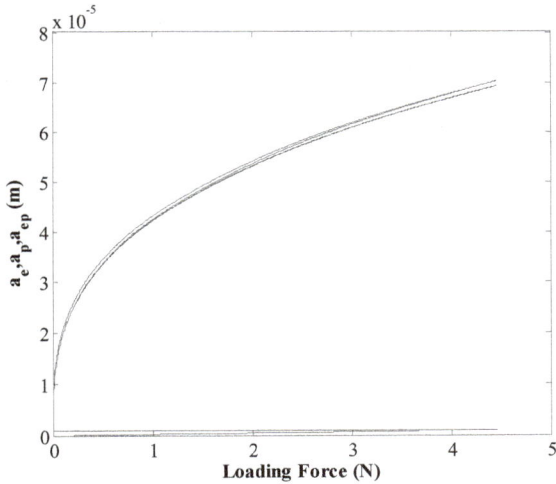

**Fig. 3.** Elastic, plastic and elastic-plastic radius in term of normal force.

### 2.3. To change the radius of the contact area

Considering the radius of the contact area from elastic-plastic contact ($a_{ep}$) and the radius of the elastic contact area ($a_e$) under force $P$, the radius of elastic-plastic is determined by Eq. (9):

$$a_{ep} = a_e + a_p \tag{9}$$

$$\begin{cases} a_{ep} = a_e & for \ P \leq P_y \\ a_{ep} > a_e & for \ P > P_y \end{cases} \tag{10}$$

Figure 4 shows the radius of plastic contact $a_p$ versus the normal force $P$ for loading and unloading to calculate

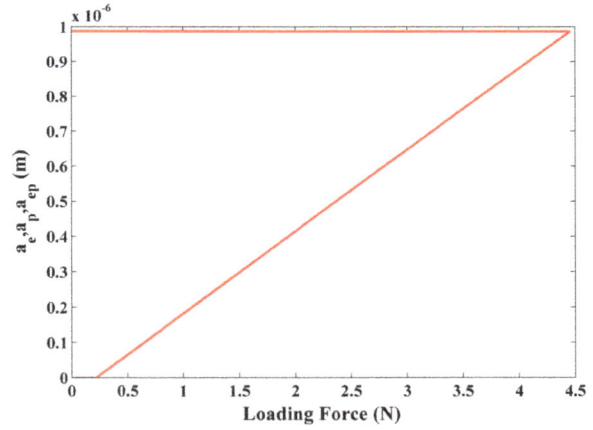

**Fig. 4.** plastic contact radius in terms of normal contact force.

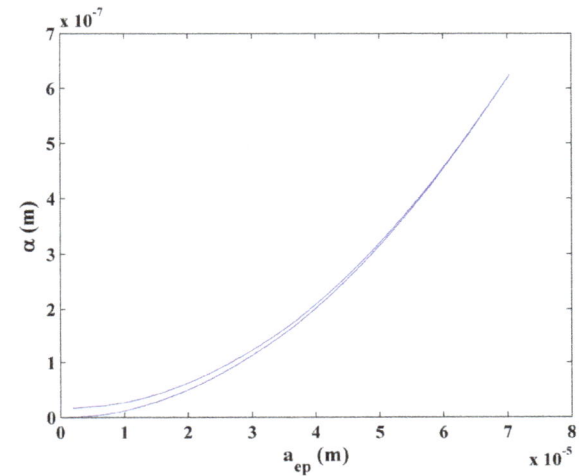

**Fig. 5.** Normal displacement versus elastic-plastic radius of collision.

calculate the ultimate force from the equation proposed by Mio et al. [24]. It is observed that during loading, plastic radius $a_p$ is increased linearly by increasing force. However, during unloading, the radius of plastic contact $a_p$ is not reduced by reducing $P$ after complete unloading. Stable deformation remains. In other words, when normal force $P$ goes to zero, the contact radius is inclined to non–zero values. Based on these observations, the area of plastic contact $a_p$ is roughly estimated in the proposed model NFD to be as follows [14]:

$$a_P = \begin{cases} C_a \left( P - P_y \right) & for \ loading \\ C_a \left( P_{max} - P_y \right) & for \ unloading \end{cases} \tag{11}$$

$C_a$ is a constant determined according to the characteristics of the contacting balls.

### 2.4. The collision frequency characteristic

The results of research conducted on distinct collisions

are suitable to determine the collision frequency, and this factor clearly influences the required time to perform the process. The volume of powder with the balls is calculated by the following equation [22]:

$$V_p = \frac{4\pi}{3} \left( \frac{R_b^3 \rho_B}{\rho_p CR} \right) \tag{12}$$

and the powder volume affected by a collision are achieved from Eq. (13):

$$V_c = \pi f_p h_0 a_p^2 \tag{13}$$

where $R_b$ is the ball radius, $\rho_b$ the ball density, $CR$ the charge ratio (mass of balls/mass of powder), $\rho_p$ the powder density, $f_p$ is the volumetric packing factor (typically = 0.6 to 0.7), $h_0$ the powder thickness coating the balls, and $a_p$ the radius of the plastic contact zone.

The powder volume affected by a collision ($V_c$, Eq. (13) [4]) is only a small fraction of the powder "associated" with a ball, so it will take several impacts by a given ball before all particles associated with it have been struck once. The number of impacts required for each particle to be struck once (on the average) is given by $V_p/V_c$, where $V_p$ is the volume of powder associated with each ball:

$$\frac{V_p}{V_c} = \frac{4}{\sqrt{3}} \left( \frac{R_b \rho_b^{(1/2)} H_v^{(1/2)}}{\rho_p CR h_0 v} \right) \tag{14}$$

The time between collisions is obtained using the following equation [21]:

$$\tau = \frac{\pi}{2k} \tag{15}$$

In above equation, coefficient $k$ is calculated as follows [21]:

$$k = \frac{1}{2R} \left[ \frac{3H_v}{\rho \left( 1 + 3C^2 tan^2 \theta_i \right)} \right]^{1/2} \tag{16}$$

Due to the force angle ($\theta$=0) in this research, the equation is simplified as follows:

$$k = \frac{1}{2R} \left[ \frac{3H_v}{\rho} \right]^{1/2} \tag{17}$$

Finally, the duration of collision is calculated from Eq. (18):

$$T = \left( \frac{1}{60 \left( \frac{1}{\tau} \right)} \right) \frac{V_p}{V_C} \tag{18}$$

where $1/\tau$ is the impact frequency. Time of each collision is the sum the of $\tau$ and T.

*2.5. Powder hardness*

Hardness is defined as the resistance of a solid material against the penetration of another harder material into its surface. This definition is not completely fulfilled by the usual hardness test methods for metallic materials as shown in Brinell, Vickers and Rockwell. This is due to the fact that the relevant measuring values for the calculation of hardness are measured after removal of test force. Depth sensing indentation gives the possibility to simultaneously record the acting force and corresponding penetration depth [25].

An important process parameter is powder hardness, in which the degree of powder is affected during impact and the normal elastic force acting is determined to separate particles during welding. There is a lack of structural relationship for metals in the strain range which is considered during BM. Thus, a simple plastic structural equation used in the range of less strain will be applied [21], where $k$ is the tenacity coefficient, $n$ is the hardness rate, $\sigma_y$ is the tension of flow in dense plastic strain and $\sigma_{y_0}$ is the initial flow. By using $H_v=3\sigma_y$, the following equation is used to determine strain:

$$\sigma_y = \sigma_{y0} + k\varepsilon^n \tag{19}$$

and the equation below is used for strain rate [21]:

$$H_v = H_{v0} + 3k\varepsilon^n \tag{20}$$

$$\varepsilon = -\ln \left( \frac{h_0 - a(r)}{h_0} \right) \tag{21}$$

where $a(r)$ is the normal displacement (approach) of two balls from each other and $h_o$ is the thickness of powder coating the ball. In this way, the determination of strain as a function of radius is simple in the collision area.

*2.6 Model accuracy*

The accuracy of forecasted values depends on how well the model fits your actual data. The value of R-squared represents the degree of fit. An R-squared value of closer to 1 indicates a better fit and more accurate model. To calculate the model accuracy in this paper we used the $R^2$ parameter. To calculate the model accuracy in this paper we used the $R^2$ parameter. The following equation is used to calculate the $R^2$ value:

$$R^2 = \frac{\sum_{i=0}^{n}(\hat{y}_i - \bar{y}_i)}{\sum_{i=0}^{n}(y_i - \bar{y}_i)} \tag{22}$$

where $y_i$ is an experimental data, $\hat{y}_i$ is a model data, and $\bar{y}_i$ is the mean of experimental data [26,27].

## 3. Powder hardness evaluation

Hardness caused by collision can be displayed by factoring certain data into the computational methods. Figure 6 shows the hardness as a function of milling time in the experimental and different calculation methods.

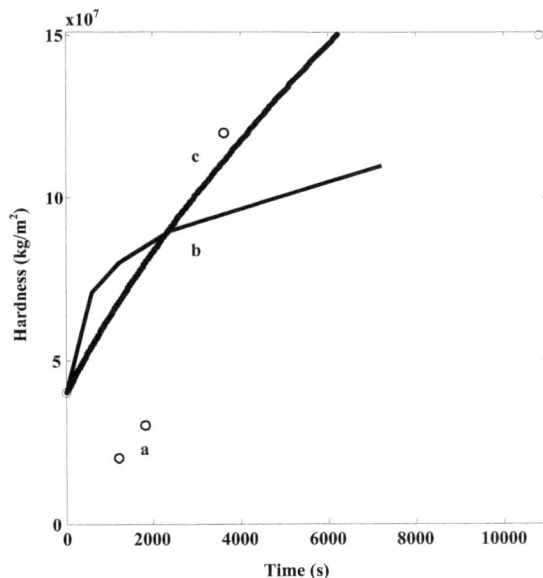

**Fig. 6.** Hardness changes versus time at ball milling process: a) Experimental data [23], b) Maurice et al. model [23], c) This paper model.

Calculation of results accuracy shows, model has a better agreement with experimental data at the beginning than Maurice et al. By calculating $R^2$ parameters for the two models (Maurice et al. and this paper), the results show these parameters are 0.5058 and 0.6849, respectively. As a result, the accuracy of our model was 35% more compared to Maurice et al. In this calculation, the point ($t$ =10800, $H_v$= 150000000) is neglected.

## 4. Conclusion

Description of aluminum nanostructure hardness

represented by a compound model was examined in this study. The two NFD models were combined with the Vo-Quoc model and the given model was used to calculate powder hardness. The results were then compared with the Maurice et al. model and experimental data. In the model developed in this study, the initial properties of powder and ball and its original dimensions were entered into the software. The first condition of collision, the rate of plastic, elastic, and elastic–plastic were then defined. The normal displacement of two balls was calculated by changing the force during the collision. After calculating the indention rate of two balls from the equation, the rate of hardness was calculated. It should be noted that calculation process time was divided in two parts during the collision and time distance between each collision. This was also used in each collision calculation and for numerical values in the model after calculating final hardness. It was observed that the compound model is more compatible with experimental results as compared to the model of Maurice et al. On the other hand, considering the plastic radius, in addition to the elastic radius, the collision will have better compatibility with the experimental data.

## Acknowledgments

The authors are grateful to the Research Council of the Quchan Islamic Azad University for financial support of this research.

## References

[1] M. Sopicka-Lizer, High-energy ball milling: Mechanochemical processing of nanopowders, Woodhead, New York, 2010.

[2] A. Nazari, M. Zakeri, Modeling the mean grain size of synthesized nanopowders produced by mechanical alloying, Ceram. Int. 39 (2013) 1587-1596.

[3] B. Nasiri-Tabrizi, A. Fahami, R. Ebrahimi-Kahrizsangi, J. Ind. Eng. Chem. 20 (2014) 245-258.

[4] M. Abdellahi, M. Bahmanpour, A novel technology for minimizing the synthesis time of nanostructured powders in planetary mills, Mater. Res. 17 (2014) 781-791.

[5] M.S. Khoshkhoo, S. Scudino, T. Gemming, J. Thomas, J. Freudenberger, M. Zehetbauer, C. Koch, , J. Eckert, Nanostructure formation mechanism

during in-situ consolidation of copper by room temperature ball milling, Mater. Desgin 65 (2015) 1083-1090.

[6] J.S. Benjamin, Mechanical alloying-A perspective, Metal Powder Report. 45 (1990) 122-127.

[7] M.S. El-Asfoury, M.N. Nasr, A. Abdel-Moneim, Effect of Friction on Material Mechanical Behaviour in Non-equal Channel Multi Angular Extrusion (NECMAE), in Book of Abstracts, 2015, pp. 364.

[8] T.P. Yadav, R.M. Yadav, D.P. Singh, Mechanical Milling: a Top Down Approach for the Synthesis of Nanomaterials and Nanocomposites, Nanosci. Nanotech. 2 (2012) 22-48.

[9] M. Zandrahimi, M.D. Chermahini, M. Mirbeik, The effect of multi-step milling and annealing treatments on microstructure and magnetic properties ofnanostructured Fe-Si powders, J. Mag. Mag. Mater. 323 (2011) 669-674.

[10] J. Ding, P. McCormick, R. Street, Structure and magnetic properties of mechanically alloyed SmxFe100-x nitride, J. Alloy. Compd. 189 (1992) 83-86.

[11] J. Ding, W.F. Miao, P.G. McCormick, R. Street, Mechanochemical synthesis of ultrafine Fe powder, Appl. Phys. lett. 67 (1995) 3804-3806.

[12] M. Abdellahi, H. Bahmanpour, M. Bahmanpour, The best conditions forminimizing the synthesis time of nanocomposites during high energy ball milling: Modeling and optimizing, Ceram. Int. 40 (2014) 9675-9692.

[13] A. Canakci, S. Ozsahin, T. Varol, Modeling the influence of a process control agent on the properties of metal matrix composite powders using artificial neural networks, Powder Technol. 228 (2012) 26-35.

[14] L. Vu-Quoc, X. Zhang, L. Lesburg, A normal force-displacement model for contacting spheres accounting for plastic deformation force-driven formulation, J. Appl. Mech. 67 (2000) 363-371.

[15] H. Hertz, Uber die Beruhrung fester elastische Korper and uber die Harts (On the contact of rigid elastic solids and on hardness), Verhandlunger des Vereins zur Beforderung des Gewerbefleisses, Leipzig, Nov. 1882.

[16] R.D. Mindlin, H. Deresiewica, Elastic spheres in contact under varying oblique forces, J. Appl. Mech. 20 (1953) 327-344.

[17] L. Vu-Quoc, X. Zhang, L. Lesburg, Contact force-displacement relations for spherical particles accounting for plastic deformation, Int. J. Solids Struct. 38 (2001) 6455-6490.

[18] O.R. Walton, R.L. Braun, Viscosity, granular-temperature, and stress calculations for shearing assemblies of inelastic, frictional disks, J. Rheol. 30 (1986) 949-980.

[19] W. Goldsmith, Impact, the theory and physical behavior of colliding solids, Edward Arnold Pub., 1960.

[20] C. Thornton, Coefficient of restitution for collinear collisions of elastic-perfectly plastic spheres, J. Appl. Mech. 64 (1997) 383-386.

[21] D. Maurice, T.H. Courtney, Modeling of mechanical alloying: Part I. deformation, coalescence, bdand fragmentation mechanisms, Metall. Mater. Trans. A, 25 (1994) 147–158.

[22] D. Maurice, T.H. Courtney, Modeling of mechanical alloying: Part II. development of computational modeling programs, Metall. Mater. Trans. A 26 (1995) 2431-2435.

[23] D. Maurice, T.H. Courtney, Modeling of mechanical alloying: Part III. Applications of computational programs, Metall. Mater. Trans. A 26 (1995) 2437-2444.

[24] H. Mio, J. Kano, F. Saito, K. Kaneko, Effects of rotational direction and rotation-to-revolution speed ratio in planetary ball milling, Mater. Sci. Eng. A 332 (2002) 75-80.

[25] E. Hryha, P. Zubko, E. Dudrova, L. Pešek, S. Bengtsson, An application of universal hardness test to metal powder particles, J. Mater. Process. Tech. 209 (2009) 2377-2385.

[26] J. Walkenbach, Excel 2013 Formulas, John Wiley & Sons Publisher, 2013, pp. 483.

[27] K. Velten, Mathematical Modeling and Simulation: Introduction for Scientists and Engineers, John Wiley & Sons Publisher, 2009, pp. 69.

# Permissions

# List of Contributors

**Arin Marcous**
Department of Food Engineering, Sari Branch, Islamic Azad University, Sari, Iran

**Susan Rasouli**
Department of Nano Materials and Nano Coating, Faculty of Surface Coating and Modern Technologies, Institute for Color Science and Technology, Tehran, Iran

**Fatemeh Ardestani**
Department of Chemical Engineering, Qaemshahr Branch, Islamic Azad University, Qaemshahr, Iran, Iran

**Ahmadreza Zahedipoor, Mehdi Faramarzi, Shahab Eslami and Asadollah Malekzadeh**
Department of Chemical Engineering, Gachsaran Branch, Islamic Azad University, Gachsaran, Iran

**Sahar S. El Sayed, Amal A. El-Naggar and Sayeda M. Ibrahim**
Department of Radiation Chemistry, National Center for Radiation Research and Technology, P. N. 13759, Cairo, Egypt

**Sadegh Beigi and Mohammad Amin Sobati**
School of Chemical Engineering, Iran University of Science and Technology (IUST), Tehran, Iran

**Amir Charkhi**
Material and Nuclear Fuel research school, Nuclear Science and Technology Research Institute, Tehran, Iran

**Ehsan Zamani Souderjani and Mohammad Ali Mousavian**
Department of Chemical Engineering, Collage of Engineering, University of Tehran, Tehran, Iran

**Ali Reza Keshtkar**
Nuclear Fuel Cycle School, Nuclear Science and Technology Research Institute, Tehran, Iran

**Ali Vesal, Ahmad Rahbar-Kelishami and Toraj Mohammadi**
Department of Chemical Engineering, Iran University of Science and Technology (IUST), Tehran, Iran

**Mansoor Kazemimoghadam**
Department of Chemical Engineering, Malek-Ashtar University of Technology, Tehran, Iran

**Zahra Amiri-Rigi**
Department of Chemical Engineering, South Tehran Branch, Islamic Azad University, Tehran, Iran

**Saeid Mohammadmahdi and Ali Reza Miroliaei**
Department of Chemical Engineering, University of Mohaghegh Ardabili, Ardabil, Iran

**Hamed Khosravi**
Department of Materials Engineering, Faculty of Engineering, University of Sistan and Baluchestan, Zahedan, Iran

**Reza Eslami-Farsani**
Faculty of Materials Science and Engineering, K.N. Toosi University of Technology, Tehran, Iran

**Mohsen Askari-Paykani**
Department of Materials Engineering, Tarbiat Modares University, Tehran, Iran

**Reza Zolfaghari and Mohammad Karamoozian**
School of Mining, Petroleum and Geophysics, Shahrood University of Technology, Shahrood, Iran

**Mohamad Alizade Pudeh, Esmaeil Rahimi and Mehran Gholinejad**
Department of Mining Engineering, Islamic Azad University, South Tehran Branch, Tehran, Iran

**Amirhossein Soeezi**
School of Mining, College of Engineering, University of Tehran, Tehran, Iran

**Bahman Hassanzadeh and Farajollah Mohanazadeh**
Department of Chemical Technology, Iranian Research Organization for Science and Technology (IROST), Tehran, Iran

**Mohammad Masoud Javidi and Sedighe Mansoury**
Faculty of Mathematics and Computer, Department of Computer Science, Shahid Bahonar University of Kerman, Kerman, Iran

**Saeed Mollaei**
Phytochemical Laboratory, Department of Chemistry, Faculty of Sciences, Azarbaijan Shahid Madani University, Tabriz, Iran

**Biuck Habibi**
Electroanalytical Chemistry Laboratory, Department of Chemistry, Faculty of Sciences, Azarbaijan Shahid Madani University, Tabriz, Iran

**Alireza Amani Ghadim**
Applied Chemistry Laboratory, Department of Chemistry, Faculty of Sciences, Azarbaijan Shahid Madani University, Tabriz, Iran

**Milad Shakouri**
Phytochemical Laboratory, Department of Chemistry, Faculty of Sciences, Azarbaijan Shahid Madani University, Tabriz, Iran
Electroanalytical Chemistry Laboratory, Department of Chemistry, Faculty of Sciences, Azarbaijan Shahid Madani University, Tabriz, Iran

**Shohreh Saffarzadeh-Matin and Majid Shahbazi**
Department of Chemical Technologies, Iranian Research Organization for Science and Technology (IROST), Tehran, Iran

**Kamran Janghorban**
Department of Chemical Engineering, Farahan Branch, Islamic Azad University, Farahan, Iran

**Payam Molla-Abbasi**
Department of Chemical Engineering, Faculty of Engineering, University of Isfahan, Isfahan, Iran

**Hossein Aghajani and Maryam Pouzesh**
Department of Materials Engineering, University of Tabriz, Tabriz, Iran

**Mohammad Hojjat**
Department of Chemical Engineering, Faculty of Engineering, University of Isfahan, Isfahan, Iran

**Elham Farzan, Mahmood Reza Rahimi and Vahid Madadi Avargani**
Department of Chemical Engineering, Yasouj University, Yasouj, Iran

**Hanieh Panahi**
Department of Mathematics and Statistics, Lahijan Branch, Islamic Azad University, Lahijan, Iran

**Saeid Asadi**
Department of Mechanical Engineering, Payame-Noor University (PNU), Tehran, Iran

**Fatemeh Masoudzadeh, Mohammad Fasihi and Masoud Jamshidi**
School of Chemical, Petroleum and Gas Engineering, Iran University of Science and Technology, Tehran, Iran

**Vahid Bidarian**
Young Researchers and Elite Club, Quchan Branch, Islamic Azad University, Quchan, Iran

**Esmaeil Koohestanian**
Department of Chemical Engineering, University of Sistan and Baluchestan, Zahedan, Iran

**Maryam Omidvar**
Department of Chemical Engineering, Quchan Branch, Islamic Azad University, Quchan, Iran

# Index

www.ingramcontent.com/pod-product-compliance
Lightning Source LLC
Chambersburg PA
CBHW080638200326
41458CB00013B/4672